# FIBER TECHNOLOGY

## INTERNATIONAL FIBER SCIENCE AND TECHNOLOGY SERIES

Volume 1: Handbook of Fiber Science and Technology: Chemical Processing of Fibers and Fabrics – Fundamentals and Preparation (in two parts), *edited by Menachem Lewin and Stephen B. Sello*

Volume 2: Handbook of Fiber Science and Technology: Chemical Processing of Fibers and Fabrics – Functional Finishes (in two parts), *edited by Menachem Lewin and Stephen B. Sello*

Volume 3: Carbon Fibers, *by Jean-Baptiste Donnet and Roop Chand Bansal*

Volume 4: Fiber Technology: From Film to Fiber, *by Hans A. Krässig, Jürgen Lenz, and Herman F. Mark*

*Other Volumes in Preparation*

# FIBER TECHNOLOGY

## FROM FILM TO FIBER

HANS A. KRÄSSIG
JÜRGEN LENZ
Chemiefaser Lenzing AG
Lenzing, Austria

HERMAN F. MARK
Polytechnic Institute of New York
Brooklyn, New York



MARCEL DEKKER, INC. New York and Basel

Library of Congress Cataloging in Publication Data
Main entry under title:

Fiber technology.

(International fiber science and technology series; v. 4)
Bibliography: p.
Includes index.
1. Textile fibers, Synthetic. 2. Plastic films. I. Krässig, H. (Hans), [date]. II. Mark, H. F. (Herman Francis, [date]. III. Lenz, Jürgen, [date]. IV. Series.
TS1548.5.F512 1984 677'.4 84-1778
ISBN 0-8247-7097-8



MARCEL DEKKER, INC.
270 Madison Avenue, New York, New York 10016

Current printing (last digit):
10 9 8 7 6 5 4 3 2 1

PRINTED IN THE UNITED STATES OF AMERICA

# Preface

The rapid growth and the high technological state of the art of the man-made fiber industry has initiated the demand for a comprehensive description of the processes practiced today in the manufacture of man-made fibers. This demand is all the more justified since the manmade fiber development—skyrocketing with a fantastic pace in the early 1950s—has now reached a certain state of consolidation. Each type of manmade fiber has now found its specific place in the textile industry, fulfilling to a reasonably high degree the needs and requirements of the textile processors and of the consumers of textile goods. This also applies to tapes, filament yarns, and fibers manufactured by way of polymer films as an initial preliminary stage. In a continuous process of selection, only the fibers best suited for a given end-use application have survived, combining specific properties with prevailing quality requirements. These fibers continued to be produced on a large scale.

The development of film-to-fiber technology (such a possibility was first considered approximately 50 years ago) actually started in the beginning of the 1960s with high expectations. In the following two decades a large number of technologies in the manufacture of film tapes, split-film yarns, and film fibers were developed by inventors and companies throughout the world, including the best known industrial giants involved in the development of conventional spinneret fiber manufacture. The initially exaggerated expectations have given way, in the meantime, to a more realistic judgment of the potential of this simplified technique for the manufacture of yarnlike and fibrous products. Film tapes, split-film yarns, and fibrillated film fibers have found their permanent place in a good number of end-use applications, such as in wide areas of the packaging field or in the manufacture of technical fabrics, household textiles, or carpets.

This book intends to offer to fiber technologists, machine and process engineers, textile specialists, textile dealers, and textile salesmen a practical guideline to become acquainted with and to deepen their knowledge of the processes for the manufacture of film tapes, split-film yarns, and fibrillated film fibers and of the potential of these products in the processing of textiles and textile-like products. The book also deals with the physicochemical basic knowledge of polymer processing, thus giving the chemists and engineers active in research and development work, the necessary background knowledge.

A comprehensive and extensive bibliography of relevant literature contained in the book is intended to help the interested reader in his or her efforts for further specialization. A large number of tables and illustrations facilitates the understanding of technological know-how and scientific knowledge in cases where more complicated physicochemical problems are involved.

The authors hope this book will be of benefit for the practical person in production and in engineering, as well as for development people in the laboratory or in pilot plants, or for the student training for future tasks in the film or fiber industry.

Hans A. Krässig
Jürgen Lenz
Herman F. Mark

# Contents

# Introduction

The direct way to produce artificial filament or staple fibers by extrusion of a polymer spinning dope or melt through small spinneret holes in well known, widely applied, and often described [see Mark et al. (1968)].

When, during the last three decades, a large number of industrial companies and research institutions began to consider and study the possibilities of producing fibrous products from polymer films, there must obviously have existed special reasons and justifications. Those are mainly the relatively low investment and processing costs for film, film tape, and film fiber manufacturing and some special properties of the film-to-fiber products, such as the suitability of film tape for the manufacture of packaging fabrics or of primary carpet backing fabrics, or of fibrillated film yarns for twines, ropes, face yarns for outdoor carpeting, etc.

The idea to produce fibers or fibrous products via the film is quite old. The first known efforts were made in the early 1930s. Jacqué and his coworkers at the I. G. Farbenindustrie AG (1938) applied, in 1936, for their first patents to produce fibers or continuous filaments from polymer films, using films of polyvinyl chloride and of polystyrene. Upon applying high stretch ratios these films could be split into fibrous products by mechanical action due to the high anisotropy of their tensile strength. Approximately at the same time Larsen (1932) described in an American patent the use of ribbons cut from paper for the manufacture of woven fabrics. Mattler (1933) suggested shortly thereafter in an American patent the use of leather cut into thin strips. However, at this time, polyvinyl chloride and polystyrene were still too expensive to compete with natural fibers. Products made from paper or leather strips were of only limited use, due either to low strength and wear resistance or to the limited length of strips cut from leather sheets.

However, when in the early 1950s the polyolefins, namely high-density polyethylene (HDPE) and the isotactic polypropylene, came into commercial reality, the situation changed completely. These polymers, being relatively inexpensive and having good crystallization and orientation ability, produced excellent films. Due to their limited tendency to develop intermolecular bonds, highly oriented films showed especially high anisotropy of strength in length and transverse direction and could easily be cut into tapes or ribbons of high longitudinal strength. This gave the development of methods to produce fibers from polymer film new momentum.

It is the purpose of this book to summarize the knowledge gathered during the recent past and to describe today's status in this relatively young field of manufacturing fiber products from polymer film.

There are a number of more or less comprehensive articles in technical journals and a few books dealing with this subject, i.e., Ford (1965, 1966, 1967, 1970), *Text. Rec.* (1966), *Mod. Plast.* (1967), *Br. Plast.* (1968), *Text. World* (1971), *Cah. Text. Cenatra* (1971), Lagowski (1967), Okabe and Chino (1967), Peuker (1968a, 1968b, 1968/1969), Condit and Johnson (1969), Hensen (1969, 1978a, 1978b), Real (1969), Schuur (1969), Van Boskirk (1969), Moorwessel and Pilz (1969), Albers and Duiser (1970), Coen and Zilio (1970), Nightingale (1970), Schubert (1970), Badrian and Choufoer (1971a, 1971b), Harms et al. (1971a, 1971b, 1973a, 1973b), Koninklijke Shell Plastics Laboratorium (1971), Ruta (1971), Thomas (1971), Campbell and Skoroszewski (1972), Hajamasy et al. (1972), Ivanyukov et al. (1972), Krässig (1972, 1975a), Polovikhina and Zverev (1972), Krejci and Stepankowa (1973), Dowgielewicz (1974), Pozleb (1974), Tweedale (1974), Schuur and Van der Vegt (1975), Clayton (1976), Droghanova and Martynova (1976), Feuerböther (1976), Michels et al. (1977), Scheiner (1977), Hensen and Braun (1978a, 1978b), and VDI-Gesellschaft Kunststofftechnik (1978).

# FIBER TECHNOLOGY

# 1
# Processes for Manufacturing Film, Film Tapes, and Film Fibers

In this chapter a general review of the processes for the manufacture of film, film tapes, and film fibers shall be given.

## I. FILM PROCESSES

The first step in the manufacture of fibrous film products such as film tapes and film fibers is the film casting. It is understandable that the manufacture of such products is only economically feasible when the film processing is relatively simple and inexpensive. For this reason, film processing by melt extrusion, particularly of polyolefins, is widely applied. Other polymers calling for more elaborate film processing methods, such as casting from solutions, are only suited for film-to-fiber technologies when they yield special products for expensive end-use applications. Furthermore, uniaxial stretching methods are most generally applied in the manufacture of fibrous products from films.

Film processing comprises the following steps: (1) extrusion of the polymer melt; (2) film solidification by cooling; (3) uniaxial stretch at elevated temperature; (4) heat relaxation or fixation; and (5) the final takeup. The cutting into tapes and the splitting or slicing into finer fibrous products is done either after the film solidification, during the stretching operation, or later in the processing before final takeup.

For the extrusion step (through a flat die or annular die) three methods are available; the extrusion through a flat die uses either water quenching or chill-roll cooling for the film solidification, and the extrusion through annular die normally uses air cooling. The three routes for casting polymer films by melt extrusion techniques as well as the processing conditions applied, such as extrusion temperature, extrusion speed, solidification conditions, etc., have a marked influence on the ease of stretching, the maximum stretch ratio achievable, and the final mechanical and thermal properties of the resulting products. The same can be said for the influence of stretching and annealing conditions applied later in the process. Details will be discussed in later chapters of this book.

## II. FILM TAPE PROCESSES

The first step towards the production of fibrous products suitable for textile and related uses by way of the film was the development of film tapes. As mentioned before, until the early 1950s the production of tapes was limited to paper, leather, or films from regenerated cellulose, polyvinyl chloride, and polystyrene. These tape products, however, were either of only limited durability or too expensive. The appearance of the polyolefins, namely high-density polyethylene and isotactic polypropylene, changed the situation completely.

As recently as the early 1960s a German mechine manufacturer developed special machinery for the production of polyolefin film tapes suitable for weaving (Becker, 1962; Poelchau, 1966). In a matter of only a few years a large number of other manufacturers throughout the world (for instance, in Austria, England, France, Germany, Italy, United States etc.) have developed improved equipment for the same purpose. Descriptions of such machinery have been given by *Kunststoffe* (1966), *Mod. Text. Mag.* (1968), *Kunststoffe* (1969), *Z. Gesamte Textilind.* (1969), *Chemiefasern* (1970), *Chemiefasern & Text.-Anwendungstech./Text. Ind.* (1972), *Mod. Text. Mag.* (1973), *Tufting Needling News Bulletin* (1973), *Chemiefasern & Text.-Anwendungstech./Text. Ind.* (1974), Hensen (1968, 1969, 1973, 1974, 1978, 1979), Mozawa (1966), Slack (1968), Nightingale (1970), Hossack (1971), Barmag (1973), Braun (1975), Windmöller & Hölscher (1981).

Principally, there are two basic ways to produce monoacially stretched film tapes:

1. Processes where tape cutting is done before the stretching operation. The polymer film being delivered from an extruder equipped with a circular or flat slit die in the form of a tubular or a flat film is being cut into tapes of 1 to 20 mm width. The resulting primary tapes are stretched to such an extent as necessary to achieve the desired dimensions and properties. In Figure 1 schemes of production lines for stretched film tape are being given. The operation can either be performed continuously or discontinuously, separating in the latter case the film formation from the tape cutting and stretching step. Further information on film tape manufacture will be given in Chapter 2. Publications dealing with this subject have been written by Ford (1965, 1966, 1967, 1970), Burggraf (1967), Hensen and Klawonn (1967a, 1967b), Schwenkedel (1967), Weber (1967), Lennox-Kerr (1968), Mayer (1969), Nau (1968), Peuker (1968a, 1968b, 1968/1969), Schrader (1968), Hensen (1968, 1979), Moorwessel and Pilz (1969), Real (1969), and Schuur (1969).

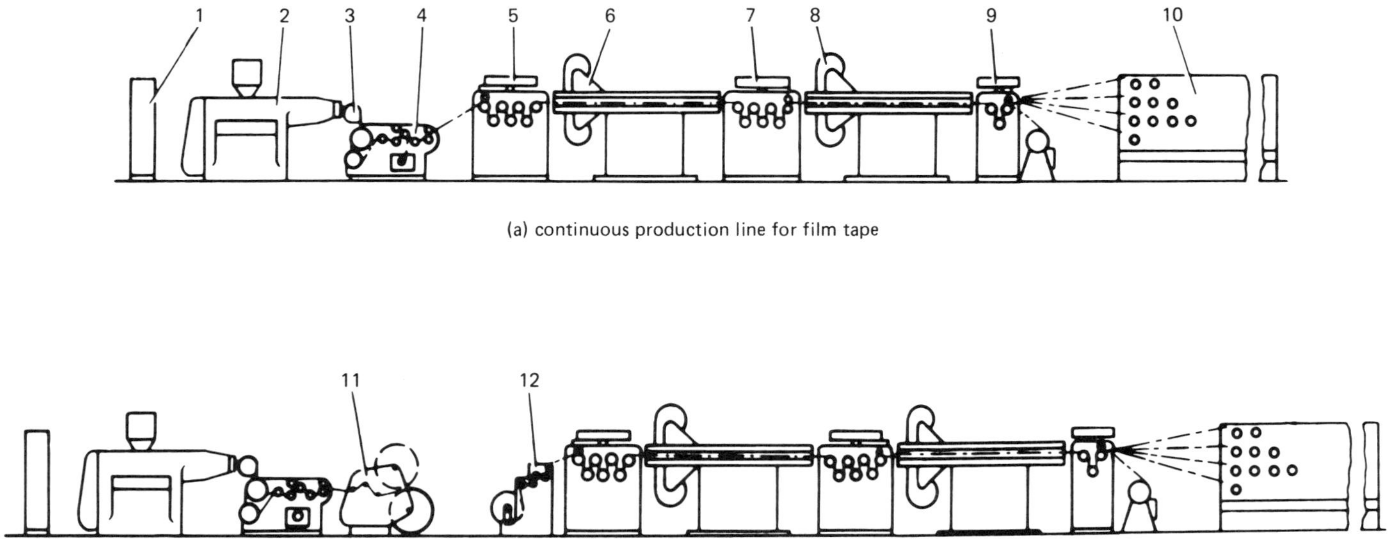


(a) continuous production line for film tape

(b) discontinuous production line for film and film tape

**Figure 1** Scheme of production lines for stretched film tapes: (1) control cabinet; (2) extruder; (3) flat die; (4) chill roll, tape cutting tool; (5) septette; (6) drawing oven; (7) septette; (8) heat setting oven; (9) trio; (10) bobbin winder; (11) film winder; (12) film unrolling stand, tape cutting tool. (From Krässig, 1977.)

2. Processes, where tape cutting is performed after the stretching operation. The polymer film extruded as described under (1) is being stretched to an extent necessary to achieve the desired thickness dimension and mechanical properties in the final tape before the separation in film tapes of desired width. In most cases the film stretching operation is done in such a way as to prevent width shrinkage of the film, such as by performing the length extension between heated rollers over a narrow distance. In addition, static charging is often used to achieve good adhesion of the film to the stretch rollers counteracting the width shrinkage tendency.

The cutting of the uniaxially stretched film is performed either in a continuous operation following the stretching step or it can be done separately on a special cutting line (see Figure 2). The Austrian fiber manufacturer Chemiefaser Lenzing AG and the German machine manufacturer Bruckner collaborated in the late 1960s in the development of this type of manufacturing of monoaxially oriented polyolefin film and film tapes. The basic idea established by Chemiefaser Lenzing AG was the direct use of monoaxially oriented film as warp sheet on weaving and knitting machines, cutting the warp tape by special blade arrangements directly on the loom (Chemiefaser Lenzing AG, 1972a, 1972b, 1973; Hossack, 1971), thus avoiding the tedious and expensive creeling and beaming operations (see Figure 2). This idea was suggested earlier in a somewhat different manner by Larsen (1932) in relation to paper, by Mattler (1933) in relation to leather, and by Pakleppa (1966) in relation to polyolefin film. Marks (1965) has described a somewhat related slit-knitting process from polymer sheet material.

For cutting operations, bars equipped with razor-bladelike cutting tools are generally used, as shown in Figure 3. The distance of the cutting tools has to be adjusted according to the desired width and thickness of the film tapes to be manufactured. When the cutting of the tapes is done before the stretching, width and thickness shrinkage have to be accounted for in the film extrusion and in the cutting step.

The structure and properties of the film tapes, especially those made from polyethylene and polypropylene, manufactured by either of the two ways outlined above, are not identical. Stretching after the tape cutting of the unstretched film, leading normally to a reduction in tape width and tape thickness, results in a more pronounced monoaxial orientation and in a higher anisotropy of strength. Tapes manufactured in this way show a higher tendency for splitting. Stretching the film before cutting it into tapes more efficiently prevents width reduction, introduces some cross orientation, and reduces the length-splitting tendency. Specific problems of tape processing from various polymer film, their properties, and applications will be discussed in detail in following chapters of this book.

In the United States, consumption of polyethylene and polypropylene for various film tape applications had already reached approximately

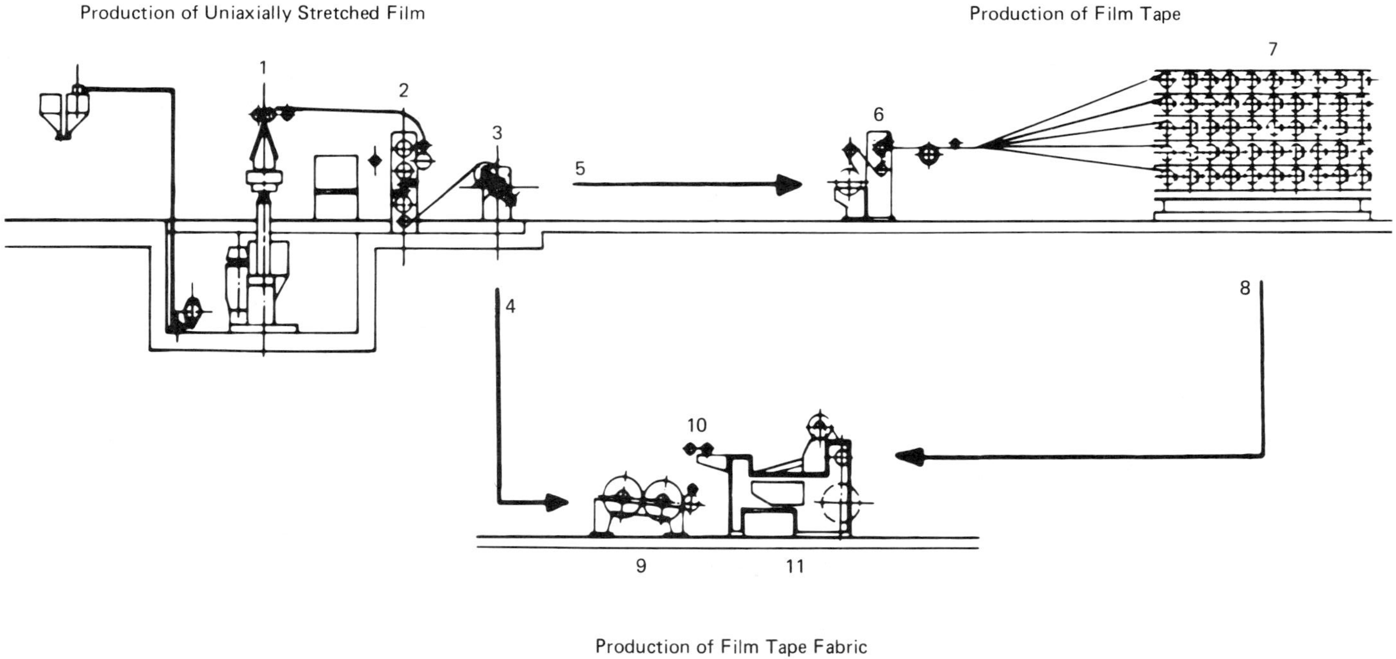


**Figure 2** Scheme of production lines for uniaxially stretched film, film tapes, and woven film tape fabrics using the "slit-weaving" process. (1) extruder ("Rotatruder," Oerlikon); (2) drawing system; (3) film winding; (4) film roll used instead of beamed tape; (5) film roll going to weft tape cutting; (6) film unrolling stand, tape cutting tool; (7) bobbin winder; (8) film tape going as weft yarn to loom; (9) film unrolling stand; (10) slit weaving attachment; (11) loom. (From Krässig, 1977.)

Figure 3 Film tape cutting tool. (From Krässig, 1977.)

200 million pounds by 1969. In 1970, nearly 500 million square yards of film tape fabrics were used in the United States alone (Kerr, 1971) for primary carpet backing and for bags of all sizes. In 1975, 270 million pounds of film tape goods were produced from polyolefins in the United States, and in 1979 this amount had risen to approximately 350 million pounds. In Europe the corresponding development started somewhat later. In 1977 Western Europe had reached a total of about 400 million pounds of film tape production, and in 1979 almost 500 million pounds.

This development was a heavy blow for the suppliers of natural fibers. The production of Manila hemp in the Philippines dropped from over 200 million pounds in 1960 to under 100 million pounds in 1970. The production of sisal in Tanzania declined similarly during the same period (Ford and Govier, 1971; Grilli, 1975).

## III. FILM YARN AND FILM FIBER PROCESSES

The pioneering work in filament and fiber processing via polymer film was performed at the beginning of the 1930s by Jacque et al. (1938)

at the I. G. Farbenindustrie in Germany. Jacque and his coworkers observed that films of polyvinyl chloride or of polystyrene, when stretched monoacially to high stretch ratios at elevated temperatures, resulted in materials having a high degree of orientation and a high tensile strength in the draw direction. At the same time, the tensile strength in the cross direction is strongly reduced, resulting in a pronounced splitting tendency. Monoacially stretched samples of these polymers could easily be split into fibrous products by mechanical action, such as brushing, rubbing, or twisting.

Somewhat later, at the beginning of the 1950s, the Dow Chemical Company developed a closely related process for the production of fibers from polyvinylidene chloride films (Dow Chemical Company, 1955).

In the early 1960s Rasmussen (1961, 1963a, 1963b, 1963c, 1964, 1965, 1966a, 1966b, 1966c, 1967a, 1967b, 1969a, 1969b, 1969c, 1970a, 1970b, 1971a, 1971b, 1972a) began his extensive work on film-to-fiber processing, which has been remarkably stimulating and which has yielded a large number of useful suggestions for the further development of textile yarn and fiber manufacturing by way of polymer films.

To the best of our knowledge, the first large-scale industrial realization of a process using film-to-fiber technology to manufacture textile yarn products was executed by Chevron Research Company in 1968. Since then, this company has produced a polypropylene carpet face yarn in their plant in Dayton, Tennessee which is sold under the trade name Polyloom.* Thereafter, many chemical companies and fiber and machine manufacturers throughout the world, such as Plasticisers, Ltd. in Great Britain, the Toshiba Company and Mitsubishi Rayon Company in Japan, the German firms Barmag-Barmer Maschinenfabrik AG and J. F. Adolff AG, the Celanese Corporation and Phillips Petroleum Company in the United States, and many others have independently developed film yarn and film fiber processes and the machinery to perform such processes [see Nozawa (1967), Slack (1968), Hensen (1969, 1974, 1978), *Chemiefasern* (1970), *Chemiefasern & Text.-Anwendungstech./Text. Ind.* (1972), *Tufting Needling News Bulletin* (1973), Nightingale (1970), Barmag (1973), and Hensen and Braun (1978a, 1978b)].

In order to process a polymer film to be subsequently converted into fibrous products, the same general techniques are applied as those used by film tape manufacturers. Specific problems of the film processing will be discussed in detail, with special attention to polyolefins, in later sections of this book. The main features of fiber manufacture via film, however, lie in the techniques of film splitting, film slicing, and film cutting. During the last 20 years many different methods of converting a film into filament yarns, fibers, or other fibrous products were developed. About 15 years ago Condit and Johnson (1969) attempted to classify the various film-to-fiber techniques. We will use their

---

* Registered trademark.

suggestions, with slight modifications. According to the principles underlying the film-to-fiber separation, one can distinguish three main types of film-to-fiber processes: (1) The processes applying random (uncontrolled) mechanical fibrillation; (2) the processes applying random (uncontrolled) chemomechanical fibrillation, and (3) the processes applying controlled mechanical film-to-fiber separation.

## A. The Processes Applying Random (Uncontrolled) Mechanical Fibrillation

In this group all the processes will be listed in which the polymer film is separated by purely mechanical action: (1) by brushing [see Phillips Petroleum Co. (1967)]; (2) by rubbing [see Rasmussen (1961, 1963b, 1966d), Shirley Institute (1964), Societe Normande de Matière Plastique (1966), Plasticisers, Ltd. (1967), Seifert (1968), Phillips Petroleum Co. (1969, 1971a), Eastman Kodak Co. (1969), Polymer Processing Research Institute, Ltd. (1970, 1972), Continental Linoleum Union (1971), Rheinstahl Henschel AG (1971a), Sekusui Chemical Co., Ltd. (1971), Lambeg Ind. Res. Assoc. (1973a)]; (3) by bending [see Rasmussen (1965b), Allied Chemical Corp. (1969a), Leykam-Josefsthal AG (1969), Hercules, Inc. (1970), Phillips Petroleum Co. (1970a, 1970b, 1971b)]; (4) by twisting or false twisting [see Dow Chemical Co. (1965), Société Rhodiaceta (1965), Shell Internationale Res. Maatschappij N.V. (1966), Friedrich (1971), Imperial Chemical Industries, Ltd. (1971a, 1971b, 1971c)]; (5) by air jet or sand blast treatment [see Société Rhodiaceta (1966), E. I. du Pont de Nemours (1966a), Monsanto Chemicals, Ltd. (1968), Allied Chemical Corp. (1969b), Dunlop Rubber Co., Ltd. (1971), Courtaulds, Ltd. (1971), Imperial Chemical Industries, Ltd. (1971c), Fiber Industries, Inc. (1972)]; (6) by special stretching techniques [see Johnson & Johnson, Ltd. (1970), Monsanto Chemicals, Ltd. (1971a, 1971b), Soko Co., Ltd. (1970, 1971, 1973)]; and (7) by ultrasonic treatments [see Branson Instruments, Inc. (1967)].

The resulting yarnlike products have, in most cases, a networklike texture of fiber segments showing a wide statistical distribution in separated fiber lengths, fiber thicknesses, and fiber cross sections. The main prerequisite for the application of the uncontrolled statistic mechanical fibrillation techniques in the manufacturing of film-to-fiber products is the creation of high tensile anisotropy in the film before fibrillation by the application of very high stretch ratios. Furthermore, polymers are used which have preferably, owing to their chemical structure, only a limited tendency to form strong secondary valence bonds between individual molecules. This is, of course, the case with polyolefins.

The uncontrolled fibrillation of highly stretched films or film tapes having a pronounced strength anisotropy was the basis of Jacque et al.'s (1938) original film-to-fiber process patents. They originally observed that rubbing a highly stretched polymer film, for example, between two countercurrently running belts, resulted in a lengthwise splitting into numerous filament segments interconnected in a networklike structure. Further splitting and parallelization could be achieved by additional treatment with rotating cylindrical brushes. Twisting highly stretched polyolefin film tapes up to 1000 turns per yard acts in a similar manner, causing the tapes to split up into fine fibrous segments of widely varying cross sections. This type of treatment is the basis for the manufacture of coarsely fibrillated yarns used as binder twine in harvesting agricultural products, in packaging applications, and in rope manufacturing. Another type of uncontrolled mechanical fibrillation treats highly stretched film tapes with high-velocity air jets. The results of such uncontrolled random fibrillation techniques applying various types of mechanical stresses on highly stretched polymer films is shown in Figure 4.

The side view shown in the upper part of Figure 4 indicates clearly that such treatments result in an incomplete separation into filament segments of widely varying width, still interconnected in a network structure. The varying thickness of the fibrillar segments is especially apparent in the cross sections shown in the bottom part of Figure 4.

## B. The Processes Applying Random (Uncontrolled) Chemomechanical Fibrillation

In this group all the processes will be listed in which additions are made to the polymer introducing randomly distributed inhomogeneities in the extruded film which, during the drawing operation, act as weak spots and enhance the tendency for lengthwise splitting during stretching and in subsequent mechanical splitting treatments.

Such additions can either be compounds decomposing near the extrusion temperature, forming gases, and leading to voids in the film, as described by Rasmussen (1963c), du Pont (1966b), Shell International Res. Maatschappij N.V. (1967a), Volans and Changani (1968), Monsanto Chemicals, Ltd. (1969, 1973), Shell Oil Co. (1970a), Celanese Corp. (1971a, 1971b, 1971c), and Farbwerke Hoechst AG (1974). Also, the use of soluble salts added to the polymer prior to or during the film extrusion has been suggested. In extracting the film, with solvents for the salt added, voids are formed which act as weak spots in stretching and fibrillation treatments, thus enhancing film-to-fiber separation. This technique has been proposed by Rasmussen (1963a, 1963b, 1963c). Another way to introduce weak spots into the film is the addition of incompatible polymers or the use of polymer blends, as claimed by Rasmussen (1965a, 1966a, 1967b, 1969c, 1970b, 1971c, 1971d), by Bakelite Xylonite, Ltd. (1968), Allied Chemical Corp. (1969c), Celanese Corp. (1969, 1971d), Mitsubishi Rayon Co. (1970a), Badische Anilin & Soda Fabrik AG (1970a), J. P. Bamberg AG (1970), Shell International Res.

Figure 4 Longitudinal view (above) and cross sections (below) of film fibers made by uncontrolled mechanical fibrillation. (From Krässig, 1977.)

Maatschappij N.V. (1971a), Fiber Industries, Inc. (1972), Reifenhäuser KG (1972), and by Rusznak et al. (1973).

If films which contain weak spots in the form of voids or material inhomogeneities are drawn uniaxially, these "flaws" will be extended in the draw direction, thus forming well defined starting points for splitting into networklike fibrous systems. The amount of voids or material inhomogeneities will be determined by the percentage of the gas-forming compound or of the incompatible material added. The splitting will, in many cases, occur during the stretching operation and can be additionally enhanced by mechanical action.

In Figure 5 an example of the texture of a split-fiber yarn made by random (uncontrolled) chemomechanical fibrillation technique is illustrated by its side view (left) and its cross sections (right). The side view demonstrates that the length splitting here is also incomplete, resulting in a more or less irregular network structure. The cross sections show that the chemomechanical splitting gives fibrous segments of a still nonuniform character, but somewhat more regular than those obtained by only uncontrolled mechanical splitting.

The fibrillar products, manufactured from films, either by random (uncontrolled) mechanical or chemomechanical fibrillation methods, as described in Sec. III.A and III.B, should be termed "Split-film fibers" or "Split fibers" in order to differentiate them from those film-to-fiber products which are produced by the controlled film separation techniques.

### C. The Processes Applying Controlled Mechanical Film-to-Fiber Separation

In this group all methods for film-to-fiber processing will be included which apply accurate slitting, cutting, or other defined separation procedures, which result in products having regular network structures,

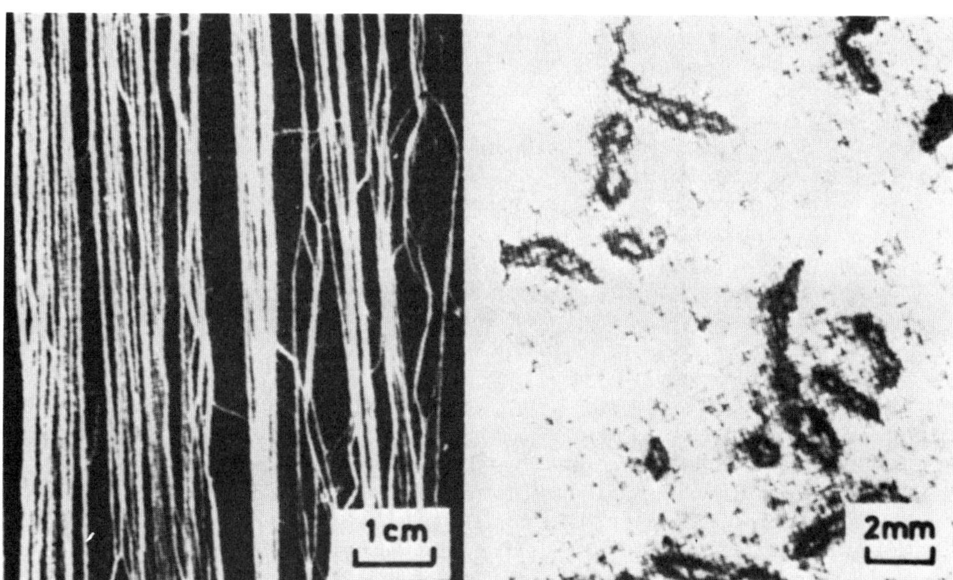


Figure 5 Longitudinal view (left-hand side) and cross sections (right-hand side) of film fibers made by uncontrolled chemomechanical fibrillation. (From Krässig, 1977.)

or preferably, continuously separated filament structure. These processes have the potential of competing with conventionally produced synthetic fiber products in fields of coarse multifilament yarn or of staple fiber applications due to their relatively good uniformity.

According to the basic principle of the film separation technique, one can distinguish the following three methods of controlled mechanical film-to-fiber separation: (1) the film separation by needle rollers, (2) the embossing techniques, and (3) the slicing or cutting techniques.

When a stretched polymer film is passed between, or treated with rotating needle rollers, more or less well defined separation into networklike fibrous structure is achieved. The structure of the resulting product is predetermined by the pattern in which the teeth or the needles are positioned on the rollers, the relative speed of the film and the needle-rollers, and whether the drawing is done before or after the needle roller treatment.

These methods are used by a number of companies throughout the world in their film-to-fiber processes, and a good number of patents were granted for processes based on needle roller film separation, such as to Phillips Petroleum Co. (1967, 1971c), Shell International Res. Maatschappij N.V. (1967b, 1970, 1972d), Plasticisers, Ltd. (1967), Chevron Research Co. (1968, 1970, 1973a, 1973b), Mitsui Petroleum Industries, Ltd. (1968), Rheinstahl Henschel AG (1971b), Scragg, Ernest & Sons, Ltd. (1970, 1971b), Friedrich (1971), J. Mackie & Sons, Ltd. (1971, 1973a, 1973b), Mitsubishi Rayon Co., Ltd. (1971a), Soko Co., Ltd. (1973), Wm. R. Steward & Sons, Ltd. (1971), Burckhardt (1972), Lambeg Industrial Research Assoc. (1971, 1973a, 1973b), McMeekin (1968), Darragh and Johnson (1968), Condit and Johnson (1969), Shell Rev. (1970), Badrian and Choufoer (1971a), Fletcher (1971), and Choufoer (1975).

Figure 6 demonstrates that the treatment of polymer films with teeth or needle rollers normally produces somewhat irregular network patterns. This is particularly the case if the teeth or needles are only punching holes or incisions into the film, and the major part of the separation occurs in the subsequent film drawing. In this case the fiber segments of the network show a pronounced rectangular shape, as shown in the lower part of Figure 6.

In some of the processes, such as the one developed by Chevron Research Co. (1968, 1970, 1973a, 1973b), the needle rollers or cutting devices insert a defined incision pattern into the film which already has undergone a major portion of the drawing. Examples of the texture of fibrous products made with such processes, and of their cross section characteristics, are shown in Figure 7.

In the upper part of Figure 7 the network texture of two Polyloom* carpet yarns made with needle-roller film separation is demonstrated. The two products are only different in the degree of slicing, the right one having a much finer texture. In the lower part of Figure 7, cross

---

*Registered trademark.

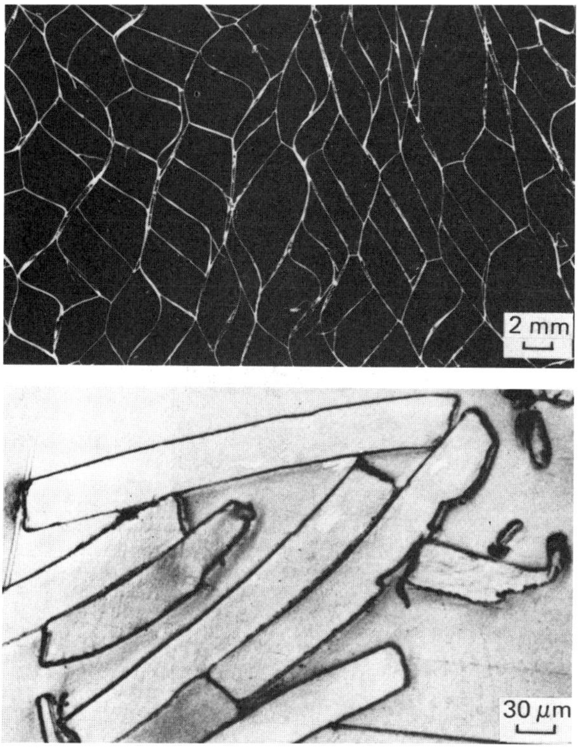


Figure 6 Longitudinal view (above) and cross sections (below) of film fiber products made by needle-roll fibrillation. (From Krässig, 1977.)

sections of these film-yarn products are shown, indicating the rectangular nature of the fibrous sections of these products.

The second method of controlled mechanical film separation uses the introduction of well defined weak areas into the film in its length direction. These weak structures can either be grooves along the length of the undrawn film, introduced during the course of the extrusion by applying dies with one-sided profiled lips [Courtaulds, Ltd. (1968), Hensen (1969, 1974, 1978), Barmag-Barmer Maschinenfabrik AG (1971a, 1972, 1973), Imperial Chemical Industries, Ltd. (1971d), Rasmussen (1971a), *Tufting Needling News Bull.* (1973). Such length grooves or weakening profiles may also be introduced into the film after the extrusion using heated profiled rollers, such as in processes described and performed by Shell International Res. Maatschappij N.V. (1969a, 1971b, 1972a, 1972b, 1972c, 1973a, 1973b), Plasticisers, Ltd. (1971), Scragg, Ernest & Sons, Ltd. (1971a, 1972, 1973a, 1973b, 1974), Smith & Nephew

Figure 7 Longitudinal view (above) and cross sections (below) of split film fiber yarns (Polyloom) made by a special needle roller technique. (From Krässig, 1977.)

Polyfabrik, Ltd. (1971), Smith & Nephew Research, Ltd. (1972), Hercules, Inc. (1971a, 1971b, 1972, 1973), Isvoranu (1972), Seymour and Dow (1968), Dow et al. (1972), Nichols (1971, 1972), Campbell and Skoroszewski (1972), Velluire (1972), Tweedale (1974), Aspin (1975), Foster (1975), and Freedman (1975).

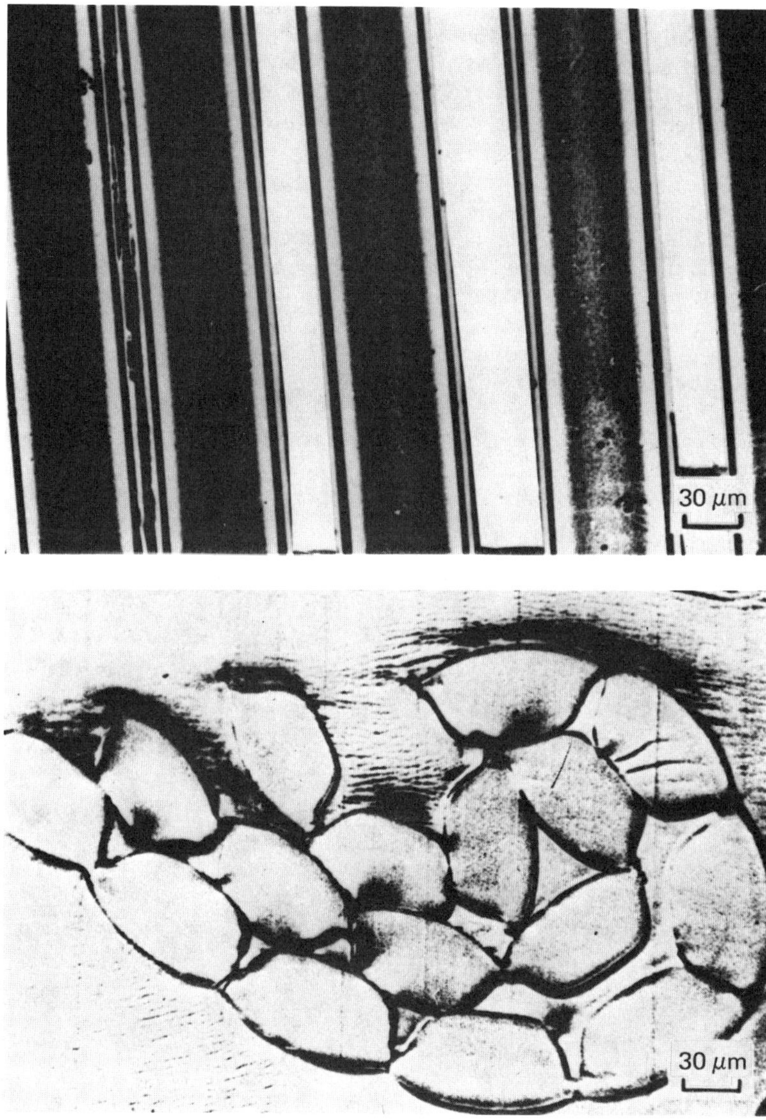


Figure 8 Longitudinal view (above) and cross sections (below) of film fibers made by a profiled die process and subsequent mechanical fibrillation. (From Krässig, 1977.)

In subsequent stretching the profiled film separates along the grooves or according to the profile pattern into more or less completely separated filaments, or into defined networks as predetermined by the embossed profile pattern.

The lower part of Figure 8 shows the split-filament cross sections predetermined by the length-grooved structure of a film extruded by using profiled dies, as suggested by Barmag-Barmer Maschinenfabrik AG for their Barfilex process. In the uniaxial drawing of the profiled film or of profiled tapes there occurs a separation into more or less completely individual filaments which have very uniform dimensions, as shown in the upper part of Figure 8. Twisting of profiled and stretched tapes further enhances the fibrillation.

By introducing more complex profile patterns into the film, network-like structures can be obtained during subsequent mono- or biaxial stretching. Such products may be used to replace lightweight non-wovens or spunbonded products. An example is given in Figure 9.

The third method of controlled film-to-fiber separation is based on the use of finely divided knife combinations or very fine sawtoothlike cutting or splicing tools. These separation tools are in a fixed position and immovable. An exception is the rotation of circular-knife combinations in order to establish continuously new cutting positions. These techniques can be used for cutting or slicing unstretched or

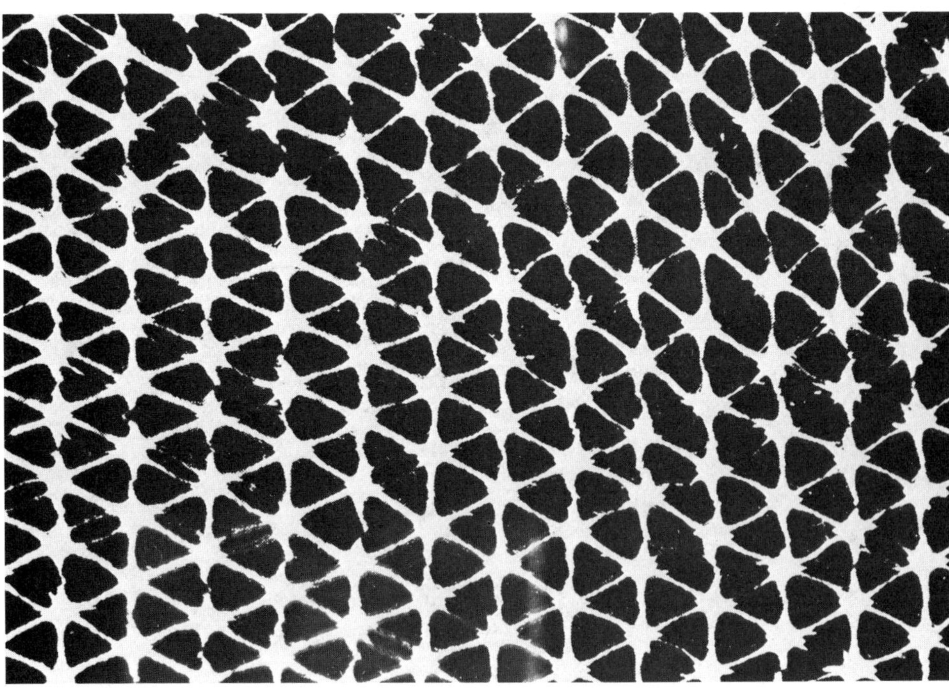

Figure 9 Nonwovenlike film fiber product made by an embossing technique. (From Krässig, 1977.)

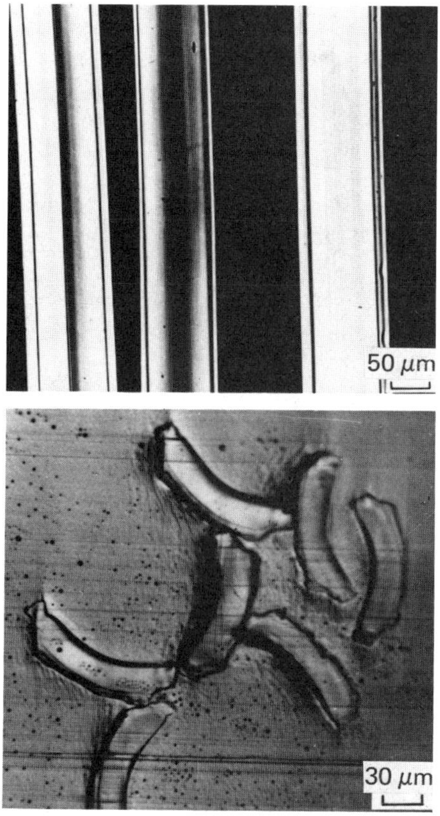


Figure 10 Longitudinal view (above) and cross sections (below) of film fibers made by a controlled cutting technique. (From Krässig, 1977.)

stretched films of almost any polymer, and result in continuously separated filaments of regular rectangular cross sections and uniform fineness down to a lower limit of approximately 6 denier (6.6 dtex).

Such processes have been developed by a number of companies, such as Dow Chemical Co. (1956), Nozawa (1966), Mirsky (1967), Mitsui Petrochemical Ind., Ltd. (1968), Hercules, Inc. (1969), Asahi Kasei Kogyo K.K. (1968, 1972), Scragg & Sons, Ltd. (1969), Shell Oil Co. (1970b), Chemiefaser Lenzing AG (1970, 1972c), Harms et al. (1971a, 1971b, 1973a, 1973b), Krässig (1972, 1975a, 1975b), Farbwerke Hoechst AG (1970), Mitsubishi Rayon Co. (1971b), and Chori Co., Ltd. (1972a, 1972b).

The texture of fibers made with such processes based on defined cutting or slitting is demonstrated in Figure 10 by the example of

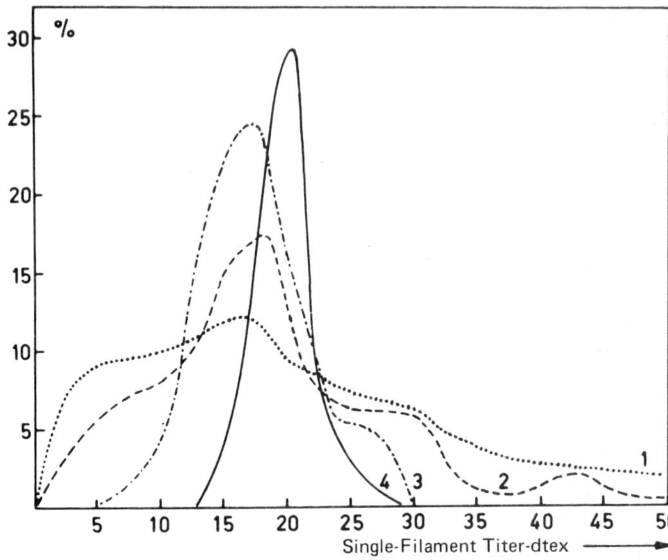


Figure 11 Distribution of the single-filament titer for film fiber yarns made by different film-to-fiber separation techniques compared with the titer distribution of a conventional spinneret-spun multifilament carpet yarn. (1), Twist fibrillation; (2), needle roller splitting; (3), sawtooth edge splitting; (4), spinneret-spun polypropylene yarn. (Modified from Krässig, 1977.)

fibers made with a process developed by Chemiefaser Lenzing AG (Austria). The upper part of Figure 10 shows the uniform length texture and the completeness of the filament separation achieved with this type

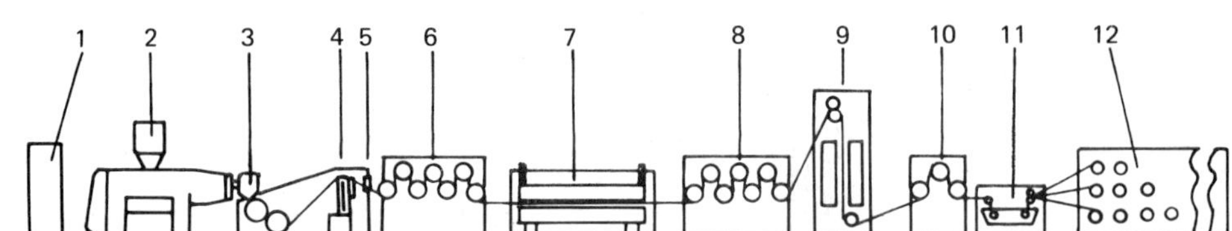


Figure 12 Scheme of production line for texturized split-fiber yarns made from film. (1), Control cabinet; (2), extruder; (3), flat die; (4), chill roll, tape cutting tool; (5), yarn separation device; (6), draw-off septette; (7), drawing oven; (8), septette; (9), heat relaxation crimp development; (10), trio; (11), applicator for antistatic spin finish; (12), bobbin winder.

of film-to-fiber process. In the lower part of Figure 10 the regular rectangular cross sections of these products are demonstrated.

One of the advantages of the defined film cutting methods is the better titer uniformity of filaments in split-film yarns and of the individual fibers in split-film fiber products. This becomes evident if products of this type are compared with film-to-fiber systems made by uncontrolled mechanical or chemomechanical fibrillation, or by the needle-roll and profiling techniques. This is illustrated in Figure 11. The simplicity of these latter processes for film-to-fiber manufacturing is illustrated by the flow sheet of a production line shown in Figure 12.

Whenever cutting or slicing of polymer film is performed using a profiled film technique, or cutting and slicing techniques with needle rollers, knife assemblies, or other separating tools, the resulting products approach in longitudinal regularity and in uniformity of cross sections that of spinneret-spun continuous filaments. This is in contrast to the products manufactured with uncontrolled, random film fibrillation techniques and one should speak here of "slit-film fibers" or "slit fibers" (in German: *Spaltfasern*).

## IV. POLYMERS FOR FILM-TO-FIBER PROCESSING

In principle, all film- or fiber-forming polymers are suited for the manufacture of film tapes or of split and slit fibers. However, as mentioned before, the manufacture of such products is only economically justifiable when the film processing is simple and inexpensive. Therefore, film-to-fiber processing by melt extrusion of suitable polymers is the most frequently applied technique. This narrows down the range of polymers applicable for this type of fibrous goods manufacturing.

In the first place, only thermoplastic polymers which are thermally stable at their processing temperature are suited for the film-to-fiber processing technique. Since the resulting products, due to the characteristics of the film-to-fiber techniques are at least presently less uniform in fineness and limited to the upper range of titers, they are until now only suited for applications with somewhat lower quality requirements. Hence, the price of the polymeric raw material is basically a decisive influence. Furthermore, as indicated earlier, only polymers with weak intermolecular forces which produce film at high stretch ratios, with a pronounced anisotropy, comply favorably with film fibrillation, film slitting, or film cutting techniques. They put up less resistance against mechanical fibrillation and exert less strain on the slitting or cutting devices during film-to-tape or film-to-fiber separation. For these reasons, the polyolefins, especially high-density polyethylene (HDPE) and isotactic polypropylene (PP), have gained growing importance in film-to-tape and film-to-fiber processing.

In addition, polyesters [polyethylene terephthalate (PET)] and polyamides (–6,6 and –6) have found a place in film-to-fiber processing.

However, their use is limited to special cases, such as for the utilization of inexpensive polymer waste or as additives to achieve certain effects in processing or in product properties (see Chapter 2).

Some basic information on high-density polyethylene and isotactic polypropylene will be given with respect to their manufacture, the possibilities of the suppliers for modifications, the various polymer types offered, and their application potential in film-to-tape and film-to-fiber processing. However, in the framework of this book we cannot attempt any sort of complete and exhaustive listing. Readers interested in more data on these polymers are referred to the following sources: Kresser (1968), Hagen and Domininghaus (1961), Raff and Doak (1964/65), Brandrup and Immergut (1966), Galanti and Mantell (1965), Raff et al. (1967), Jezl and Honeycutt (1969), Solomon (1967), Brydson (1970), Roff and Scott (1971), Lenz and Stein (1973), Briston and Katan (1974), and Gropper et al. (1980).

### A. Polyethylene

There are two general types of polyethylene: (1) the low-density polyethylene (LDPE), and (2) the high-density polyethylene (HDPE).

The development of low-density polyethylene, which is mainly used for film processing (approx. 75%), for injection molding (approx. 10%), and for coating applications (approx. 5%), but is rarely used in film-to-fiber or film-to-tape processing, started in 1936 with the pioneer work of Imperial Chemical Industries, Ltd. (ICI) in England on the high-pressure (2000 to 2500 bar) polymerization of ethylene (see Imperial Chemical Industries, Ltd. 1936). Figure 13 gives a flow sheet of a high-pressure polymerization unit for low-density polyethylene.

Due to the high pressure and the elevated temperatures (up to 300°C), the commercial realization of the ICI process met with numerous technical difficulties which had to be solved to make this process economically feasible. The reaction mechanism is that of a radical initiated process. The initiation is caused by oxygen and occurs in a complex and still not completely clarified manner. The polymerization is strongly exothermic, freeing approximately 3500 KJ/kg of polyethylene formed. Today's production units use either autoclaves or tubular reactors.

The low-density polyethylenes resulting from high-pressure polymerization processes have densities between 0.915 to 0.935, relatively high molecular weights, broad molecular weight distributions, are highly branched, show only moderate crystallinities, and melt in the range between 105 to 120°C (compared with 130 – 140°C for HDPE). The lower crystallinity and melting temperature for low-density polyethylenes is caused by short branches which interfere with the formation of lateral order. They are mainly ethyl or butyl groups. The content of such short side chains lies in the range of one branch per 25 to 30 ethylene base units. Generally there are more branches the higher the molecular

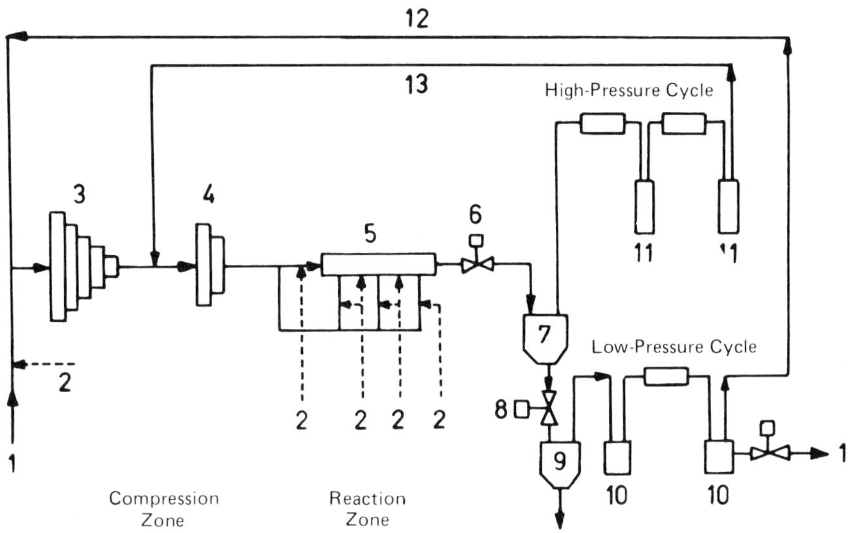


Figure 13 Scheme of the LDPE process: (1), reaction gas; (2), catalyst injection; (3), precompressor; (4), after-compressor; (5), reactor; (6), pressure control valve; (7), high-pressure separator; (8), product valve; (9), low-pressure separator; (10), low-pressure segregation; (11), high-pressure segregation; (12), low-pressure gas recycling; (13), high-pressure gas recycling; (14), flue gas. (From Gropper et al., 1980.)

weight. The end groups in low-density polyethylene are mainly unsaturated and possess vinyl or vinylidene structure. Infrared spectroscopy is used for the determination of branching (intensity of the 1375 $cm^{-1}$ band) and of end groups (extinction coefficients of the 888, 908, and 964 $cm^{-1}$ bands). Examples of typical infrared spectra are given in Figure 14.

Ethylene can also be copolymerized at high pressures with a number of more polar polymerizable unsatured compounds, such as acrylic or methacrylic acid and their esters, and with vinyl acetate to yield polymers of modified properties and extended application potential, especially suited as melt adhesives, coating materials, etc. Recently, some polyethylene manufacturers such as Mitsui in Japan and Du Pont and Union Carbide in the United states have been producing low-density copolymers of ethylene with one-butene or 4-methylpentene, respectively, with low-pressure processes.

Presently, there are approximately 10 mullion metric tons of low-density polyethylene being produced per year. There are more than 100 production plants for low-density polyethylene operating throughout

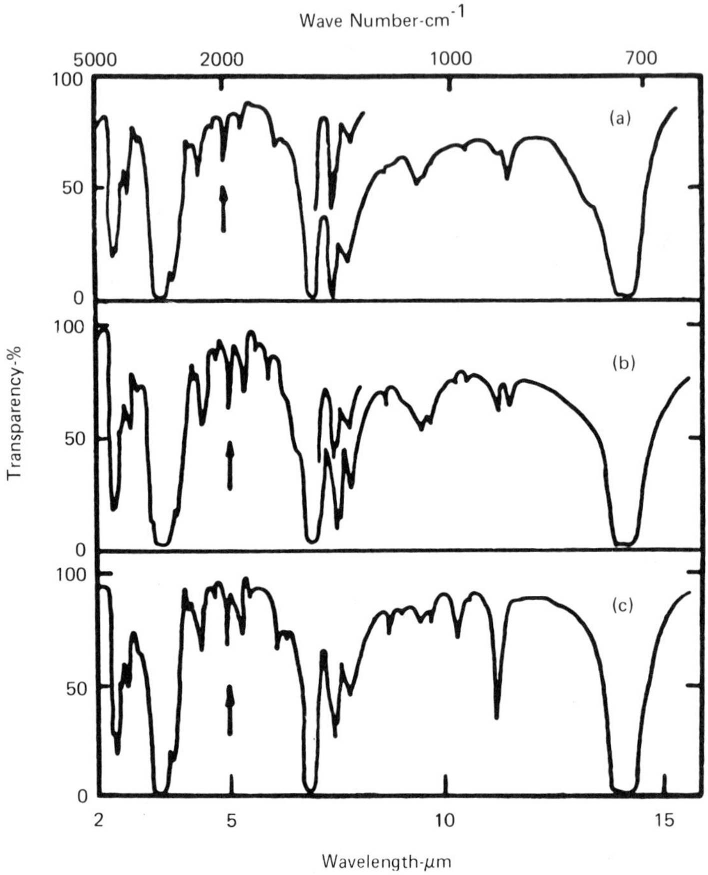


**Figure 14** Infrared spectra of polyethylenes made by different manufacturing processes: (a), high-pressure process; (b), Ziegler low-pressure process; (c), Phillips low-pressure process. (From Hagen and Domininghaus, 1961.)

the world. The production capacities (based on Figures for 1979) in European Economic Community countries, Austria and Switzerland, is about 5.2 million metric tons, in the United States about 3.9 million metric tons, and in Japan about 1.5 million metric tons per year. Table 1 lists the major low-density polyethylene producers in the major European countries, in the United States, and in Japan.

Much more important for film tape and split- or slit-fiber products are the high-density polyethylenes. In 1953 Ziegler et al. (1955) at the Max-Planck-Institut für Kohleforschung in Mühlheim/Ruhr found

**Table 1** Major Low-Density Polyethylene (LDPE) Producers in Western Europe, North America, and Japan and their Respective Nameplate Capacities (in 1000 metric tons/year)

| Country/company | Location | Nameplate capacity | Trade Name |
|---|---|---|---|
| Western Europe (1981) | | | |
| *Austria* | | | |
| Danubia (BASF/Linz/ÖMV) | Schwechat | 200 | Daplen |
| *Belgium* | | | |
| BASF | Antwerp | 40 | Lupolen |
| Essochem | Antwerp | 500 | Esso-PE |
| BP Chemicals | Antwerp | 150 | Rigitex |
| *Finland* | | | |
| Pekema Oy | Uusimaa | 165 | |
| *France* | | | |
| ATO Chimie | Balan | 120 | Lacqtene |
| | Mont | 80 | |
| | Gonfreville | 150 | |
| CDF Chimie | Carling | 205 | Lotrene |
| | Lillebonne | 160 | |
| Cochime (BASF/Shell) | Berre | 100 | Lupolen |
| Copenor (CDF/OPGC) | Dunkerque | 170 | |
| ICI | Fos | 100 | Alkathene |
| *West Germany (FRG)* | | | |
| BASF AG | Ludwigshafen | 18 | Lupolen |
| Erdoel Chemie (Bayer/BP) | Dormagen | 300 | Baylon |
| Rheinische Olefinwerke (BASF/Shell) | Wesseling | 690 | Lupolen |
| Ruhrchemie (Hoechst) | Oberhausen | 160 | Hostalen LD |
| *Italy* | | | |
| ANIG Fibre SPA | Gela | 100 | Eraclene |
| | Ragusa | 125 | |
| Montedison SPA | Brindisi | 180 | Fertene |
| | Ferrara | 90 | |
| | Priolo | 160 | |
| Rumianca SPA | Cagliari | 40 | Rumitan |
| SIR | Porto Torres | 120 | Sirtene |

Table 1 (Continued)

| Country/company | Location | Nameplate capacity | Trade name |
|---|---|---|---|
| *Netherlands* | | | |
| Dow Chemical Co. | Terneuzen | 200 | Dow-LDPE |
| DSM | Beek | 430 | Stamylan |
| ICI | Rozenburg | 150 | Alkathene |
| *Norway* | | | |
| Norolefin (Saga/ Norsk Hydro) | Bamble | 110 | |
| *Spain* | | | |
| Alcudia (ICI/Enpetrol) | Puertollano | 150 | Alkathene |
| | Tarragona | 100 | |
| Dow Chemical Co. | Tarragona | 125 | Dow-LDPE |
| UERT | Tarragona | 140 | |
| *Sweden* | | | |
| Unifos Kemi (Union Carbide/Kema Nord) | Stenungsund | 140 | Unifos-PE |
| *United Kingdom* | | | |
| ICI | Wilton | 160 | Alkathene |
| Monsanto | Fawley | 55 | Monsanto-PE |
| Shell Chemical UK | Carrington | 160 | Carlona |
| BP Chemicals | Grangemouth | 100 | Rigitex |
| Total nameplate capacities (West Europe) 1981 year end (1000 metric tons/year) | | 6143 | |
| North America (1981) | | | |
| *United States* | | | |
| Arco | Port Arthur, TX | 180 | Dylan |
| Chemplex | Clinton, IA | 185 | Chemplex |
| Cities Service | Lake Charles, LA | 240 | |
| Dow Chemical Co. | Freeport, TX | 300 | Dow-LDPE |
| | Plaquemine, LA | 165 | |
| du Pont | Orange, TX | 250 | Sclair |
| | Victoria, TX | 110 | |
| El Paso (Rexene Polym.) | Bayport, TX | 70 | Rexene |
| | Odessa, TX | 180 | |

Table 1 (Continued)

| Country/company | Location | Nameplate capacity | Trade name |
|---|---|---|---|
| Exxon Chemicals | Baton Rouge, TX | | Dexon |
| Gulf Oil | Cedar Bayou, TX | 260 | Poly-Eth |
| | Orange, TX | 130 | |
| Mobil Chemical | Beaumont, TX | 135 | |
| Northern Petrochemical | Joliet, IL | 250 | Norchem |
| Texas Eastman | Longview, TX | 165 | Epolen |
| Union Carbide | Seadrift, TX | 335 | |
| | Penuelas, PR | 135 | |
| U.S. Industrial Chemicals | Tuscola, IL | 80 | |
| | Houston, TX | 250 | |
| *Canada* | | | |
| Canadian Industries | Edmonto, Alb. | 75 | |
| Dow Chemical Co. | Sarnia, Ont. | 70 | Dow-LDPE |
| du Pont | Sarnia, Ont. | 235 | Sclair |
| Union Carbide | Montreal, Que. | 140 | Zendel |
| | Sarnia | 90 | |
| Total nameplate capacities (North America) 1981 year end (1000 metric tons/year) | | 4030 | |
| Japan (1981) | | | |
| Asahi-Dow | Kawasaki | 35 | |
| | Mizushima | 80 | |
| Mitsubishi Chemical | Mizushima | 110 | Novatec |
| Mitsubishi Petrochemical | Kashima | 50 | Beauron/Yikalon |
| | Yokkaichi | 200 | |
| Mitsui Polychemicals | Chiba | 105 | Mirason/Evaflex |
| | Ohtake | 70 | |
| Nippon Petrochemical | Kawasaki | 95 | Rexlon |
| Nippon Unicar | Kawasaki | 185 | Uniclene |
| Showa Denka | Oita | 125 | Sho-Allomer |
| Sumitomo Chemical | Chiba | 200 | Sumikathene |
| | Niihama | 80 | |
| Toyo Soda Manufacturing | Shin-Nanyo | 85 | Nipolon |
| | Yokkaichi | 80 | |
| Ube Industries | Chiba | 145 | |
| Total nameplate capacities (Japan) 1981 year end (1000 metric tons/year) | | 1645 | |

that diethyl aluminum chloride, in combination with titanium tetrachloride, initiated ethylene polymerization at room temperature and atmospheric pressure. The resulting polyethylene was of a stiffer nature than the high-pressure polymerized products. Natta and his co-workers (1955) and others showed that, in general, combinations of compounds from transition elements of group IV and V of the periodical system with organometal compounds of group II and III were able to polymerize nonpolar monomers with double bonds in end position.

Further studies revealed that the stiffer nature of high-density polyethylene resulting from low-temperature polymerization and its other characteristic properties were due to a much lower degree of branching (only approximately one side branch per 1000 ethylene base units), allowing a remarkably better crystalline arrangement of the linear polyethylene molecules. In 1954, patents for this novel class of catalysts for olefin polymerization were applied for by Ziegler and by Natta and later assigned to Montecatini Societa Generale (1955, 1959).

About the time of Ziegler and Natta's findings, two additional low-temperature and low-pressure polymerization methods for ethylene were found by Phillips Petroleum Company (Clark et al., 1956) and by Standard Oil Company of Indiana (Peters et al., 1957). The Phillips process uses chromium oxide as a catalyst supported on silica alumina. The Standard Oil process utilizes an alumina support-impregnated molybdena ($MoO_3$) or vanadium oxide for the initiation. These processes also yield unbranched polyethylene of high density, improved crystallinity, and markedly elevated melting temperature. In Figure 15 the property differences of low- and high-density polyethylenes obtained by the various polymerization processes are demonstrated. The polymerization of ethylene by the Ziegler-Natta, the Phillips, and the Standard Oil processes can be performed either in solution, in suspension, or in the gaseous phase.

Solution polymerization processes are used, for example, by Amoco, Dow Chemical, du Pont, Phillips Petroleum, and Dutch State Mines (DSM). Figure 16 gives a flow sheet of the DSM-solution polymerization process (*Hydrocarbon Process.*, 1974) which uses $TiCl_4/AlR_2Cl/R\text{-}MgX$ as catalyst, hexane as solvent, and 130°C as reaction temperature. The short reaction time of only about 10 min makes the use of relatively small reactors (5 $m^3$ vol for 5 tons/hr) possible.

Suspension processes are practiced by Chemische Werke Hüls, Hoechst, Mitsubishi, Montedison, Phillips, Solvay, and others. A flow sheet of the suspension polymerization process of Hoechst AG is given in Figure 17. In this process Ti/support + $AlR_3$ is applied as catalyst, hexane as suspension medium, and 80–90°C as reaction temperature. The polymerization occurs at pressures between 8 and 10 bar (Kreuter and Diedrich, 1974; Diedrich, 1974; *Hydrocarbon Process.*, 1979).

Polymerization processes in the gaseous phase were developed in the late 1960s and first applied in 1968 by Union Carbide (Union Carbide

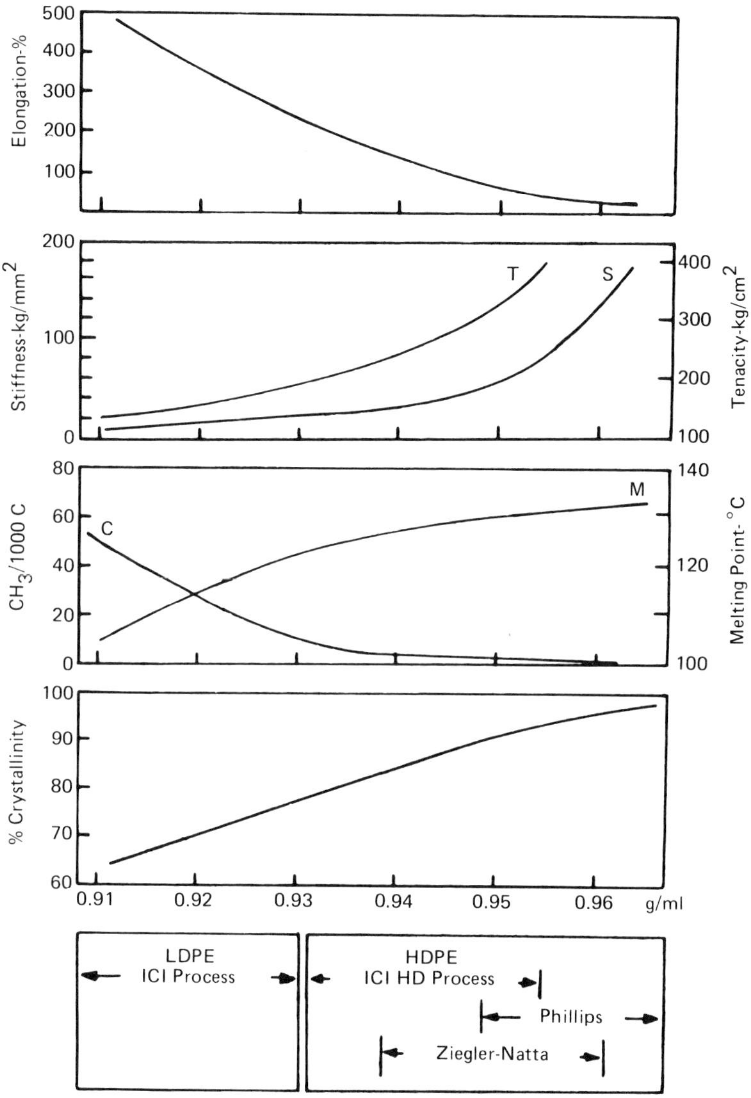


**Figure 15** Interrelations of the density of polyethylenes, their process of manufacturer, and their chemical and physical properties.

Company, 1970) and later by Naphthachimie (Naphthachimie, 1974, 1975). Figure 18 illustrates schematically the Union Carbide process. Here triphenyl-silyl-chromate on silica support saturated with polyisobutyl-alumina oxide is used as catalyst. The polymerization occurs at about

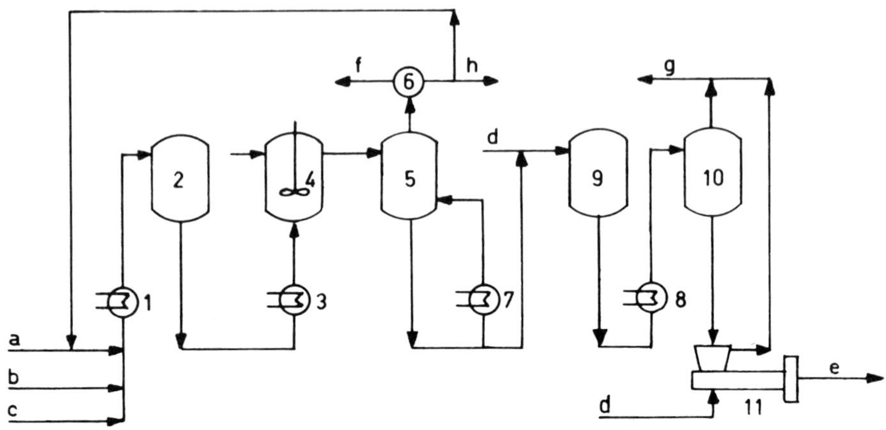


Figure 16 Scheme of the DSM solution polymerization process.

| | |
|---|---|
| 1 = cooler | a = ethylene |
| 2 = absorber | b = comonomer |
| 3 = deep freezer | c = solvent |
| 4 = reactor | d = additives |
| 5 = flash tank | e = granulated HDPE |
| 6 = condenser | f = solvent recycling |
| 7 = heater | g = solvent purification |
| 8 = heater | h = flue gas |
| 9 = mixing tank | (From Gropper et al., 1980.) |
| 10 = flash tank | |
| 11 = extruder | |

20 bar pressure and at 85–100°C. The growing demand for high-density polyethylene has called for a broad spectrum of polymers differing in properties suited for specific end uses. This can be achieved mainly by changing the crystallinity or crystallization tendency, the molecular weight, and the molecular weight distribution. These characteristics can be altered by choosing appropriate polymerization conditions.

The crystallinity, for which the density is an indirect measure, can be influenced mainly by copolymerization with α-olefins, such as 1-butene or hexene, leading to side branches interfering with crystal structure formation. By this modification, polymers with densities in the range of 0.93 g/cm$^3$ have been obtained. Novel developments of polyethylene types, such as the linear, low-density polyethylenes (LLDPE), showing densities similar to LDPE grades, are produced by copolymerization with $C_3$ to $C_6$ olefins. The nature of the comonomer used influences the effect on density of the resulting polymer markedly (Hogan and

Witt, 1979). Figure 19 illustrates this graphically for a number of examples.

The molecular weight can generally be influenced by the polymerization temperature. In the case of polymerization with Ziegler-Natta catalysts, the molecular weight of the resulting polyethylene can be controlled by the addition of hydrogen, causing chain termination by radical recombination. Also, changing the Al/Ti ratio in Ziegler-Natta

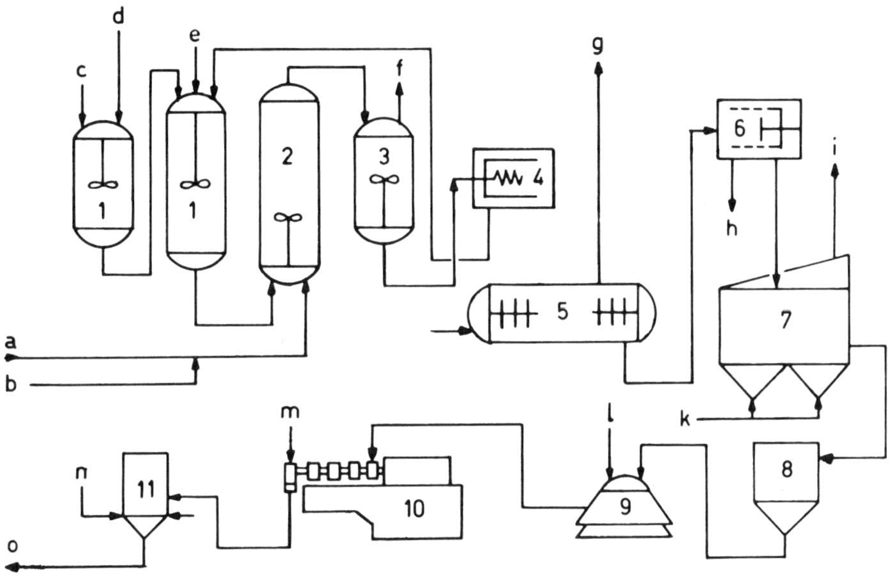


Figure 17 Scheme of the Hoechst suspension polymerization process.

1 = mixing tanks
2 = reactor
3 = expansion tank
4 = centrifuge
5 = stripper
6 = centrifuge
7 = dryer
8 = polymer silo
9 = mixer
10 = extruder
11 = dryer

a = ethylene
b = comonomer $H_2$
c = titanium compound
d = aluminum organic compound
e = hexane
f = flue gas
g = steam/hexane to purification
h = to sewage
i = exhaust air
k = hot air
l = additives
m = water
n = hot air
o = granulated HDPE

(From Gropper et al., 1980.)

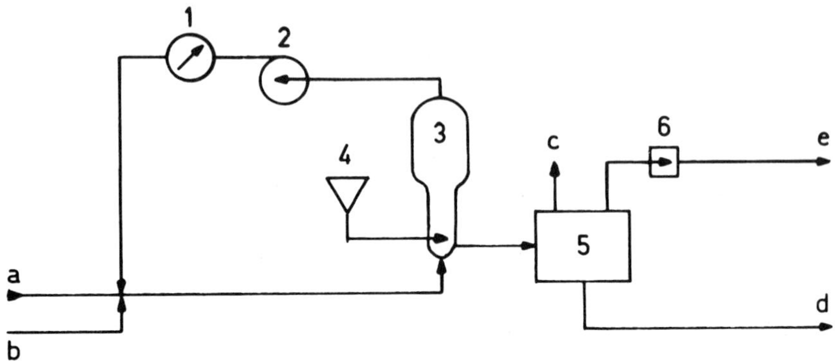


Figure 18 Scheme of the gas-phase "Unipol" polymerization process (UCC).

1 = cooler for the ethylene recycling
2 = compressor
3 = reactor
4 = catalyst feed
5 = polymer discharge system
6 = compressor

a = ethylene
b = comonomer + hydrogen
c = flue gas
d = HDPE powder
e = ethylene recovery
(From Gropper et al., 1980.)

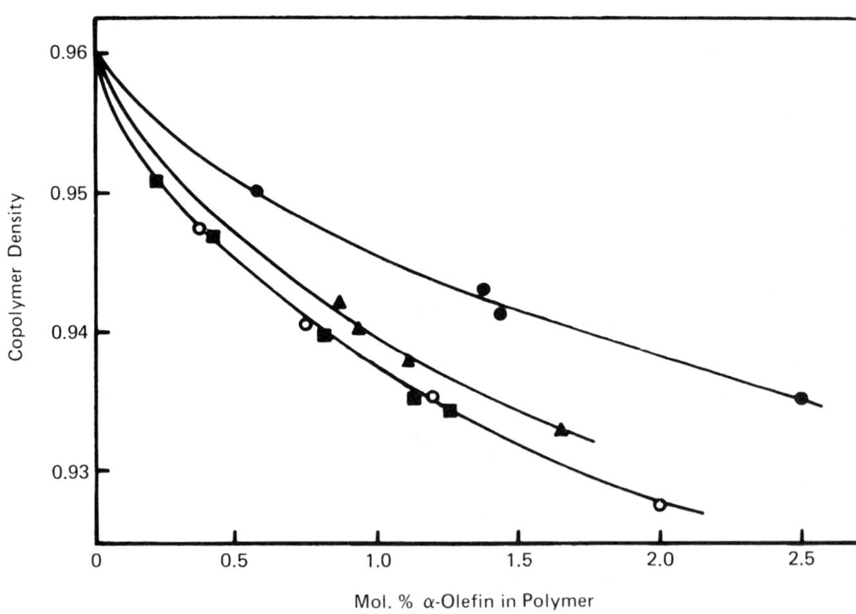


Figure 19 Dependence of HDPE density on the nature of the comonomer applied. ●, Propylene; ▲, 1-butene; ○, 1-pentene; ■, 1-octene. (From Gropper et al., 1980.)

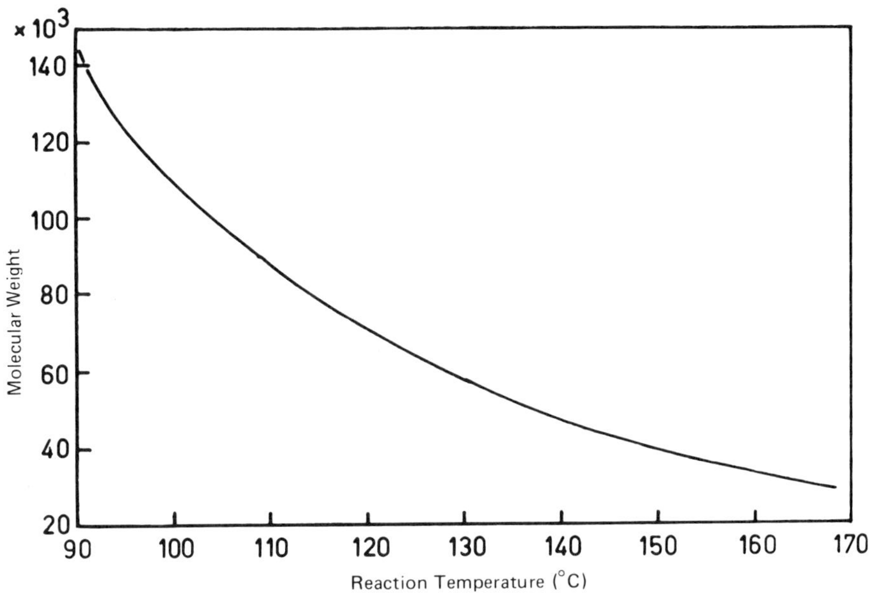


Figure 20 Dependence of HDPE molecular weight on polymerization temperature. (From Gropper et al., 1980.)

catalysts can affect the molecular weight. An increased Al/Ti ratio leads to higher molecular weights of the resulting polyethylene. Furthermore, certain compounds, such as diethyl zinc, titanium tetrachloride, or low molecular weight alcohols can be used in Ziegler-Natta polymerizations to regulate the molecular weight of the polymer. Figure 20 demonstrates the effect of the polymerization temperature on molecular weight in ethylene polymerization by the Phillips process (Clark et al., 1976).

Generally ethylene polymers with relatively broad molecular weight distribution are obtained by the use of metal-organic catalysts. The specific type of catalyst is of strong influence on the molecular weight distribution. The use of catalysts containing esters of titanic acid instead of titanium chloride leads to polymers of remarkably narrower molecular weight distributions. The most powerful means to alter molecular weight distribution is the polymerization temperature. Polymers resulting from polymerizations at lower temperatures always show broader molecular weight distributions.

Today's manufacturers of high-density polyethylene offer a rather wide selection of grades differing in density, molecular weight [or melt-flow index (MFI); inversely related to the molecular weight], and molecular weight distribution. Table 2 lists the major uses of high-density polyethylene in relation to the melt-flow index of the most suitable grades and to the applied processing method.

Table 2 Processing Methods for High-Density Polyethylene (HDPE), Suitable Polymer Melt-Flow Indices (MFI), and Typical Resulting Products

| MFI 190°/ 5 kg 10 min | Processing method | Typical applications |
|---|---|---|
| 0.05–0.15 | Compression molding extrusion | Profile, preformed blocks |
| 0.1–1.3 | Extrusion | Pipes, round bars |
| 0.1–0.4 | Blown film extrusion | Films |
| 0.4–0.7 | Extrusion blow molding | Fuel oil tanks |
| 1.3–3 | Extrusion blow molding | Hollow bodies (i.e., bottles |
| 3–13 | Extrusion blow molding Injection molding | Toys, household articles, screw caps |
| 13–25 | Injection molding | Beer cases |
| 25 | Injection molding | Mass-produced articles for household uses, non-deposit goods |

For extrusion processing, high-density polyethylene grades of medium-to-high molecular weights (low MFI between 0.1–1.5 mg/10 min at 190°C/5 kg) are being used. The processing temperatures are usually between 180 and 240°C; in certain cases, such as fiber spinning, extrusion temperatures up to 300°C are sometimes applied. Extrusion methods are used in the manufacture of pipes, rods, and other profiled articles. and in the manufacture of fibers, monofils, film, and film tapes. Blow extrusion techniques are used in the processing of bottles, gasoline containers, automobile tanks, oil storage tanks, etc.

For injection molding, polyethylene grades of low molecular weight (high MFI between 13–30 g/10 min at 190°C/15 kg) and relatively narrow molecular weight distribution are generally best suited. Injection molding is applied in the manufacture of beer cases, garbage cans, transport containers, storage bins, etc. The distribution of the use of high-density polyethylene for the manufacture of various products in Western Europe during 1978 and 1979 is listed in Table 3 (*Mod. Plast. Int.*, 1980).

The main applications for high-density polyethylenes are the manufacture of hollow bodies by blow extrusion (approx. 40–45% of all uses), and the processing of injection molded articles (approx. 35–40%). The use for film, film tapes, monofils, and fibers is still substantially lower (approx. 10–12%); however, it is growing rapidly. The present production capacity for high-density polyethylene (*Hydrocarbon Process.*, 1979) was approximately 7 million metric tons worldwide in 1978. From this total 2.6 million metric tons capacity (37% of the capacity) was located in North America, 2.4 million metric tons capacity (34%) was available in Western Europe, 1.0 million metric tons capacity (14%) in Japan, and 0.55 million metric tons yearly capacity (8%) in Eastern Europe. One estimates is that in the next few years the world production capacity will grow to over 10 million metric tons per year. The highest growth rate is to be expected for East European countries and for the less industrialized Asian, African, and South American countries. Their share in the world production capacity for high-density polyethylene will increase from 14% in 1978 to approximately 28–30% in a few years. The major producers of high-density polyethylene in North America, Western Europe, and Japan, together with their respective capacities and the trade names of their products are listed in Table 4.

**Table 3** Distribution of the High-Density Polyethylene (HDPE) Consumption for Various End Use Applications, 1978/79 in Western Europe

| | Consumption | | | |
|---|---|---|---|---|
| | 1978 | | 1979 | |
| Application | 1000 tons | % | 1000 tons | % |
| Hollow bodies (i.e., bottles) | 565 | 43.9 | 622 | 43.7 |
| Films, sheets | 95 | 7.3 | 115 | 8.1 |
| Injection-molded articles | 480 | 37.3 | 513 | 36.0 |
| Filaments, tapes | 39 | 3.0 | 43 | 3.0 |
| Pipes | 76 | 5.9 | 89 | 6.2 |
| Wire and cable sheathing | 10 | 0.8 | 10 | 0.7 |
| Others | 23 | 1.8 | 32 | 2.3 |

**Table 4** Major High-Density Polyethylene (HDPE) Producers in Western Europe, North America, and Japan and Their Respective Nameplate Capacities (in 1000 metric tons/year)

| Country/company | Location | Nameplate capacity | Tradename[a] |
|---|---|---|---|
| Western Europe (1981) | | | |
| *Belgium* | | | |
| Dow Chemical Co. | Tessenderloo | 85 | Dow-PE |
| Polyolefins (Phillips/ Petrofina) | Antwerp | 120 | Manolene |
| *France* | | | |
| Soc. Ind. Polyolefins | Gonfreville | 60 | Manolene |
| Naphthachimie (BP Chimie) | Laver | 80 | Natene |
| CdF Chimie | Lillebonne | 30 | Lotrene HD |
| Solvay | Sarralbe | 125 | Eltex |
| *West Germany (FRG)* | | | |
| Hoechst | Hoechst | 220 | Hostalen G |
| | Knapsack | 100 | |
| | Munchmuenster | 80 | |
| Rheinische Olefinwerke (BASF/Shell) | Wesseling | 215 | Lupolen |
| Ruhrchemie (Hoechst/ UKW) | Oberhausen | 130 | Hostalen G |
| Chemische Werke Hüls | Gelsenkirchen | 140 | Vestolen |
| Wacker | Koeln | 30 | Wacker Polyäthylen |
| *Italy* | | | |
| Montedison | Brindisi | 180 | Moplen |
| Solvay | Rosignano | 90 | Eltex |
| Rumianca | Cagliari | 35 | Rumiten |
| ANIC | Gela | 20 | Eraclene HD |
| *Netherlands* | | | |
| DSM | Beek | 150 | Stamylan |
| *Norway* | | | |
| Norpolefin | Bamble | 60 | |

Table 4 (Continued)

| Country/company | Location | Nameplate capacity | Tradename[a] |
|---|---|---|---|
| *Portugal* | | | |
| EPSI (CNP/CdF) | Sines | 60 | |
| *Spain* | | | |
| Calatrava (Entropol/ Phillips) | Puertollano | 65 | Marlex |
| Dow | Tarragona | 50 | Dow-HDPE |
| TAQSA (Hoechst) | Tarragona | 80 | Hostalen G |
| *Sweden* | | | |
| Unifos (UCC/Kema Nobel) | Stenungsund | 100 | Polysten |
| *United Kingdom* | | | |
| BP Chemicals | Grangemouth | 170 | Rigitex |
| Total nameplate capacities (West. Europe 1981 year end (1000 metric tons/year) | | 2475 | |
| North America (1981) | | | |
| *United States* | | | |
| Allied Chemical | Baton Rouge, LA | 300 | Paxon |
| American Hoechst | Bayport, TX | 100 | Hostalen G |
| Amoco Chemicals | Chocolate Bayou, TX | 160 | |
| Arco Polymers | Port Arthur, TX | 150 | Durethene |
| Chemplex | Clinton, IA | 135 | Chemplex |
| Cities Service | Texas City, TX | 90 | |
| Dow Chemical Co. | Plaquemine, LA | 150 | Dow-HDPE |
| | Freeport, TX | 55 | |
| du Pont | Orange, TX | 100 | Sclair |
| | Victoria, TX | 100 | |
| Gulf Oil | Orange, TX | 190 | Poly-Eth |
| National Petrochemicals (USI) | La Porte, TX | 270 | Petrothene |
| Phillips Petroleum | Pasadena, TX | 460 | Marlex |
| Soltex Polymer | Deer Park, TX | 340 | |
| Union Carbide | Seadrift, TX | 155 | |

Table 4 (Continued)

| Country/company | Location | Nameplate capacity | Tradename[a] |
|---|---|---|---|
| *Canada* | | | |
| Dow Chemical, Canada | Sarnia | 40 | Dow-HDPE |
| du Pont Canada | Sarnia | 125 | Sclair |
| Union Carbide | Sarnia | 90 | |
| Total nameplate capacities (North America) 1981 year end (1000 metric tons/year) | | 3010 | |
| Actual capacity use (85% in 1981) | | 2560 | |
| Consumption (United States + Canada) | | 2130 | |
| Japan (1981) | | | |
| Asahi Chemical | Hizushima | 110 | |
| Chisso Petrochemical | Goi, Chiba Pref. | 45 | |
| Idemitsu Petrochemical | Anegasaki, Chiba Pref. | 65 | |
| Mitsubishi Chemical Ind. | Mizushima | 65 | Novatec |
| Mitsubishi Petrochemcial | Yokkaichi | 35 | Beauron |
| Mitsui Petrochemical | Ichihara, Chiba Pref. | 170 | Hi-Zex |
| | Ohtake | 55 | |
| Nissan Chemical Ind. | Goi, Chiba Pref. | 55 | |
| Nisseki Plactic Chem. | Kawasaki | 100 | |
| Shin-Daikyowa Petrochem. | Yokkaichi | 60 | |
| Showa Yuka | Tsurusaki, Oita Pref. | 120 | Sholex |
| Tonen Sekiyukagaku | Kawasaki | 40 | |
| Total nameplate capacities (Japan) 1981 year end (1000 metric tons/year) | | 920 | |
| Actual capacity use (70% in 1981) | | 635 | |
| Consumption (Japan) | | 525 | |

[a]Most of these tradenames are registered trademarks.

## B. Polypropylene

The fact that propylene yields an oily-to-waxy polymeric compound on treatment with acid catalysts, such as concentrated sulfuric acid or tin tetrachloride, melting around 70°C, has been known for some time. As early as 1869, Berthelot described its formation. The waxy polypropylenes, even having reasonably high molecular weights, had no ability to crystallize into a well ordered solid state and therefore did not generate any technical interest.

The discovery of the metal organic catalysts by Ziegler (1955) and Natta (1955) changed the situation with respect to polypropylene decisively. The use of these catalysts yielded for the first time a propylene polymer of high molecular weight capable of crystallizing into a well ordered solid state of a high degree of crystallinity. The credit for clarification of the reasons for this drastic change in properties goes to Natta. He showed that the coordinative polymerization on the metal organic titanium/aluminum catalysts he and Ziegler had found produce propylene polymers of high chemical and sterical regularity ("tacticity"). He introduced for this polymerization technique the term "stereospecific" polymerization.

In the polypropylene species known until then, the methyl side groups had a random distribution in sequence along the carbon-carbon chain as well as in their sterical placing along the polymer backbone (see Figure 21A). Natta named this configuration "atactic." This random placing of the methyl groups prevents the formation of a well ordered stable crystalline solid structure (see Figure 22A). In contrast, the novel class of catalysts forms polypropylenes of high regularity in methyl group sequence and steric placing along and around the polymer chain. X-ray studies of the crystalline polypropylenes obtained by stereospecific polymerization revealed that two forms exist, namely the "syndiotactic" and the "isotactic" arrangements, both giving rise to a high crystallization tendency.

In the syndiotactic arrangement the methyl groups alternate regularly on both sides of the polymer chain, being positioned more or less in one plain (see Figure 21B). In the isotactic form all the methyl groups are situated on one side and in the same plain along and around the polymer backbone chain (see Figure 21C). The formula in Figure 21 illustrates the differences between the three steric forms of polypropylene.

Besides the steric configuration the heat treatment during processing is of great influence on the crystalline structure of the solidifying polymer. This can best be shown by x-ray diffraction analysis. Upon rapid cooling, isotactic polypropylene will solidify in a less crystalline supercooled smectic (paracrystalline) form (see Figure 22B). If the cooling rate is slower, the polymer crystallizes in the stable monoclinic form (see Figure 22C). The paracrystalline smectic structure can be

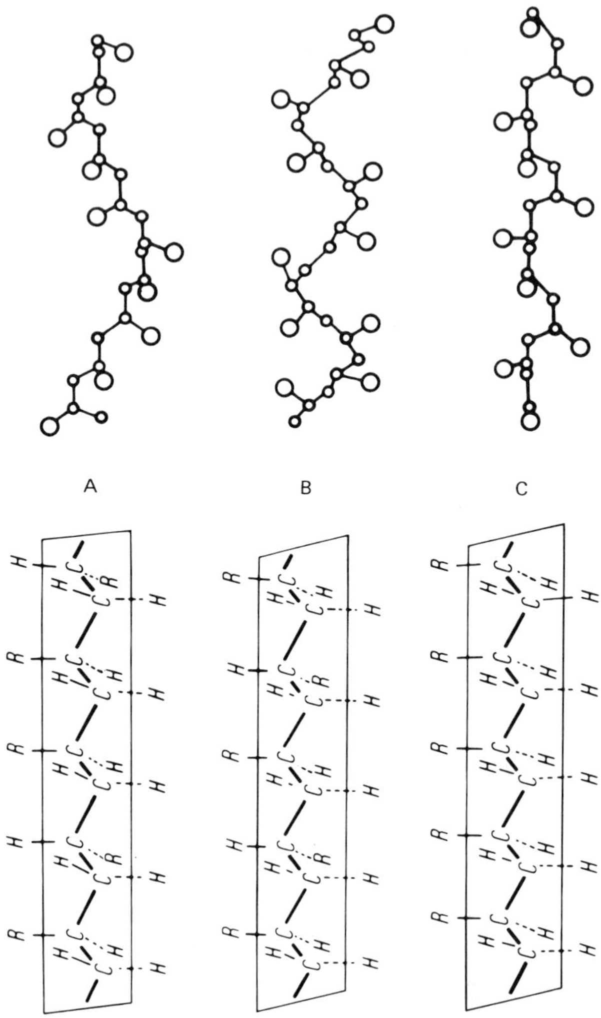


**Figure 21** Molecular configuration of various polypropylenes. A, atactic polypropylene; B, syndiotactic polypropylene; C, isotactic polypropylene (left-handed helix).

transformed into the more stable monoclinic form by annealing through heat. Figure 22 shows the x-ray diffraction patterns of atactic and of the two forms of isotactic polypropylene.

Technically, isotactic polypropylene is produced by polymerization in suspension, in bulk liquid propylene, in the gas phase, or in solution. In all four cases the process involves four processing steps:

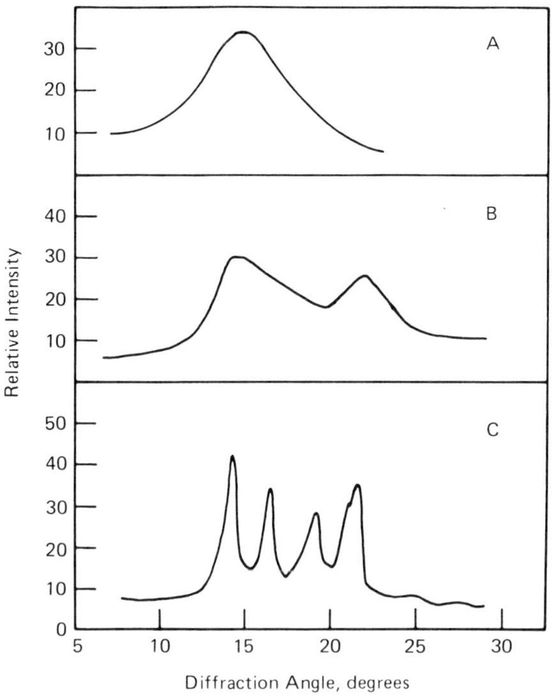


**Figure 22** X-ray diffraction patterns for various polypropylenes. A, Atactic polypropylene; B, isotactic "smectic" polypropylene; C, isotactic crystalline (monoclinic) polypropylene. (From Galanti and Mantell, 1965.)

(1) the polymerization itself, (2) the decomposition of the residual catalyst and its separation from the polymer, (3) the purification during which the diluent and/or residual monomer (in some cases also atactic portions) are removed, and (4) the final granulation.

The actual polymerization is often preceded by the catalyst preparation. The technically used Ziegler-Natta catalyst preparations of the early days (still being used by some manufacturers) contain titanium trichloride of the $\alpha$-, $\gamma$-, and $\delta$-modification as heavy metal component in a 1:1 mole ratio with deithyl aluminum chloride as cocatalyst. The typical layer lattice of these crystal modifications of titanium trichloride is essential for obtaining a high stereospecificity of the resulting polymer. For the stereospecificity of the catalyst system the cocatalyst is also of importance. Chlorine-free aluminum alkyles are much less stereoregulating. The same is true for aluminum alkyles of higher chlorine content. These older type catalysts have an "isotactic index" of 85–90%, whereas the new generation of catalysts have an improved isotactic index between 93–97%. Their improved activity is based on a modification with compounds exhibiting electron donor action. Quite a number of different chemicals can be used as donors, such as olefinic compounds (Farbwerke Hoechst AG, 1971), phosphorus containing compounds (Badische Anilin & Soda Fabrik, 1976), carbonic acids and carbonic acid esters (Toyo Stauffer Chemical Co., 1978), etc. These compounds are complexed favorably with the nonstereospecific sites of the catalyst; the creation of a larger specific surface also enhances the stereospecific activity. Even more recent developments are based on the embedding of diethyl aluminum monochloride into the layers of the crystalline $TiCl_3$ and upon the use of $TiCl_4$ absorbed on water-free magnesium chloride in the presence of benzoic acid ethyl ester as an electron donor (Montecatini Edison SA, 1974; Mitsui Petrochemical Ind., Ltd., 1975). These catalysts produce such high stereospecificities that in technical polypropylene production the separation of the small atactic portions is not necessary.

The process most frequently applied in practice is the suspension polymerization. Figure 23 illustrates schematically the principle of the manufacturing process for isotactic polypropylene by this polymerization technique. The reaction is normally performed continuously in a series of stirred vessels. The temperature is kept between 40–80°C; pressures are between 5–20 bar. The polymer yield reaches 90% or more. In most cases the residual monomer is recovered. The purity requirements for the propylene and for the suspension medium are very high, since a large number of impurities, such as oxygen, water, sulfur compounds, acetylene, dienes, etc., are drastically interfering with catalyst activity. Molecular sieve columns are used for purification. As a suspension medium, saturated alkyls with 6 to 11 carbon atoms are normally used.

Residues of the catalyst are usually removed by the use of alcohols. They transfer the metal halogenides and the alkyl metal halogenides into soluble esters. Elevation of the temperature and the addition of dry hydrochloric acid enhances the ester formation. For the removal of the esterified metal component of the catalyst and of the atactic portions dissolved in the suspension medium, various methods are possible. Hydrolysis of the metal esters and extraction of the oxide hydrates with water, followed by separation of the powdery polymer from the organic suspension conclude the process. Finally, the polypropylene is dried, compounded with stabilizing and other additives, and granulated.

Polymerization in the liquid state (bulk polymerization) allows remarkable increases in the speed of the reaction. However, this technique always requires recovery and recycling of large portions of the monomer since conversions higher than 50% are not possible. The polymerization temperatures applied are similar to those of the suspension polymerization. The reaction pressures are somewhat higher, namely

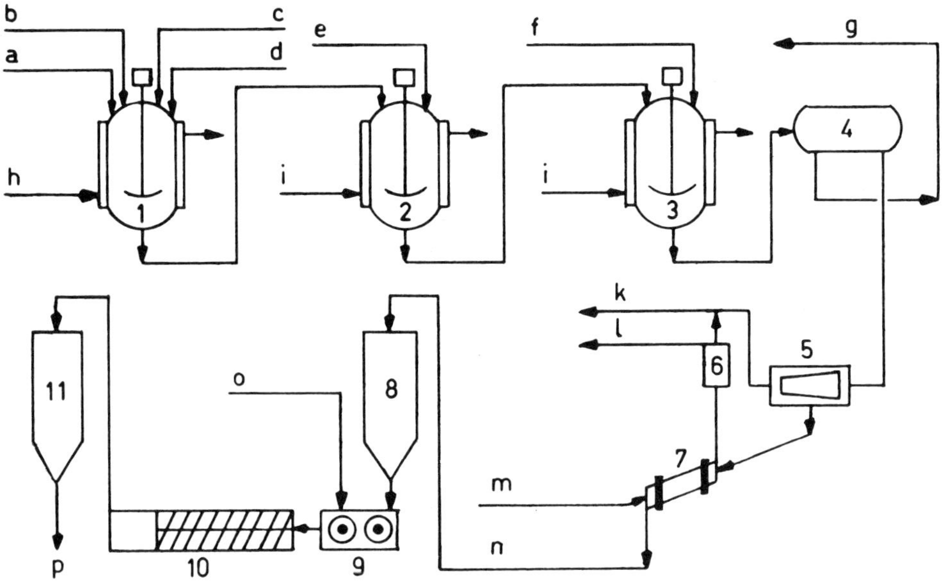


Figure 23 Scheme of the suspension polymerization process for propene.

1 = polymerization reactor
2 = catalyst desactivation
3 = rinsing operation
4 = separator
5 = centrifuge
6 = cooler
7 = dryer
8 = silo for PP powder
9 = mixer
10 = extruder
11 = silo for granulated PP

a = propene
b = diluent
c = hydrogen
d = catalyst system
e = decomposition agent
f = washing agent
g = residual catalyst + washing agent
h = cooling water
i = hot water
k = diluent + dissolved atactic PP
l = nitrogen
m = hot nitrogen
n = PP powder
o = additives
p = granulated PP for shipment

(From Gropper et al., 1980.)

20–30 bar. The finishing of the polymer resulting from bulk polymerization is analogous to the procedures described for the suspension polymerization.

Polymerization in the gaseous phase is for propylene technically quite different from the other processes (Badische Anilin & Soda Fabrik, 1970). The reaction is performed in a fluidized bed in which the catalyst components are introduced together with the gaseous propylene. With the newer type mixed catalysts polymers can be produced in this way at relatively low temperatures, having low residual catalyst contents, and low levels of atactic portions. In the finishing treatment, propylene oxide and steam are used for the catalyst destruction and the removal of bonded chlorine.

In solution polymerization, the monomer and the resulting polymer are dissolved in a diluent medium (Eastman Kodak Co., 1967). Since temperatures above 100°C are necessary to achieve solubility of the formed polypropylene, only temperature-stable catalysts can be used. This limits the choice of suitable catalysts drastically. Due to the high viscosity of the polymer solution the polymerization can only be advanced to approximately 10% by weight of polymer content. At the end of polymerization the solution is degassed from residual monomer. The precipitation of polypropylene is achieved by cooling the solution, and the powdery precipitate is filtered and dried. The atactic portion remains under these conditions in solution.

Besides polypropylene homopolymer grades, the industry also produces copolymers. The production of copolymers of propylene with other polymerizable compounds is in all respects more difficult than the processing of homopolymers. Therefore, the suspension technique is most suitable. Normally copolymerization reduces production capacity of existing facilities and increases production costs due to the necessity of more elaborate polymerization and finishing procedures. Copolymers are produced in praxis both with random distribution of the comonomer in the resulting polypropylene chains or in the form of block copolymers. Random copolymers with ethylene (which is usually used) can only be produced with approximately 4% ethylene content. The use of 1-butene as comonomer allows random incorporation of up to 10% butene. For the production of block copolymers, ethylene is almost exclusively being used. Usually the polyethylene blocks contribute between 5 and 30% to the total polymer. There are techniques to produce block copolymers consisting of two or of multiple block sections in the polymer chains (Heggs, 1973). The polymerization has to be performed stepwise, first producing a block of desired average molecular length from one monomer and thereafter adding the other monomer. Since at the time of the addition of the second monomer not all of the polymer molecules of the first monomer have any more active (living) ends, in most cases the resulting product contains block polymer molecules together with homopolymer molecules of both comonomers in an intimate mixture.

The polypropylenes are generally marketed in granulate form. Due to the tertiary carbon atoms in the molecular chain, polypropylene is rather susceptible to heat and light (UV) degradation. It has to be stabilized by the addition of stabilizers added in amounts of 0.1 to 1.0% of weight, usually shortly before granulation. Good stabilizers against

**Table 5** Melt-Flow Indices of Various Polypropylene Grades and Their End Use Applications

| End use application | Melt-flow index (MFI) 230°/5 kg (g/10 min) |
|---|---|
| Compression moldings, pipes | <2 |
| Extrusion blow moldings | 1-5 |
| Biaxially oriented film | 5-15 |
| Film tapes | 5-15 |
| Monofilaments | 5-15 |
| General injection moldings | 5-20 |
| High-speed injection molding (i.e., disposable bags) | >30 |
| Flat film | 30-40 |
| Staple fibers | 40-80 |
| Spun-bonded fabrics | >60 |

thermal degradation are mixtures of sterically hindered phenols together with sulfides or phosphites acting as antioxidants, through destruction of peroxidic groups. Metal salt desactivators, such as hydrazones or bis-hydrazones of aliphatic or aromatic aldehydes prevent metal catalyzed heat degradation. Stabilizers against photodegradation are alkoxy-hydroxy benzophenones, hydroxyphenol benzotriazols, piperidine derivatives, or hexamethyl phosphorous acid triamides.

The various grades of polypropylene are defined by their melt-flow index (MFI), either determined at 190 or 230°C under 2.16 or 5 kg loading. Standard polypropylene grades offered on the market have melt-flow indices between 0.5 and 100 g/10 min MFI (230/5). Table 5 gives the MFI values of common polypropylene grades in relation to their respective end uses. Isotactic polypropylenes crystallize very well due to their regular structure. Their crystallite melting temperature ranges from 160 to 168°C. Until about 100°C polypropylene has reasonably good long-term heat stability. At more elevated temperature, even when stabilized, service life goes down significantly. Therefore, long residence times at higher temperature, such as the normal processing temperatures between 230 and 300°C, should be avoided in order to prevent degradation.

The first company producing isotactic polypropylene industrially was Montecatini Edison Sp. in Italy starting in 1955. Interest in this new polymeric material grew very fast thereafter and a good number of companies throughout the world acquired licenses from Montecatini. The first producer in the United States was the Hercules Powder Company which erected a 10,000 metric ton per year plant in 1957.

In the meantime, isotactic polypropylene has grown to one of the large volume plastics. Today there are approximately 70–80 producers of polypropylene throughout the world. In Western Europe alone there are more than 20 companies producing this interesting material. In Table 6 the major producers in North America, Western Europe, and Japan are listed together with their respective capacities, their trade names, and their major grades. The worldwide consumption of polypropylene has reached about 5 million metric tons in 1980; the production capacity is estimated to be in the range of nearly 7 million metric tons. Table 7 gives the development in the polypropylene consumption from 1960 to 1980 (estimated) in Western Europe, North America, Japan, and in other parts of the world.

The application of isotactic polypropylene covers a very wide spectrum. It is mainly processed by injection molding and by extrusion. In the large variety of applications, the largest amount of polymer is consumed by extrusion processing to films, film tapes, textile fibers, hollow bodies, and pipes. In Western Europe about 55% of all polypropylene went into such applications in 1978. The distribution over the various end uses of polypropylene in Western Europe during 1978 is given in Table 8.

By far the most important end use for the moment is the use for the manufacture of film, film tape, and film tape-derived products. In Western Europe approximately 450,000 metric tons were used in this field in 1978. From this total, 95,000 metric tons were film applications in the packaging field; 210,000 metric tons were used for film tape and slit- or split-film applications, such as for the manufacture of woven carpet backings, wall coverings, screens, upholstery fabrics, artificial grass, etc. In the same year, 125,000 metric tons went into the manufacture of polypropylene staple fibers and filament yarns for textile applications, especially for face yarns, tufted carpets, and needle-felt floor coverings. In 1978 about 20,000 metric tons of propylene polymer was used in Western Europe for the manufacture of nonwovens, mainly in the form of spun-bonded nonwovens for geotextile application. This end use is a relatively fast growing business.

In the United States in 1978, approximately 350,000 metric tons of polypropylene were used for film, film tape, and film tape-derived products. About 120,000 metric tons went to the manufacture of carpets, mainly for primary carpet backing (around 100,000 metric tons), about 125,000 metric tons were consumed in the manufacture of slit- and

**Table 6** Major Polypropylene Producers in Western Europe, North America, and Japan and Their Respective Nameplate Capacities (in 1000 metric tons)

| Country/company | Location | Nameplate capacity | Tradename[a] |
|---|---|---|---|
| Western Europe (1981) | | | |
| *Austria* | | | |
| Petrochemie Schwechat | Schwechat | 105 | Daplen-P |
| *Belgium* | | | |
| Amoco | Geel | 130 | Amoco PP |
| Hercules | Paal | 130 | Pro-fax |
| Montefina (Montedison/ Petrofina) | Feluy | 100 | Moplen |
| *France* | | | |
| BP Chimie | Lavera | 70 | |
| SNMP (Hoechst) | Lillebonne | 80 | Hostalen P |
| ATO Chimie | Gonfreville | 70 | Lacqtene P |
| Solvay | Sarralbe | 100 | Eltex P |
| Shell | Berre | 70 | Shell PP |
| *West Germany* | | | |
| BASF | Ludwigshafen | 10 | Novolen |
| Hoechst | Kelsterbach | 130 | Hostalen P |
| | Knapsack | 70 | |
| Rheinische Olefinwerke (Shell/BASF) | Wesseling | 75 | Novolen |
| Hüls (Vestolen) | Gelsenkirchen | 80 | Vestolen-P |
| | Marl | 30 | |
| *Italy* | | | |
| Montedison | Ferrara | 75 | Moplen |
| | Brindisi | 150 | |
| | Terni | 120 | |
| ANIC | Gela | 65 | Kastilene |
| SIR | Porto Torres | 60 | Sirtene P |
| *Netherlands* | | | |
| Shell | Pernis | 50 | Shell PP |
| DSM | Beek | 110 | Stamylan P |
| ICI | Rozenburg | 50 | Propathene |

Table 6 (Continued)

| Country/company | Location | Nameplate capacity | Tradename[a] |
|---|---|---|---|
| *Norway* | | | |
| Norpolefin (Norsk Hydro/ (Statoil/Saga) | Bamble | 65 | |
| *Spain* | | | |
| Alcudia (Entropol) | Puertollano | 40 | |
| Taqsa (Hoechst) | Tarragona | 40 | |
| *United Kingdom* | | | |
| ICI | Wilton | 215 | Propathene |
| Shell | Carrington | 100 | Shell PP |
| *Portugal* | | | |
| EPSI (CNP/CdF) | Sines | 95 | |
| Total nameplate capacities (West. Europe 1981 year end (1000 metric tons/year) | | 2485 | |
| North America (1981) | | | |
| *United States* | | | |
| Amoco Chemicals | Chocolate Bayou, TX | 240 | Oleflo |
| Arco Polymers | La Porte, TX | 180 | |
| El Paso Products | Bayport, TX | 70 | |
| | Odessa, TX | 70 | |
| Eastman Chemicals | Longview, TX | 70 | Epolene |
| Exxon Chemical | Baytown, TX | 200 | |
| Gulf Oil | Cedar Bayou, TX | 180 | Poly-pro |
| Hercules | Bayport, TX | 180 | Pro-fax |
| | Lake Charles, LA | 430 | |
| Northern Petrochemical | Morris, ILL | 90 | |
| Novamont (USS Chemicals) | Neal, W. VA | 195 | |
| Phillips Petroleum | Pasadena, TX | 85 | Marlex |
| Shell Chemical | Norco, LA | 135 | Shell PP |
| | Woodbury, NJ | 135 | |
| Soltex Polymer | Deer Park, TX | 90 | |

Table 6 (Continued)

| Country/company | Location | Nameplate capacity | Tradename[a] |
|---|---|---|---|
| *Canada* | | | |
| Hercules | Varennes | 70 | Pro-fax |
| Shell | Sarnia | 70 | Shell PP |
| Total nameplate capacities (North America) 1981 year end (1000 metric tons/year) | | 2490 | |
| Actual capacity use (87% in 1981) | | 2165 | |
| Consumption (United States + Canada) | | 1610 | |
| Japan (1981) | | | |
| Chisso Petrochemical | Goi, Chiba Pref. | 155 | |
| Idemitsu Petrochemical | Chiba Pref. | 75 | |
| Mitsubishi Chemical Ind. | Mizushima | 30 | |
| Mitsubishi Petrochemical | Kashima | 100 | |
| | Yokkaichi | 120 | |
| Mitsui Petrochemical | Chiba Pref. | 120 | |
| Mitsui Toatsu Chemicals | Ohtake | 95 | |
| | Sakai | 60 | |
| Showa Denko | Oita | 75 | Bifan |
| Sumitomo Chemical | Chiba Pref. | 110 | Noblen |
| | Niihama | 25 | |
| Tokuyama Soda | Tokuyama | 80 | |
| Tonen Petrochemical | Kawasaki | 75 | |
| Ube Industries | Sakai | 105 | |
| Total nameplate capacities (Japan) 1981 year end (1000 metric tons/year) | | 1225 | |
| Actual capacity use (78% in 1981) | | 960 | |
| Consumption (Japan) | | 920 | |

[a]Most of these tradenames are registered trademarks.

and split-film fibers or yarns, and 90,000 metric tons for the production of filament yarns and staple fibers. Table 9 gives a survey of the development of the consumption of polypropylene fiber products in the United States since 1965 and lists the capacities which will be available in the near future.

**Table 7** Development of Polypropylene Consumption in North America, Japan, Western Europe, and in Other Parts of the World Between 1960 and 1980 (in 1000 metric tons/year)

| Area | Consumption in 1000 metric tons | | | | | Annual average growth rate 1966-1980 in % |
|---|---|---|---|---|---|---|
| | 1960 | 1966 | 1970 | 1976 | 1980 | |
| North America | 14 | 239 | 445 | 1120 | 1301 | 12.9 |
| Japan | – | 100 | 400 | 580 | 854 | 16.6 |
| Western Europe | 5 | 102 | 300 | 824 | 1450 | 20.9 |
| Other countries | – | 17 | 120 | 525 | 850 | 32.2 |

**Table 8** Distribution of Polypropylene Consumption for Various End Use Applications (in % of Total Polypropylene Consumption in 1978)

| Application area | Federal Republic Germany | Western Europe |
|---|---|---|
| Injection-molded article | 44 | 45 |
| Films | 16 | 10 |
| Film tapes, twines | 19 | 25 |
| Fibers, monofils | 6 | 12 |
| Hollow bodies, bottles | 2 | 1 |
| Pipes, profiles | 7 | 3 |
| Sheets | 4 | 1 |
| Miscellaneous applications | 2 | 3 |

The most important field of application of injection molded articles is in the automobile industry. In Western Europe approximately 75,000 metric tons were applied in 1978 in this particular field. For the United States the application of polypropylene in the automotive industry amounted in 1978 to about 155,000 metric tons. Until 1985 the use of polypropylene in the American car industry is supposed to triple and and reach up to 450,000 to 500,000 metric tons per year. The applications in car manufacturing include battery casings, air filter housings,

**Table 9** United States Textile Polyolefin Production and Capacity in 1000 Metric Tons, 1975 to 1979 and Estimate for 1982

| Product type | 1975 | 1976 | 1977 | 1978 | 1979 | Estimated 1982 |
|---|---|---|---|---|---|---|
| Filament yarn | 44 | 42 | 48 | 51 | 52 | 108 |
| Monofilament | 71 | 75 | 84 | 83 | 95 | 158 |
| Film fiber | 86 | 117 | 120 | 136 | 143 | 205 |
| Staple fiber | 23 | 26 | 35 | 42 | 53 | 125 |
| Total | 224 | 260 | 287 | 312 | 343 | 596 |

air ducts, fan blades, dashboards, gas pedals, bumpers, shock absorbers, etc. Another technical application is injection molded textile spools made from polypropylene. In this field, polypropylene holds a market share of approximately 85% in Europe, leading in 1978 to a consumption of about 15,000 metric tons. Polypropylene is also widely accepted in the household appliance industry.

# 2
# Tape Yarn and Film Fiber Processing

The properties of products made from thermoplastic polymers are dependent on the molecular characteristics and structure of the material used and on the processing conditions under which the product has been made, such as the thermal conditions and the flow characteristics during extrusion, drawing, and heat setting, the extent and the ratio of stretch, etc.

## I. CHOICE OF POLYMER AND POLYMER CHARACTERISTICS

The proper choice of polymer or copolymer type, respectively, or of blends of several thermoplastic polymers or copolymers is essential for the processibility and the desired property combination, as well as for the applicability of the derived product to the intended end use.

Following the selection of a certain polymer material, one has to decide what molecular weight the polymer ought to possess for the specific application. The molecular weight distribution is also of importance in many cases. Furthermore, one has to consider the glass transition temperature and the melting point of the selected polymer with respect to temperature exposure of the end product in actual use. In the following sections, the main polymer types used in film, film tape, and film fiber manufacture will be discussed and compared. In doing so, we begin with polypropylene because of its wide versatility and its great economic importance.

### A. Polypropylene

Polypropylene generally yields film tapes and film fibers of higher tenacity along with lower elongations at break in comparison to

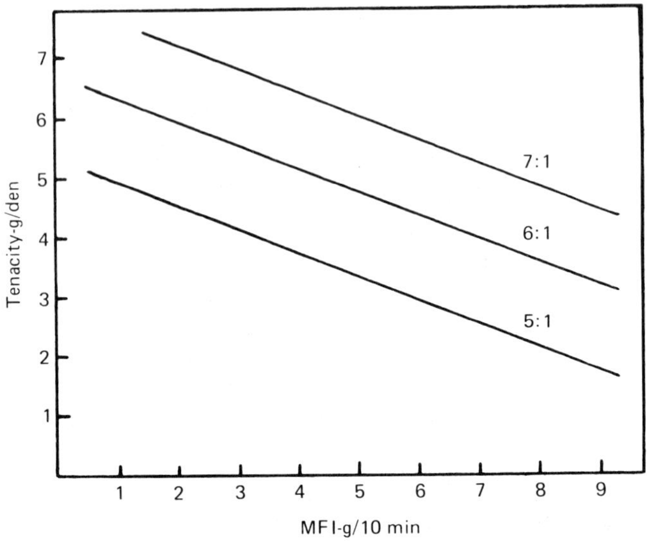


**Figure 24** Relationship between film melt-flow index and film-tape tenacity at different draw ratios. (From Krässig, 1977.)

high-density polyethylene. The main characteristics of polypropylene film-related products are the higher softening point, the greater stiffness, and the markedly higher fibrillation tendency.

The effect of molecular weight on the properties of polypropylene film products is very pronounced and important. In practice the molecular weight is measured by the melt-flow index (MFI); it is inversely proportional to the molecular weight. One obtains the MFI as the amount of polymer melt flowing through a defined standardized dye hole at a temperature of 230°C under a load of 5.0 or 2.16 kg during a test time of 10 min (g/10 min). For polypropylene the rule applies that a lower molecular weight (i.e., a higher MFI) leads to films or film-related products with decreased tenacity. This is illustrated by the experimental results given in Figure 24.

When compared with high-density polyethylene the correlation between melt-flow index and tenacity is less pronounced for polypropylene. Therefore, grades with relatively high melt-flow indices, which are easier to process, can be used without fear of significant losses in strength. The application of somewhat higher draw ratios also helps to balance possible losses. Only in the manufacture of strappings, where maximum strength is required, should polymers with low MFI of 0.3 to 0.5 g/10 min be used. Melt-flow index values of 2 to 3 g/10 min are favorably taken in the manufacture of polypropylene film tapes, and 4 to 6 g/10 min for slit- or split-film fibers.

In Figure 25, taken from Skoroszewski (1971), the general experience in the application of high-density polyethylene and isotactic polypropylene in the manufacture of film tapes and of film fibers is illustrated. A high-MFI polypropylene is preferred in the manufacture of flexible film products with high fibrillation tendency. However, when fibrillation tendency is disadvantageous, as in the case of film tapes for sacks or fabrics for packaging purposes, it is preferable to use a polypropylene grade with lower melt-flow index, since high fibrillation tendency would disturb the weaving performance. Further information about the effect of molecular weight on film product properties will be found in publications by Sheehan and Cole (1964), Sheehan et al., (1964), Sheehan (1965), Gouw and Skoroszewski (1968), and by Evans (1971).

Figure 25 also reflects the much wider range of application accessible to isotactic polypropylene. Due to the lower intermolecular cohesion of polyethylene melts, high-density polyethylene is offered on the market in a much narrower range of melt-flow indices. Also, because

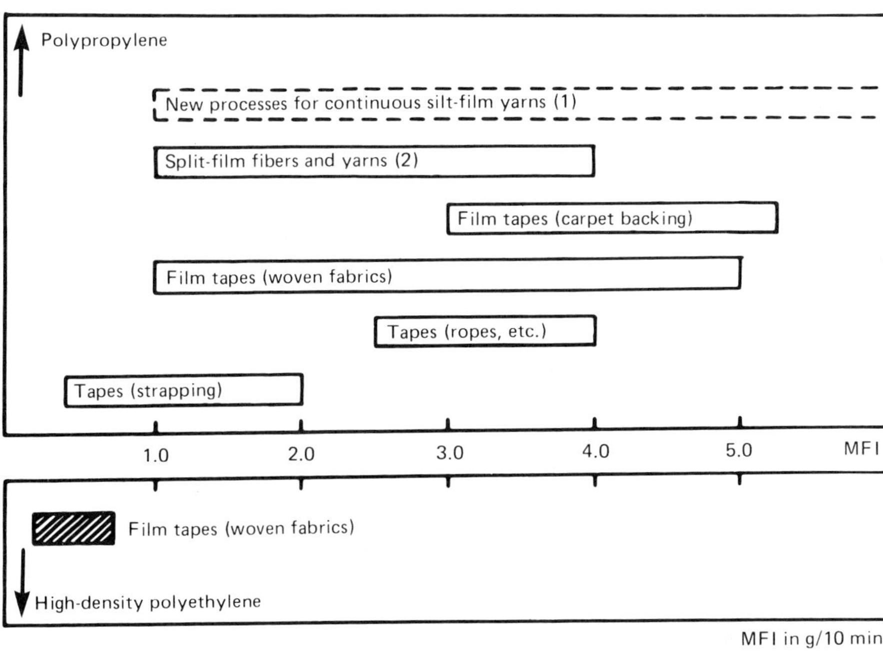


**Figure 25** Melt-flow index (MFI) ranges of polyolefins used in different areas of application: (1) the controlled film-to-fiber separation processes; (2) the uncontrolled mechanical fibrillation processes. (From Skoroszewski, 1971.)

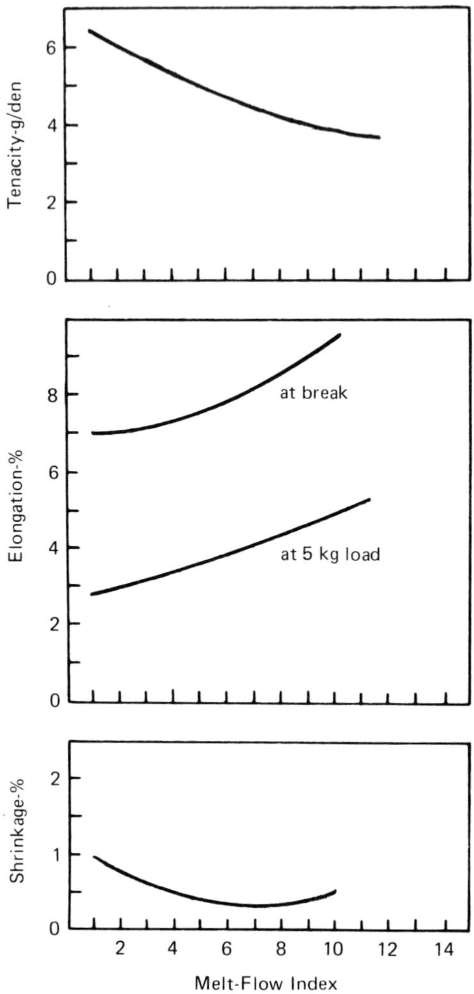


**Figure 26** The effect of melt-flow index (MFI; 230°C; 2.16 kg) on tenacity, elongation, and shrinkage (130°C) of polypropylene film tapes. (Courtesy of ICI.)

of much lower fibrillation tendency of highly stretched high-density polyethylene film products, this material is the better choice whenever the end-use can tolerate the lower softening or melting point and when high strength and low fibrillation tendency are essential. This is the case for many fields of packaging materials (Real, 1969).

Figure 26 shows the effect of MFI (or of molecular weight) on tensile strength, elongation at break, and on thermoshrinkage behavior of polypropylene films. With increasing melt-flow index (i.e., with decreasing molecular weight), tensile strength drops markedly and elongation at break increases simultaneously. The thermoshrinkage is enhanced for films made from low MFI (i.e., high molecular weight material).

Compared with other polyolefins polypropylene has a relatively high crystalline melting point at 160 to 170°C, which benefits many of its applications in the manufacture of films, film fibers, and conventionally spun fibers. A further advantage is its low creep tendency. Compared with high-density polyethylene, polypropylene possesses a markedly higher coefficient of friction. This is favorable when piling sacks made from this material; however, it may pose a problem in weaving operations. Special additives in polypropylene during processing reduces this friction to a certain extent, minimizing the so-called blocking effect.

Besides isotactic polypropylene grades with low fractions of atactic polymer (3 to 4%), there are polypropylene types on the market containing between 5 to 10% of the atactic species. The average molecular weight of the atactic component is identical to that of the main isotactic portion. From such materials film tapes and split- or slit-film fibers with good and specific properties can be obtained. The more amorphous atactic component functions as an internal lubricator, improving properties such as stretchability, ductility, tenacity, impact strength, soft handle, and better compatibility with fillers. The fibrillation tendency of films made from such polypropylene grades lies between that of products made from highly isotactic grades and that of products with low fibrillation tendency made from high-density polyethylene (Moorwessel and Pilz, 1969).

## B. High-Density Polyethylene

There are narrower limits in the melt-flow index range for high-density polyethylene suitable for extrusion processing of film tapes and film-derived products than for isotactic polypropylene. In order to attain good processibility and stretchability for achieving acceptable levels of tensile strength, polymer grades with MFI values lower than 0.5 to 1.0 g/10 min at 190°C and a load of 2.16 kg have to be used. The necessity of using polymer grades with such low melt indices results in an inferior extruder melting capacity compared with the processing of suitable polypropylene grades.

For the production of weaving tapes the melt-flow index of high-density polyethylene should be at 0.4 to 0.5 g/min. If high impact strength is required for the end product, a MFI of 0.1 to 0.2 g/min will be necessary. In spite of such low melt-flow index values (or in other words, such high molecular weights), the attainable tenacities do not amount to those received with isotactic polypropylene. However, the products obtained show higher elongation, greater flexibility, and lower fibrillation tendency (i.e., properties which are desired in packaging applications).

High-density polyethylene grades with melt-flow indices above 1.0 g/10 min are generally not suited for the production of film tapes because of their low melt strength and their insufficient stretchability. Besides the influence of the average molecular weight on properties, the effect of the molecular weight distribution has to be mentioned, especially in the case of high-density polyethylene. Polymer grades with a narrower molecular weight distribution lead to films with increased tensile and impact strength. On the other hand, polymer grades with broader molecular weight distribution are easier to process. In this context we refer to publications by Gouw and Skoroszewski (1968), and by Skoroszweski (1971) showing that polymers with a narrow molecular weight distribution yield film tapes with higher tenacity. This is shown in Figure 27.

The reduced fibrillation tendency and the soft hand of high-density polyethylene film tapes has proven to be advantageous in the production of woven tape fabrics, especially of raschels. Of special importance for many applications is the good resistance of high-density polyethylene film products against UV radiation, giving polyethylene, in many cases, preference over polypropylene. On the other hand, the low crystalline melting range of 127 to 132°C reduces the chances of

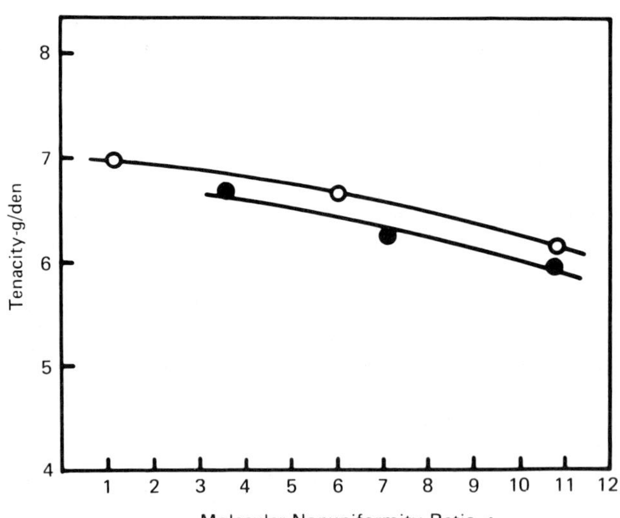


Figure 27 Effect of molecular nonuniformity on film-tape tenacity. (From Krässig, 1977.)

high-density polyethylene to qualify for many end uses. For instance, it cannot be applied for the manufacture of backing fabrics for tufted carpets. Further details on the application of high-density polyethylene film tapes are presented by Real (1969).

## C. Copolymers of Ethylene and Propylene

For special applications copolymers of ethylene with propylene and other olefinic monomers can be of interest. Characteristics of such copolymers are, in general: (1) lower degree of crystallinity, (2) lower density, and (3) lower crystalline melting range. They are easier to process and result in film tapes of reasonable tensile strength, somewhat higher elongation at break, greater flexibility, softer hand, and above all a far lower fibrillation tendency. This illustrates that especially the property profile of polypropylene film products can be altered and improved by the application of copolymers containing other olefinic monomers besides propylene. This may be demonstrated by a copolymer consisting of 96.4% propylene and 3.6% ethylene. Its MFI at 230°C is 1.6 g/10 min and it melts in the range of 147 to 149°C. The fibrillation tendency of a film made from this copolymer is significantly lower than that of a film derived from polypropylene homopolymer, however, not quite as low as that of a film processed from high-denisty polyethylene. Polymer suppliers are often marketing such polymer types as "statistical copolymers." In the recent past these copolymers have achieved increasing technical interest. Further information on the use of copolymers from propylene with higher olefins can be found in articles published by Gouw and Skoroszewski (1968), and by Skoroszewski (1971).

The copolymerization of ethylene with a certain percentage of propylene to a high-density copolymer yields a product which in the melt extrusion of films results in a markedly increased extruder efficiency, as shown in Table 10 taken from a publication by Skoroszweski (1971). The data in this table also show that at the same draw ratio the tensile strength decreases only slightly for film made from the copolymer, whereas the elongation at break is approximately three times higher than for film made from the ethylene homopolymer.

## D. Polymer Blends

The simplest and cheapest way to modify given polymers is without doubt the physical blending of various polymers in the molten stage. For this purpose the mixture of chips or pellets is thermally plasticised in a kneader or extruder and thereafter directly processed into film or film products or intermediately granulated for subsequent film extrusion.

Table 10 Influence of Polymer Type and Film Thickness on Polyethylene Tape Processing

| Polymer type | Film thickness µm | Maximum inlet speed for even stretch m/min | Maximum stretch ratio, tenacity, and elongation at maximum inlet speed | | |
|---|---|---|---|---|---|
| | | | Stretch ratio | Tenacity gf/dtex | Elongation % |
| Homopolymer (melt-flow index: 0.4 g/10 min; density: 0.962 g/ml) | 50 | 20 | 8.5:1 | 5.4 | 14 |
| | 75 | 13 | 10:1 | 5.7 | 12 |
| | 100 | 8 | 11:1 | 5.3 | 10 |
| Copolymer[a] (melt-flow index: 0.5 g/10 min; density: 0.946 g/ml) | 50 | 28 | 7:1 | 4.5 | 26 |
| | 75 | 20 | 8:1 | 4.7 | 37 |
| | 100 | 13 | 9:1 | 5.4 | 33 |

[a]The comonomer is propylene.

### *1. Blends from Polypropylene and High-Density Polyethylene*

Blends from 80 to 90% polypropylene and 10 to 20% high-density polyethylene are often used. Film or film products made from such blends have, in many respects, improved properties over products made from either homopolymer. They show higher elongation at break, have better impact strength, a softer hand, and distinctly better abrasion resistance. The latter property is of particular importance for twisted and fibrillated film tapes or yarns used in the manufacture of woven belts and of cordage (see Skoroszewski, 1971; Nojiri et al., 1967). The high abrasion resistance of split-film fibers made from a polymer blend of 92.5% isotactic polypropylene with 7.5% high-density polyethylene is specifically reported in a patent by Mitsubishi Rayon Co., Ltd. (1976).

The effect of the ratio of components in the polymer blend on splitting resistance is relatively complicated, as shown in Figure 28 (Skoroszewski, 1971). References dealing with the same subject include: Phillips Petroleum Company (1970a), Moorwessel and Pilz (1969), and Fitton and Gray (1971). Thermoshrinkage of films or film products

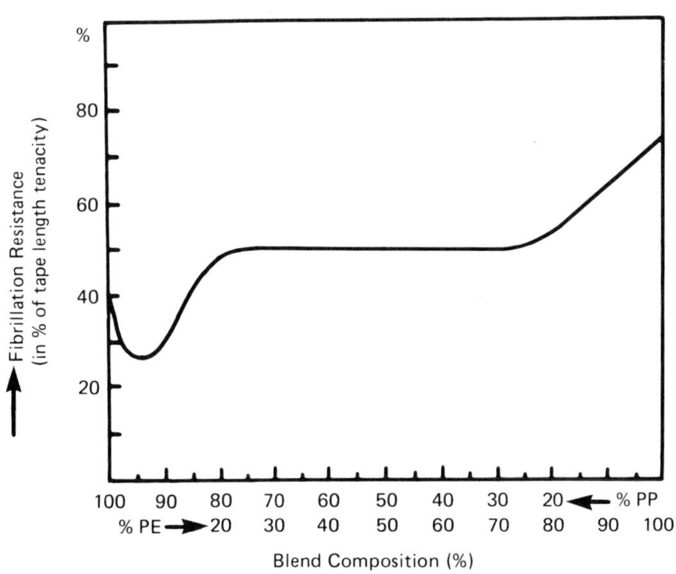


Figure 28 Fibrillation resistance of 1150 dtex weaving tapes produced from blends of polypropylene and high-density polyethylene (stretching temperature 120°C; tenacity 5.0 cN/dtex. From Skoroszewski, 1971).

made from blends of polypropylene with additions of high-density polyethylene up to 35% is not much higher than that of films from propylene homopolymer.

Badische Anilin & Soda Fabrik AG (1974) describes a procedure for the manufacture of film yarns from a polymer blend containing polypropylene, high-density polyethylene, and atactic polypropylene. From this blend films can be obtained which do not show after-splitting in the mechanical fibrillation. However, such film yarns have reduced resistance to mechanical stress and to abrasion. J. P. Bemberg AG (1970) describes for a similar use a blend of 100 parts of a polymer consisting of 80% isotactic and 20% atactic polypropylene with 20 parts of high-density polyethylene. Other blends of this kind are claimed in several patents by Badische Anilin & Soda Fabrik AG (1970a, 1970b, 1970c, 1970d). Further studies on the use of such blends will be found in a publication by Jakob and Michels (1974).

### 2. *Blends from Polypropylene and Low-Density Polyethylene*

In contrast to polypropylene and high-density polyethylene, which are compatible polymers and may be mixed homogeneously over a wide range of blending ratios, most other combinations do not harmonize. A certain degree of compatibility has been found in the combination polypropylene/low-density polyethylene. Even though the thermodynamic compatibility of these two polymers is insufficient to produce homogeneous melts in every ratio of the two components, their structural similarity guarantees good adherence between both polymers even though phase separation occurs on a microscale during solidification.

Very good end use properties show film tapes made from blends of polypropylene with low-density polyethylene in certain mixing ratios. Deanin and Sansone (1978) report that a blend of 77% isotactic polypropylene with 23% low-density polyethylene shows a maximum tensile strength with only slight reduction of the elongation at break. With an increased addition of low-density polyethylene, however, the textile-mechanical properties show a sharp deterioration, as can be seen from Figure 29.

In our own experiments we could establish that blends of 80 to 85% polypropylene with 15 to 20% low-density polyethylene produced films or film tapes with a markedly reduced fibrillation tendency. With more low-density polyethylene in the blend the possibility for fibrillation increases sharply, a tendency which may be caused by phase separation occurring between the two components during solidification. Polovikhina and Zverev (1977) have obtained similar results. According to a patent of Imperial Chemical Industries, Ltd. (1970), film tapes can be manufactured from a blend of 80% polypropylene with 20% low-density polyethylene having a tensile strength of 6.5 cN/dtex (= 7.2 g/denier) and showing no fibrillation tendency.

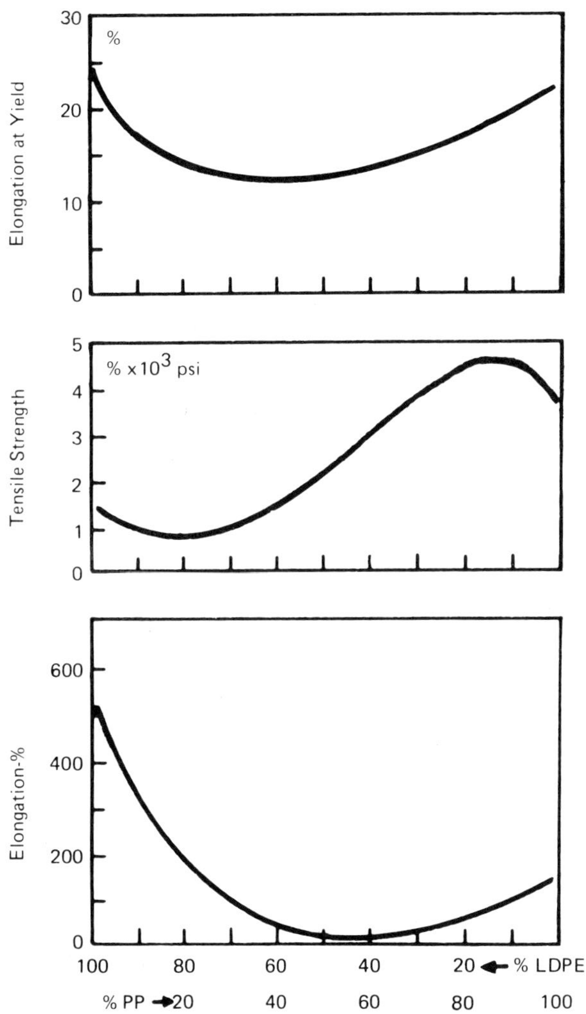


**Figure 29** Effect of polypropylene/low-density polyethylene blend compositions on elongation at yield, tensile strength, and breaking elongation of film tapes. (From Deanin and Samsone, 1978.)

Undoubtedly, low-density polyethylene in blends with polypropylene allows a more effective reduction of the fibrillation tendency than high-density polyethylene, especially when applied in the proper ratio and in good correspondence with the melt index of the blending partner (see Skoroszewski, 1971).

### 3. *Blends of Polypropylene and Polystyrene*

On account of its low price and its good processibility, polystyrene is eligible as a partner in blends for film and film product processing. According to a patent held by Badische Anilin & Soda Fabrik AG (1970a), an addition of 10 to 20% of polystyrene to polypropylene results in a higher fibrillation tendency of film tapes made from it. Skoroszewski (1971), however, states a reduction in fibrillation tendency for a blend of 90% polypropylene with 10% polystyrene. Our experience indicates that the results of Badische Anilin & Soda Fabrik AG are more reliable. By addition of only a small percentage of polystyrene to polypropylene the splitting tendency increases sharply as a result of the insufficient compatibility of the two polymers. It seems that there are existing weak spots on the interfaces formed during solidification which set off fibrillation when under transverse mechanical stress.

The practical significance of this effect lies not only in the area of split-fiber yarns. In film tapes, increased fibrillation tendency is, in some instances, a desired property, particularly in carpet backing fabrics. Here the loss of strength during tufting in warp direction is reduced when film tapes are used as warp material having a somewhat higher fibrillation tendency.

### 4. *Blends of Polypropylene and High-Density Polyethylene with Rubber-Elastic Polymers*

The main disadvantage of film products made from isotactic polypropylene are stiffness, brittleness, low resistance to abrasion, and fibrillation tendency. A very effective method for overcoming stiffness and brittleness and improving flexibility consists in using blends with rubber-elastic copolymers containing between 5 and 30% of the latter.

*Block Copolymers of Butadiene and Styrene.* Blends of polypropylene with block copolymers of butadiene and styrene yield films with excellent flexibility and fibrillation resistance. Unfortunately, it is difficult to mix these two polymer components in the extruder. Moreover, the butadiene-styrene copolymer has a tendency to stick to the screw in the compression zone of the extruder, effecting a breakdown of the conveyance of the melt after some processing time.

Split-film fiber products with a smooth wool-like hand and high fineness are obtained through needle-roller fibrillation at a high fibrillation ratio of monoaxially stretched films made from a blend containing 80% isotactic polypropylene and 20% of butadiene-styrene copolymer (Shell Internationale Research Maatschappij N.V., 1975b, 1978). Asahi Chem. Ind. (1976) confirms the fibrillation reducing effect of butadiene-styrene block copolymers in blends with high-density polyethylene. From a blend of 50% high-density polyethylene and 50% butadiene-styrene, copolymer film tapes with a tenacity of 4.5 cN/dtex (= 5.0 g/denier) have been produced, whereby the fibrillation index was reduced from 22 for tapes made from 100% high-density polyethylene, down to only 3.

*Polyisobutylene.* The use of blends of polypropylene with polyisobutylene has been mentioned by Rusznak and Huszar (1975). They report on the effect of drawing on molecular orientation and on the development of the textile mechanical properties of film products made from such blends.

*Polybutadiene.* Through the addition of 1 to 7% of polybutadiene to high-density polyethylene in the production on film tapes, good resistance against undesired fibrillation can be achieved according to Showa Yuka K.K. (1975a, 1975b). On the other hand, slit-film fibers with very good textile mechanical properties are obtained via film made from polypropylene/high-density polyethylene/polybutadiene blends applying needle-roller fibrillation (Shell Internationale Research Maatschappij N.V., 1977).

Our own experience has led us to find that terpolymers containing ethylene, propylene, and butadiene base units (generally known under the acronym EPDM; so-called thermoplastic elastomers) have strong fibrillation reducing effects and show excellent compatibility with polypropylene and with high-density polyethylene in extrusion processing.

Such blends are also useful in the production of split-film fibers. According to a patent by Shell Internationale Research Maatschappij N.V. (1975), it is possible, under special conditions to produce fibers through needle-roller fibrillation of films made from a blend consisting of 80% polypropylene and 20% polybutadiene with much lower titer (8 dtex vs. 16 dtex, i.e., approx. 7.2 and 14.5 denier, respectively) than achievable with corresponding films made from propylene homopolymer alone. According to a patent of Phillips Petroleum Co. (1970), films with good resistance against fibrillation can be obtained from a blend of 95% isotactic polypropylene and 5% 1-butene ethylene copolymer when processed at a temperature between the melting points of the two components. From these examples it can be recognized how many possibilities are open for the application of blends of polypropylene or polyethylene, respectively, with a wide variety of elastomeric materials.

### 5. *Blends of High-Density Polyethylene and Polypropylene with Polyesters or Polyamides*

According to Skoroszewski (1971) it is possible to obtain twisted film tape yarns showing a 30% increased tensile strength over yarns made from pure propylene homopolymer by applying blends containing a 10% addition of polyamid-12. As illustrated in Figure 30, the effect of increasing strength rises with increased stretch ratio.

Berger and co-workers (1974, 1975, 1976, 1978, 1979) have examined blends of polyolefins with polyamides mainly with respect to the effect of the polyamide addition on the fibrillation tendency of films made from such blends. They observed that the frequency of inhomogeneities is most distinct in the phase inversion range and that the ability to fibrillate is particularly promoted in the mixing ratio relevant for phase inversion (see Berger and Schmack, 1974). "Phase inversion range" is that blending ratio at which two polymers in a blended solid state exchange their role in the "island/matrix" structure. The fibrillation tendency was tested by these authors by milling short-cut film tapes in a laboratory refiner. In the case of high-density polyethylene/polyamide film tapes, best fibrillation was obtained with a mixing ratio of 50:50 of the components.

Aside from the interface energy between polymers of a blend, the ability to fibrillate is determined also by the elastic moduli (Berger and Kammer, 1974). Formulas for the calculation of the fibrillation

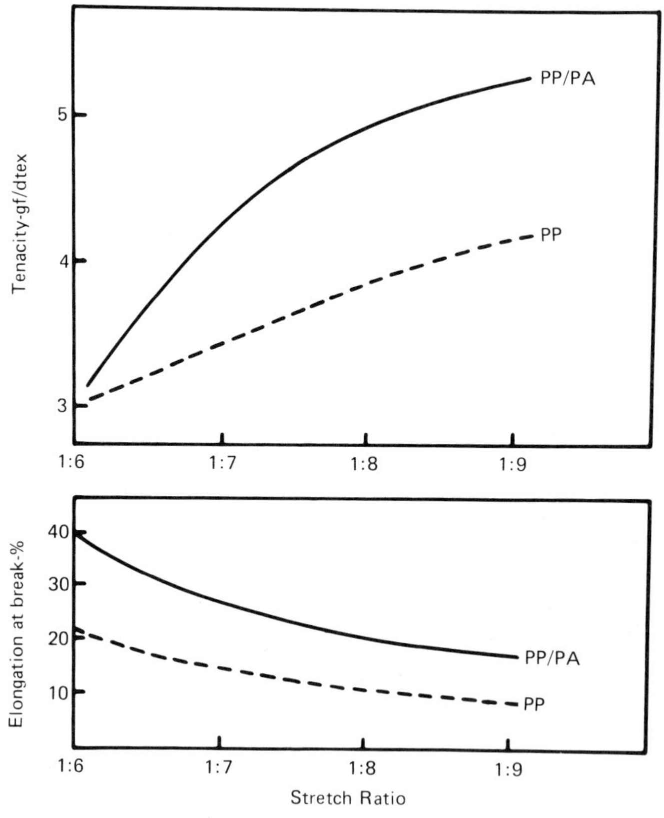


**Figure 30** Effect of blending 10% Nylon-12 with polypropylene on film-tape tenacity and elongation. (From Skoroszewski, 1971.)

**Table 11** Phase Inversion Range of Various Polymer Blends

| Blending system | Phase inversion range |
|---|---|
| Polyester / low-density polyethylene | 50/50–45/55% |
| Polyester / high-density polyethylene | 40/60–35/65% |
| Polyester / polyamide | 60/40–40/60% |
| Polyamide / low-density polyethylene | 45/55–35/65% |
| Polyamide / high-density polyethylene | 40/60–30/70% |
| Polyamide regranulated / high-density polyethylene | 35/65–25/75% |

ability of film from polymer blends were developed by Schmack et al. (1975). In their studies, Berger et al. further found that the ability of film to fibrillate in relation to various polymers increases in the following order: polyamide < polyester < low-density polyethylene < high-density polyethylene < polypropylene. Berger et al. (1976) have also experimentally examined in which blending ratio phase inversion between the coherent and the dispersed phase occurs for various blends. Such blending stages are recognizable microscopically when matrix phases of the components are observed lying side by side, as well as in "island" phases. An excerpt of their results is given in Table 11.

Concerning the splitting resistance of films from polyethylene and polyester, studies exist from Michels (1976). Among other findings, Michels found that the ability of films from polymer blends to fibrillate changes with the ratio of the components, the boundary energies, and the strength anisotropy in machine and transverse direction. Results concerning the flow characteristics of polymer blends were presented by Berger and Michels (1975). Under extrusion conditions polyamide and polyethyleneterephthalate can be considered as viscous; polyethylene and polypropylene as viscoelastic materials. Examination of blends shows that the elasticity of polymer blends is always greater than that of the matrix-forming homopolymer. Moreover, the matrix-forming homopolymer undergoes, through the addition of the dispersed phase, a reduction of the melt viscosity. Information on the measurement of the surface tension of polymers and its significance for the boundary phenomenon is found in a publication by Berger and Kammer (1978). A method for the approximate calculation of the degree of dispersion is described by Michels and Franz (1977). The morphology of the polymer blends has also been examined. The polymer particles of the dispersed phase are present in the extruder in the form of small pellets. They are transformed into cylinders in the die and turned into thread-shaped structures in the stretching process. Kammer (1978) has developed a thermodynamic theory for the surface tension in the boundary layers between immiscible polymers whereby the thickness of the boundary layer proves to be proportional to the average chain extension in the boundary layer.

### E. General Criteria of Polymer Quality

A satisfactory processing of thermoplastic polymers is only guaranteed when certain minimum requirements with respect to quality are fulfilled. In practice, the polymers are applied in the form of granulates whose grains must be formed regularly and must be free of gaseous inclusions. The grains should not show any abnormal deformations which could possibly lead to abrasions. Some of the most essential quality criteria are:

- Uniform melt-flow index which should not change during processing.
- Narrow molecular weight distribution.
- Bulk density of the granulate of at least 0.4 $g/cm^3$.
- Free of polymer dust and of impurities.
- Good brightness and homogeneous fluorescence under UV radiation.
- Sufficient stabilization against heat deterioration to prevent the occurrence of grains and bubbles.
- If pigmented master granulates are used, the pigment particles have to be distributed very finely and homogeneously; there should be no agglomerations with cross sections larger than 5 $\mu m$.

However, experience indicates that to date there has not been a reliable chemical or physical test method found which specifies and guarantees the processibility of a thermoplastic on the basis of test results with absolute certainty.

## II. EXTRUSION AND FILM COOLING

The methods for film extrusion can be classified as follows:

1. Extrusion through a flat die. Here one has to differentiate between cooling and solidification of the polymer melt by either water immersion or chill-roll treatment (see Figure 31).
2. Extrusion through an annular die. Using this technique, three possibilities for cooling and solidifying the extruded polymer exist:

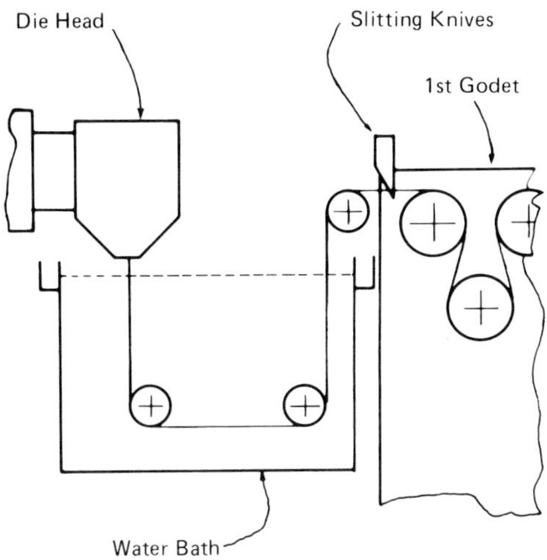


A

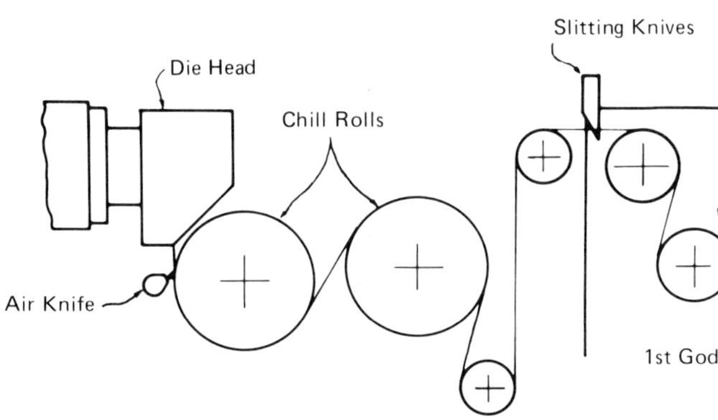


B

**Figure 31** Film production using water bath quenching system (A) and chill-roll casting system (B). (From Krässig, 1977.)

(1) with cooled air, (2) with cooled air combined with a contact cooling ring operated with cold air or cold water (see Figure 32), and (3) thin layer water cooling in case of downward extrusion.

Whereas the plastification and extrusion has only limited significance for the final properties of the resulting film products, the means of solidification is of extreme importance for further processing and for the ultimate characteristics of the end products.

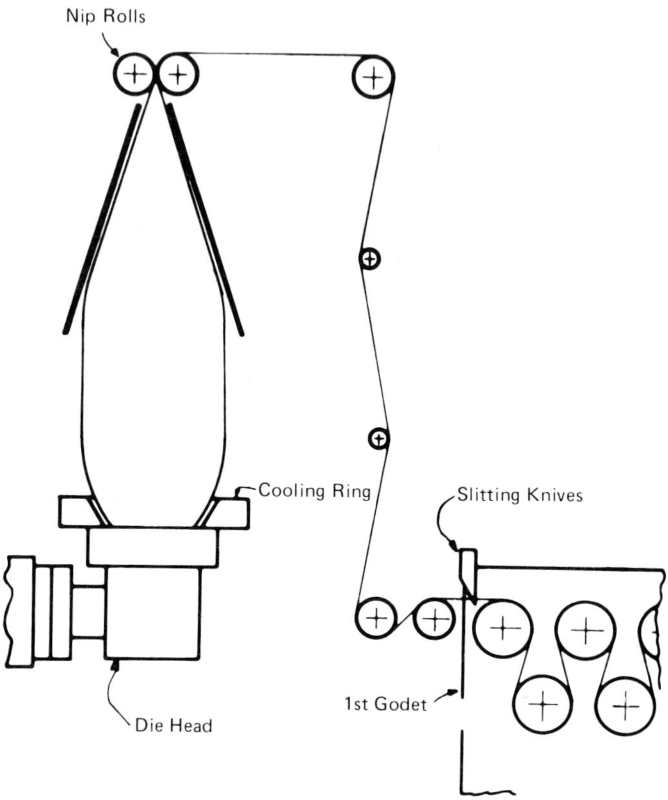


Figure 32 Film production using blowing technique. (From Krässig, 1977.)

## A. The Extrusion

The melt extrusion of thermoplastics to film requires the choice of an extruder which has a suitable screw geometry and an adjusted temperature profile of its heating zones. While high-density polyethylene is chemically relatively insensitive against heat exposure, polypropylene has a great tendency to decompose during melting and at more elevated temperature. For this reason all commercially available polypropylene grades contain thermostabilizers. However, it is advisable to keep the processing temperature in the metering zone as low as possible in order to minimize losses in the mechanical properties of the end product due to thermal decomposition of the polymer. As a reason for the decrease of the mechanical properties due to extrusion processing, molecular degradation is considered the main cause. The decline of the molecular weight is demonstrated by a rise in the melt-flow index. This increase becomes more pronounced at more elevated extrusion temperature. Examples of the increase of the melt-flow index in correlation with the extrusion temperature for various isotactic polypropylene grades stabilized to a different degree are given in Figure 33.

The decrease of molecular weight is evident; also, its effect on the further processing steps, especially the drawing operation. Figure 34, presenting results of Evans (1971), shows that an undrawn film reaches a higher tensile strength at a 5:1 stretch ratio when extruded at lower temperature. Figure 34 shows further that the yield zone, represented by the horizontal lines *n* in the graph, becomes shorter if the extrusion temperature is lower. This means that at lower extrusion temperature (i.e., at higher melt viscosity) the material responds earlier to the drawing force, resulting in earlier development of molecular orientation. In the extrusion at the lower temperature when the polymer retains a higher molecular weight.

The preorientation of the molecules occurs partly when the melt passes under high shearing forces through the die slot (preorientation) and partly thereafter under the influence of the speed difference between the die exit speed of the melt and the speed of the

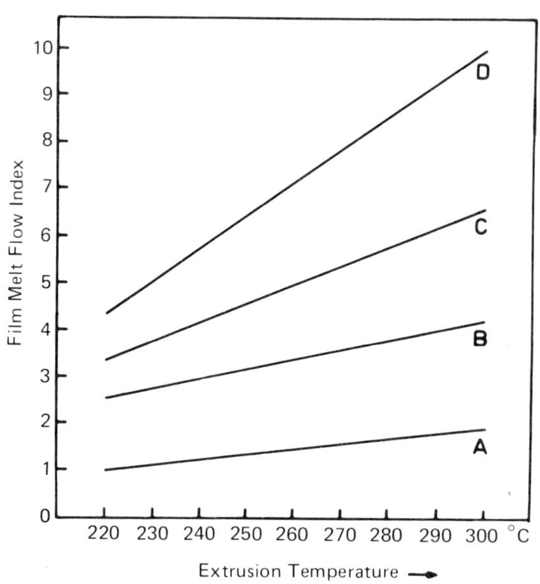


Figure 33 Effect of extrusion temperature on the film melt-flow index for polypropylene of different melt stability. (From Evans, 1971.)

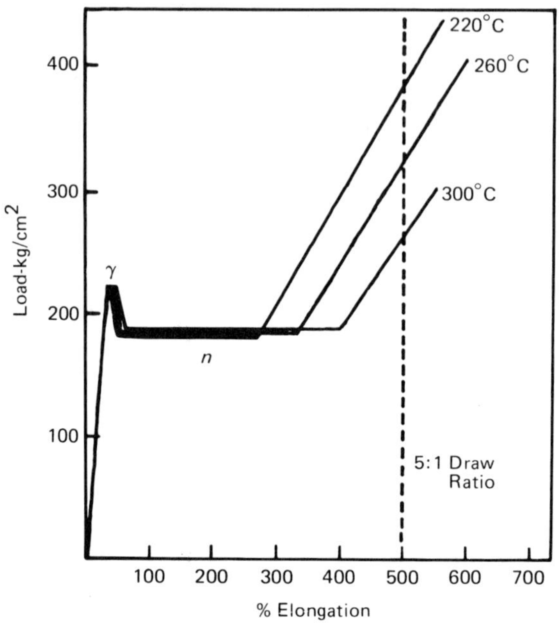


Figure 34 Effect of extrusion temperature on the tensile strength of undrawn film and its response to drawing (γ = yield point; *n* = necking zone). (From Evans, 1971.)

drawoff of the film from the die when the melt is not yet completely solidified. This effect is generally known as "melt orientation."

For the achievement of high film tenacity the extrusion conditions must therefore be adjusted under two aspects: (1) low temperatures in the compression and metering zones to avoid molecular degradation, and (2) high temperature in the die slot to reduce preorientation of the film through melt orientation. The reason for the second condition is the prevention of a high degree of melt orientation in combination with high crystallinity. It prevents the achievement of maximum orientation in the stretching operation. The alignment of the molecules or the crystallites is best achieved in the stretching operation, especially when the crystallites and the degree of orientation are not yet too pronounced.

Figure 35 shows that the highest tenacities are obtained with low melt-flow index (i.e., high molecular weight) and under low extrusion temperatures. Stretching difficulties which are caused by high melt orientation can be overcome by high temperatures at the die slot, and also by high stretching temperatures. By such measures the tendency to rupture during the stretching operation can often be overcome unless it is caused by inhomogeneities due to solid impurities.

An important influence on the molecular orientation during extrusion is the speed with which the primary film is being cast and drawn off the die slot. Figure 36 (see Evans, 1971) shows that the maximum achievable stretch ratio becomes higher if the extrusion speed of the undrawn film is reduced. Since the stretch ratio also determines the maximum attainable tensile properties, low extrusion speed is advantageous for good quality. This fact, however, is contradictory to the efforts of most producers to operate extrusion lines at maximum capacity. The draw ratio necessary to achieve optimum strength can be reduced through reduction of the width of the die slot. However, one has to take into account that a small orifice opening increases the degree of melt orientation at the higher velocity. Nevertheless, one may suppress melt orientation in the die slot by higher melt temperature and by slower film cooling thereafter.

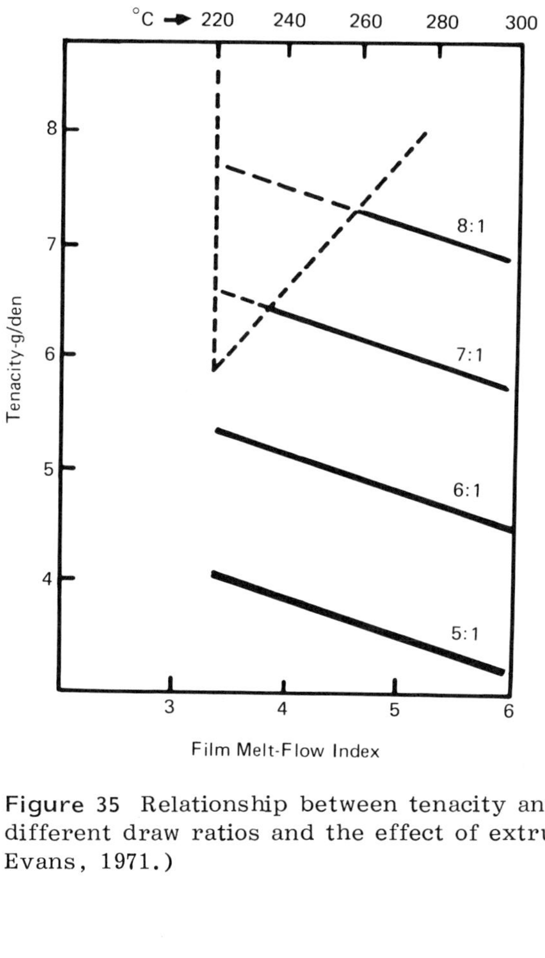


**Figure 35** Relationship between tenacity and film melt-flow index at different draw ratios and the effect of extrusion temperature. (From Evans, 1971.)

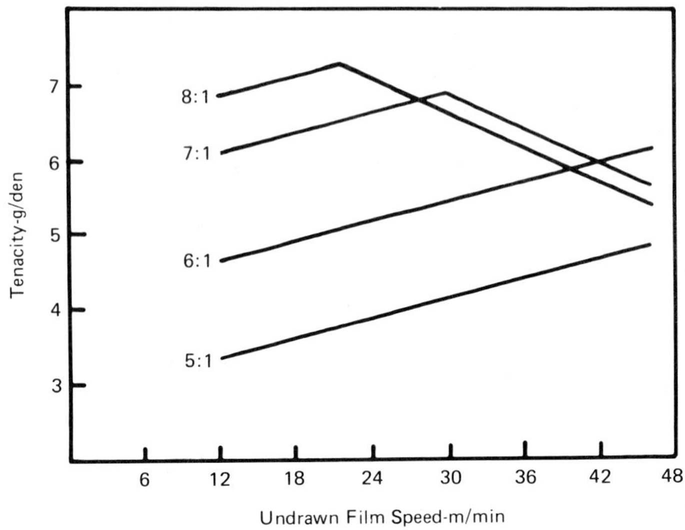


Figure 36 Effect of undrawn film speed on tenacity at various draw ratios. (From Evans, 1971.)

On top of this, the extrusion temperature profile must also take into account how the film will be cooled and solidified. In the production of blown tubular film it is necessary to work with lower temperatures than in flat die extrusion, since the film tube can only stand steady if the melt is sufficiently viscous as it leaves the die slot and is cooled more rapidly. Typical extruder temperature profiles for the extrusion of polypropylene are given in Table 12. Since high-density polyethylene has a lower crystalline melting temperature than polypropylene, the stabilization of the annular film tube is somewhat easier due to the faster solidification of this polymer.

Closely connected with the extruder temperature profile is the screw geometry. The screw must be designed in such a way that it

Table 12 Typical Temperature Profiles in the Extrusion of Polypropylene

| Processing Method | Heating zones | Head | Die | Melt |
|---|---|---|---|---|
| Cast film | 220/240/260°C | 260°C | 260°C | 260°C |
| Tubular film | 220/210/220°C | 220°C | 220°C | 220°C |

prevents too high a frictional heat development in the compression and metering zones. It is also possible to reduce frictional heat development by cooling the extruder cylinder with cold air or oil. If the cooling is not sufficient it is impossible to control the temperature in the different extruder zones. In this case the temperature will steadily increase, leading to thermal decomposition and damage of the polymer. When the so-called "adiabatic operation" method is used, where the polymer melt is heated practically solely through the energy of motion of the screw, temperature control is not possible by controlling the operation in heating zones, but only through defined cooling of the compression and metering zones.

Fundamentally an extruder screw is no mixing aggregate. Nevertheless, a mixing effect during the melting or plastification of a thermoplastic material is highly desired, especially when granulate mixtures of different polymers are processed. Only if one succeeds in intimately mixing the components in the molten state can optimum properties be achieved. Incomplete mixing in processing polymer blends leads, in film tape processing for example, to increased fibrillation tendency. For this reason, extensive efforts were made to construct screws of different designs for achieving better mixing action. Special mixing sections were incorporated in the screws. Unfortunately, such mixing segments create increased frictional heat, interfering with the temperature regulation. Good results were experienced with extruder screws which carry "naps" at the end of the metering zone. Further improvement of the mixing performance can also be achieved by the installation of static mixers into the die adapter. Here attention is drawn to mixing systems offered by the firms Barmag, Sulzer, and Kenex.

For polypropylene the following screw design can be recommended: pump zone: 3 D; first compression zone: 8 D; compression ratio: 1:1.7; second compression zone: 1 D, compression ratio: 1:2.5; metering zone: 12 D. Universally applicable for polyamides and polyesters, respectively, are screws with the following characteristics: pump zone: 6 D; compression zone: 6 D, compression ratio 1:3 to 1:4; metering zone: 12 D. If the mixing zone is situated immediately behind the compression zone, a release zone has to follow in order to avoid overheating. The danger of uncontrolled temperature increases with the number of revolutions of the extruder screw. Therefore, it is advisable to choose a larger screw pitch depth in the metering zone. Table 13 shows the screw pitch depth for the extrusion of polypropylene at different cylinder diameters.

Finally, the importance of filter packings must be mentioned. With the aid of several wire screens having different fineness the melt pressure in the extruder can be regulated and simultaneously an improvement in the homogenizing effect be achieved. Moreover, by way of filter screens, impurities are removed which could cause ruptures of

**Table 13** Extruder Screw Design for the Extrusion of Polypropylene

| Screw zones | Pumping zone | 1st Compression zone | 2nd Compression zone | Metering zone |
|---|---|---|---|---|
| Zone length in multiples of screw diameter | 3 | 8 | 1 | 12 |
| Cylinder diameter | Screw pitch depth (mm) | | | |
| 45 mm | 8.4 | 8.4–6.3 | 6.3–2.2 | 2.2 |
| 60 mm | 10.0 | 10.0–7.6 | 7.6–2.3 | 2.3 |
| 90 mm | 12.7 | 12.7–7.6 | 7.6–3.0 | 3.0 |
| 120 mm | 14.0 | 14.0–10.0 | 10.0–3.3 | 3.3 |

the film or the film tapes during stretching. Of special importance is the arrangement of the filter packings in the production of pigmented films by the removal of pigment agglomerates. The adjustment of the melt pressure in the extruder can also be achieved with the help of a throttle valve in the die adapter.

### B. Flat Die Extrusion

There exists extensive literature on film forming technology, on the maximum attainable stretch ratio, and on the mechanical properties of film products. We will mention in the following only publications which deal with this question in connection with the production of film tapes and of split- or slit-film fibers, such as Burggraf (1967), Okabe and Chino (1967), *Br. Plast.* (1968), Gouw and Skoroszewski (1968), Slack (1968), Condit and Johnson (1969), Moorwessel and Pilz (1969), Schuur (1969), Ford (1970), Balslev (1971), Evans (1971), Fitton and Gray (1971), Hossak (1971), Nott (1971), Skoroszewski (1971), Campbell and Skoroszewski (1972), McIllhagger (1975), and Akay (1979). The material published by these authors will be discussed together with our own technical knowledge.

During the solidification of a polypropylene or high-density polyethylene melt at rapid cooling in a water bath or on cool chill rolls, a fine spherulitic or "smectic" structure is created. On the other hand, slow cooling during solidification gives a coarse spherulitic texture. This effect can be explained, according to Reinshagen and Dunlap (1975), by an increase in number and growth rate of nuclei at quick cooling of the melt. This causes a reduction of spherulite size, lamella thickness, and of the degree of crystallinity.

With very thick films the conditions are more complicated, particularly if one-sided cooling is applied during solidification. When fast cooling occurs smectic and coarse spherulitic structures are formed side by side in the cross section of the film. In the extrusion through a flat die either water or chill rolls may be used for the film cooling. In these cases generally fast solidification of the polymer occurs, resulting in the formation of a smectic structure, as shown by Figure 37a.

Contrary to this, in the slow cooling occurring during tubular film extrusion a coarse spherulitic structure is formed, as indicated in Figure 37b. By adjustment of the temperature of the cooling mean the crystal texture can be influenced within certain limits. The differences in fine structure caused by various methods and conditions of cooling can be recognized and determined by optical birefringence or x-ray

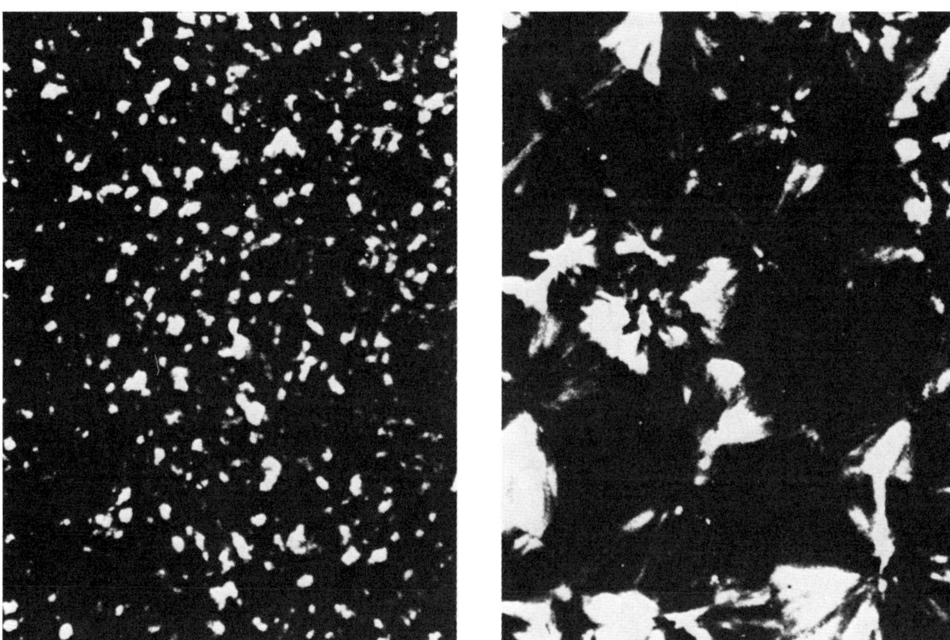

(a) (b)

Figure 37 Smectic structure of (a) rapidly cooled water-quenched film and (b) coarse spherolytic structure of slowly cooled blown film. (From Krässig, 1977.)

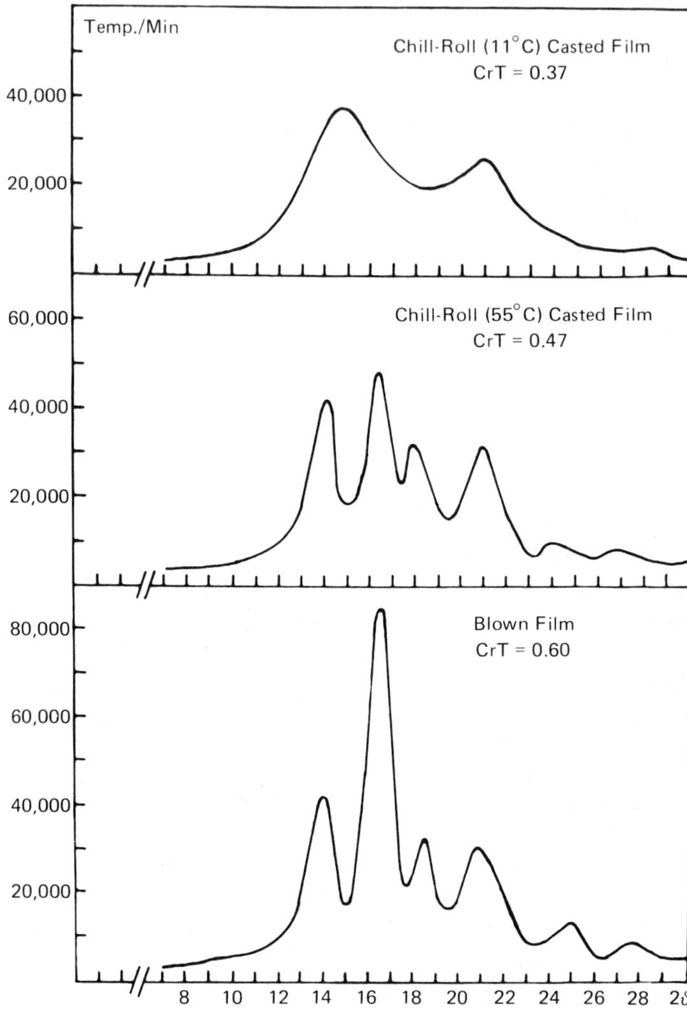


**Figure 38** X-ray wide-angle diffractograms of polypropylene films solidified at various temperatures. (From Krässig, 1977.)

diffraction, as shown by Harms et al. (1971a, 1971b, 1973a, 1973b) and by Krässig (1972, 1975a) and illustrated by Figures 37 and 38.

Figure 38 shows that a polypropylene film which has been cooled at a chill-roll temperature of 11°C yields an x-ray wide angle diffractogram with diffuse equatorial bands. This indicates small regions of crystalline order. A higher degree of crystallinity has a polypropylene film which was solidified at a chill-roll temperature of 55°C. The highest degree of crystallinity recognizable by the sharp equatorial diffractions shows air-cooled blown tubular film.

Not only the degree of crystallinity, but the molecular orientation as well is greatly influenced by the technique of primary film formation. In the blown tubular film process a certain degree of lateral orientation of the polymer chains is unavoidable. As will be shown later in greater detail, this lateral orientation can not be reversed completely in the subsequent monoaxial stretching operation. In the extrusion through a flat die and cooling on chill rolls or by water immersion, orientation in machine direction occurs almost exclusively, varying by degree in accordance with the extrusion conditions. Consequently this results in, among other things, a higher fibrillation tendency of the film or film tapes after stretching. Additionally, a quick solidification of the melt leaving the die causes a freeze-in of the melt orientation. As has been shown in the preceding chapter, high melt orientation is, to a certain degree, disadvantageous for the mechanical properties of the end product. Therefore one should always take care that the film cools off and does not solidify too rapidly. The degree of crystallinity and the molecular orientation achieved in film forming and film solidification have a great effect on the stretchability of the primary film. Water quenching of the freshly extruded film generally limits the maximum stretch ratio of polypropylene film to values between 1:10 and 1:15. The blown tubular film formation combined with air cooling, however, allows stretch ratios up to 1:20. The chill-roll process allows a higher variability in the cooling and film solidification by controlling the cooling-off speed exactly through proper adjustment of the cooling medium temperature between 5 and 95°C. At elevated temperatures it is thus possible to let the freshly extruded film solidify even more slowly than with air cooling, so that with polypropylene even stretch ratios up to 1:25 are said to be achievable.

With the water-quenching technique an analogous procedure is not possible. The utilization of hot water leads to local boiling in the neighborhood of the film immersion, resulting in a coarse film surface. The maximum stretch ratio is strongly dependent on film thickness. The thicker the primary film the lower is the cooling-off speed and the higher the maximum possible stretch ratio. The maximum final film gauge, lying in the range of 300 to 350 μm, can be best reached with the aid of water quenching. The outlined relations between the film forming conditions on processing performance and on properties are in principle also valid for high-density polyethylene, especially with respect to the influence of the cooling conditions on maximum possible stretch ratio: the quicker the film cooling and solidification, the lower the maximum achievable stretch ratio.

Of great importance for the final film properties is the crystallite size. Flat die extrusion combined with water or chill-roll cooling at low temperatures leads, as mentioned before, to a higher degree of preorientation and to the formation of a "smectic" structure composed

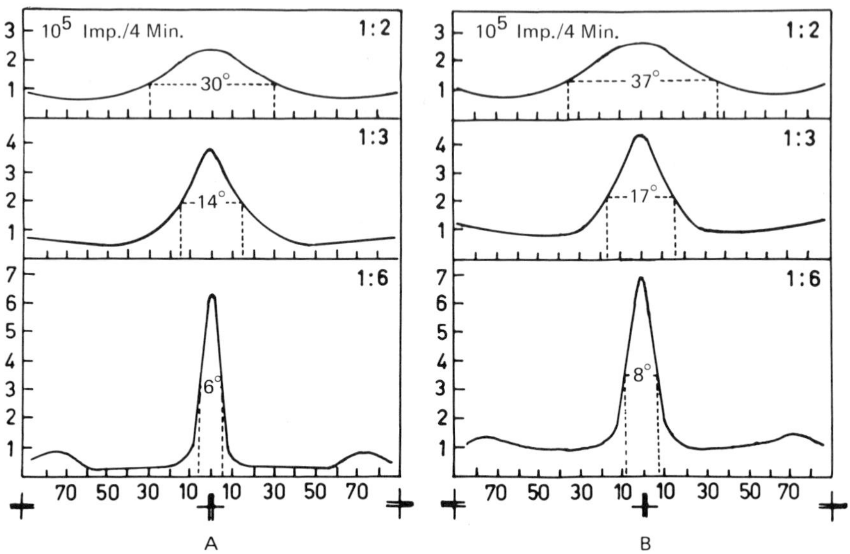


Figure 39 Effect of various stretch ratios on the development of crystallite orientation: (A) in chill-roll cast film (11°C), and (B) in blown polypropylene film (azimuthal scans through the x-ray diffraction at $2\theta = 14°$). (From Krässig, 1977.)

of relatively small crystallites. Such a system responds more quickly to the orienting forces in the stretch operation (Manson and Reed 1975; Ward et al., 1975) than the coarse crystalline structure obtained in the tubular, blown film technique. The higher degree of orientation obtained at a given stretch ratio is probably the reason for the higher fibrillation tendency. On the other hand, residues of the larger crystallite lamellas formed by air cooling in the film blowing which are not completely unfolded may also constitute an obstacle for the lengthwise splitting.

Figure 39 illustrates the effect of various stretch ratios on the development of the crystalline orientation in a flat die-cast film solidified on a chill roll water-cooled at 11°C and a tubular blown film cooled with air. Comparing the azimuthal intensity distribution in the x-ray deffraction maximum at $2\theta = 14°$, it can be seen that at all stretch ratios the cast and chill-roll solidified film shows a distinctly narrower intensity distribution indicative of better crystalline orientation.

Apparently, the stiffer network of the larger ordered regions in the blown tubular film impedes the orientation of the crystallites effected by a given stretch ratio to a greater extent than in the case of the smectic flat die extruded film (see Krässig, 1975b; Fitton and Gray, 1971; Galanti and Mantell, 1965). Similar observations were made by Gouw and Skoroszewski (1968) and by Moorwessel and Pilz (1969). The slower orientation of tubular film during stretching may also have its cause in the inflation of the film tube by the cooling air to a widening ratio of 1:1.3 to 1:1.5. A certain degree of lateral orientation is thereby introduced in the film, which has to be compensated for in the subsequent monoaxial stretch before a complete and ideal orientation in machine direction is attained.

Figure 40 gives a comparison of the tensile strength values obtained at different stretch ratios with film produced with the various extrusion and film solidification techniques. At a given stretch ratio the tensile strength increases in the following sequences: blown tubular film < chill-roll quenched flat die extruded film < water-quenched flat die extruded film. Apart from the possibility of obtaining higher tenacities, thick film tapes are more advantageously produced by the combination of flat die extrusion with water bath cooling. The application of this technique also provides a better capacity for a given production line. With this film-forming technique, the upper limit of primary film gauge lies at about 500 μm, and drawn film with a final thickness of 90 to 100 μm can be favorably produced.

Flat die extrusion also offers the opportunity of a very precise adjustment of the die slot and therefore the advantage of very accurate

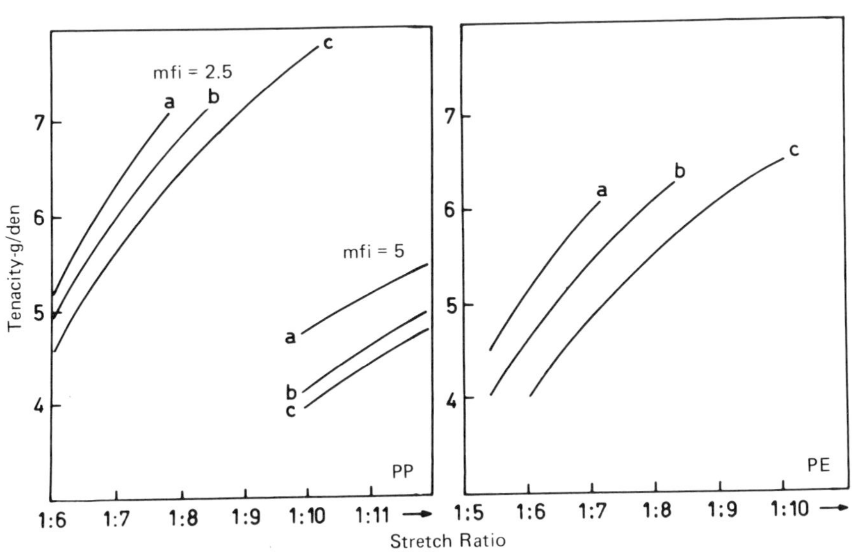


**Figure 40** Effect of film manufacturing technique on tenacity development in polypropylene (left-hand side) and polyethylene (right-hand side) film tapes: a = water quenched; b = chill-roll cast; c = blown film. (From Gouw and Skoroszewski, 1968; and Moorwessel and Pilz, 1969.)

compliance with thickness tolerances. Variation of the die slot width also permits variation of the melt pressure in adapting to the flow characteristics of the polymer melt. A very exact adjustment of the slot width is possible with dies equipped with the flex-lip system. A further advantage of flat dies consists of equipment with a so-called "blocking bar," which enables an even discharge of the melt across the whole die width. The design of the melt conduit in the shape of a coat hanger serves the same purpose. Through the application of all design possibilities the flat die extrusion gives a good titer constancy of the film tapes produced.

The good heat transfer in the quenching step, especially by the application of water bath immersion, makes high production speeds and high output possible. Furthermore, the utilization of water-bath or chill-roll quenching gives a better independence of the flow characteristics and the crystalline melting point of the polymer. The application of water baths or chill rolls in film solidification also allows thermoplastics with low melt viscosity to be produced, such as polyesters, which can not be extruded through annular dies. Finally, it should be mentioned that water bath solidification requires the lowest equipment investment costs (Balk, 1978).

Aside from these advantages and the universal applicability, the flat die extrusion procedure has some drawbacks, particularly when used in combination with water-bath quenching. In the first place the restriction in film width should be noted. Even flat die extrusion lines with an operating width of 3 m (9.83 ft) have been propagated by some machine manufacturers; the achievement of such widths is connected with great technical expenditures for the die and the chill-roll aggregates. The available width of the primary film in water-bath quenching is further restricted by the so-called "neck-in effect" (i.e., the tendency of the melt emerging from the die to reduce its surface contracting from both edges of the still plastic film). This results in an undesired increase in the edge thickness. This effect is the more pronounced the higher the draw-off speed in relation to the speed of the emerging melt. The level of the draw-off ratio, however, is not free of choice since it is predetermined by the desired film or film tape properties. Before the undrawn film enters the stretching or if the undrawn film is slit into tapes at the beginning of the stretching unit, edge trimming is necessary since the thicker edges interfere with the film windup and the edge strips would result in tape with too much titer. Generally the edge trimmings are directly fed back into the extruder after compacting and cutting (see Chemiefaser Lenzing AG, 1979).

The chill-roll technique, on the other hand, offers technical possibilities to reduce the neck-in effect and thus the amount of edge trimmings by the following steps:

1. The distance between the die slot and the chill-roll can be held very small.

2. The contraction of the film can be reduced with the help of an "air knife," a flat die from which an air stream infringes on the whole width of the emerging film. This air jet is adjusted so that it hits the film immediately above the line of contact of the melt with the chill roll and presses the solidifying film closely onto the roll surface, avoiding air locks between the former and the latter.
3. Even more effective than the air knife is the reverse procedure, namely the application of reduced pressure between the chill roll and the molten film emerging from the die with a "suction box." This aggregate prevents the formation of an air layer or of air locks between the solidifying film and the chill-roll surface, which would reduce adhesion of the film to the roller surface, thus favoring the necking-in or resulting in uneven film cooling.

Figure 41 demonstrates the action of an air knife (left) and also of a suction box attachment (right). The disadvantage of the chill roll technique is the one-sided cooling off of the film, which limits the attainable film thickness or tape titer, respectively. On the other hand, the film solidification by water bath immersion offers no solution for preventing the neck-in, leading to edge thickening and enhanced edge trimmings. The only feasible option as mentioned before, lies in reducing the distance between the die lips and the water surface and reducing the ratio between melt exit speed and draw-off speed. However, there exist practical limits for both measures, namely the danger of water boiling and undulatory motion of the water surface.

A further disadvantage of water bath quenching is the dragging along of adhering water. In order to guarantee trouble-free further processing in subsequent production steps, it is absolutely necessary

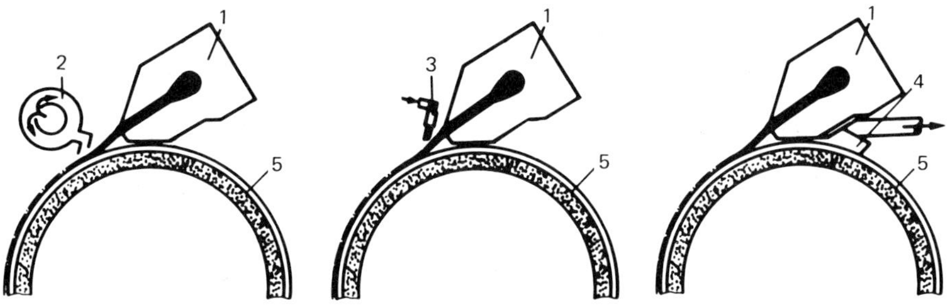


**Figure 41** Application of air knives (left), air nozzles (middle), and suction devices (right) in chill-roll casting. (From VDI-Gesellschaft für Kunststofftechnik, 1978.)

to remove the adhering water and to dry the film after it has left the water bath. This is mostly done by blowing compressed air on the film or by installation of a drier between the water bath and the process equipment (stretch intake rollers or blade bar).

For an assessment of the suitability of either water bath cooling or chill-roll cooling in a given process, a comparative study of their effects on film gauge tolerances and surface quality should be made. If the molten film is solidified by water bath immersion, there is the risk that the film surface will be creased and scared by turbulence in the water. As already indicated before, rough film surfaces result whenever elevated water temperatures are being used for slower cooling. In this case there is a danger of water boiling at the point where the hot molten film enters the bath. It is therefore advisable to arrange the water flow in such a way that the water can flow off on both sides of the film. In the case of chill-roll application one must always consider the danger of mechanical damage to the roller surfaces. Each small cut in the chromium plated chill rolls leads to damage in the film surface and causes possible rupture of the film or the film tapes in the subsequent drawing operation. A further cause of surface roughness and gauge variations of the film in machine direction is the phenomenon of "melt fracture" and the resulting draw resonance. They occur when a critical melt flow velocity is surpassed in the die slot, as illustrated by Figure 42.

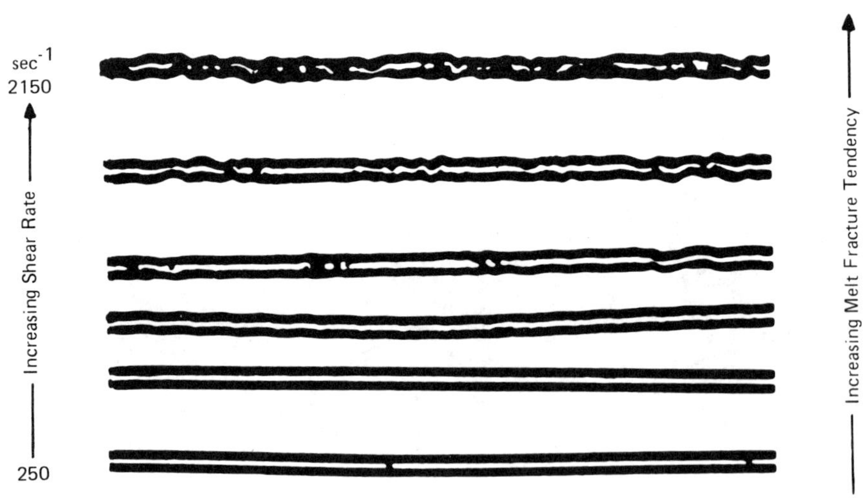


**Figure 42** Occurence of melt fracture in film extrusion demonstrated on lengthwise film cross sections.

The melt fracture phenomenon occurs especially when the ratio of the melt exit speed to the draw-off speed of the primary film exceeds the value of 1:4. It manifests itself in regularly distanced thickness variations along the machine direction. Bergonzoni and Discresce (1965; see Schuur and Van de Vegt, 1975) explain this phenomenon as follows. By internal friction the melt is heated up additionally between die and water bath surface through the longitudinal shear forces of the draw-off. The viscosity of the melt is decreased, thereby causing a higher draw-down ratio. This in turn leads to a reduction of thickness resulting in a quicker cooling. The resulting increase in draw-off tension again causes overheating of the melt and the process repeats itself. According to Schuur (1966), the draw-down energy is not sufficient to explain a rise in temperature worth mentioning, so he gave another explanation for this phenomenon. According to this author, the orientation of the molecules at high draw-down ratios becomes so high that the film extends under solidification and crystallization of the polymer. Therefore the draw-down ratio declines and the film grows thicker. As the orientation now decreases, the degree of crystallinity and the elongation of the film in machine direction decreases. The draw-down ratio increases again and the process repeats itself. The author has proven his theory by orientation measurements in the thick and thin areas of the film. In the thin areas a higher orientation of the molecules in the flow direction has been found.

For the practice of film or film tape production the following conclusions can be drawn from the above. To avoid thickness or titer variations at high line speeds the melt exit temperature has to be adjusted at the die, as high as possible using polymer grades with a moderate average molecular weight and narrow molecular weight distribution.

## C. Extrusion Through an Annular Slot Die

The principle of film formation by extrusion through an annular slot die is illustrated in Figure 32, shown in Sec. II.

The polymer melt emerges from a circular slot having opening widths of 0.5 to 1.0 mm (0.2 to 0.4 in.) and is drawn off vertically upwards from the die. The die consists of an outer ring and a core. The outer ring is adjusted so that both elements can be centered equidistant from the lips. The exact centering of the outer ring and of the core is a prerequisite for satisfying uniformity of the film over its whole width and of the titer constancy of film tapes made therefrom. In the production of film, however, the die slot centering is not sufficient in order to obtain perfectly wound cylindrical film bobbins. In order to compensate for the variations of gauge in the film the die has to be set in permanent back and forth rotating motion. There are vertical extruders on the market where the whole extruder rotates back and forth.

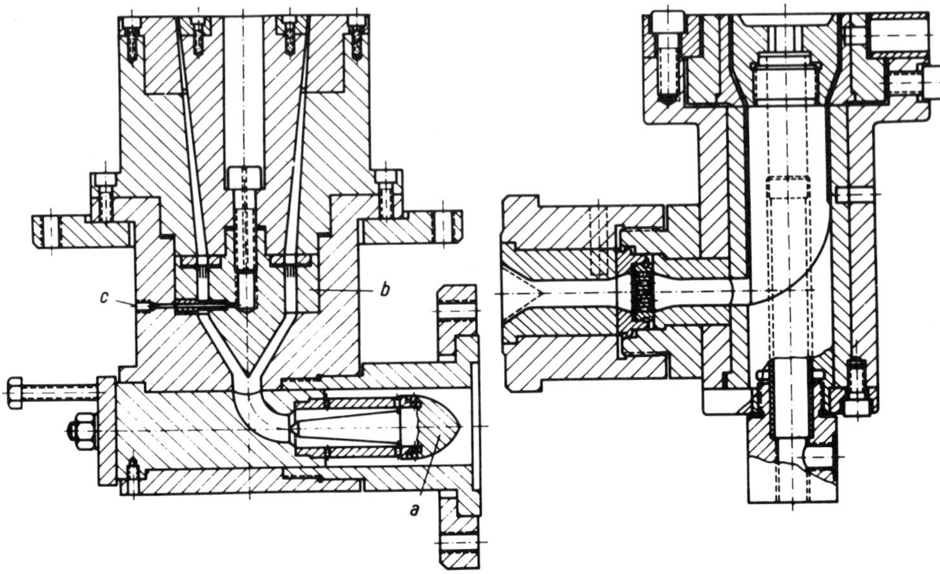


Figure 43 Tubular film blowing head with central polymer melt feed (left side) and with polymer melt side feed. a, Screen package; b, core-mounting support; c, blow air feed. (From Kunstoff-Handbuch, edited by R. Vieweg, A. Schley, and A. Schwarz, Carl Hanser Verlag, Munchen, 1969.)

Constructive problems are caused by the fixation of the die core in the film blowing head. Between the core and the outer ring of the annular die a rigid connection has to exist, which inevitably divides the melt stream. At places where the melt stream is interrupted, areas with reduced film thickness, "bar marks," are created in flow direction. The effect of the interruption caused by the holding bars for the inner die core can be eliminated only with great technical difficulties. Many suggestions for die constructions aiming for the elimination of bar marks have been made in the past. One relevant principle consists of equipping the core in the area of the circular slot between the mounting bars and the die lip with spiral grooves. With these grooves the melt is forced to follow the spiral path by which the part streams bypassing the bars are forced to flow together, eliminating the thickness reduction. If the removal of the bar marks is not successful, the film or the film tapes show an increased fibrillation and splitting tendency along these weak spots. Figure 43 shows two film blowing head designs.

After the melt has left the circular die slot it is cooled to the solidification temperature by a stream of cool air. The air stream is applied to the inside of the tube and to the outside. The air stream applied to the inside of the film tube has to be sufficient to build up a certain excess pressure, blowing up the film tube to a "balloon", thus stabilizing it statically. Normally in film or film tape production one operates with inflation ratios between 1:1.1 to 1:1.5. The cool air stream applied to the outside of the film tube additionally helps the film solidification. The area in which the film solidification occurs is called the "frost line." It is clearly to be seen on the film tube and is of practical help to the operator in establishing stable film extrusion conditions.

The solidified film tube is flattened by a pair of rolls under an angle of no more than 20° and drawn off by a pair of nip rolls. The roller combination serves as draw-off unit and at the same time prevents the blowing air from escaping the inside of the flattened film tube. The film tube has to be flattened when it is still warm. For this reason the distance between the die and the nip rolls should be no more than 4 to 5 times the tube diameter. As initial film thickness, 40 to 70 μm are recommended in the application of the film blowing technique. The attainable draw-off speed lies between 5 to 50 m/min (approx. 15 to 150 feet/min) depending on the primary film thickness. The capacity of a blown tubular film extruding line is limited mainly by the relatively poor cooling effect of air.

The insufficient heat transfer of cooling air, however, does not only limit the line speed. It also impairs the static stability of the still plastic film tube. At a high frost line due to slow cooling or because of too high a melting point of the film-forming polymer, the film tube tends to "breathe" and sway. As a result creases are being formed which cannot be removed during the stretching of the film or of the tapes, resulting in thick spots in the final product. A measure to increase operational stability consists of decreasing the distance between the annular die and the nip rolls as far as cooling speed or efficiency of the air cooling, respectively, allow.

As a further means for the stabilization of the frost line, cooling rings were developed which can be used to support the air cooling. They consist of a tubular ring through which cooling liquid is being circulated and whose diameter should be somewhat wider than that of the annular die accommodating the inflated film. On the polished inside of the cooling ring, exactly in its center, a radial groove is situated to which a weak vacuum is applied. Forced by the negative pressure the film tube fits tightly to the inside of the cooling ring and is cooled off to its solidification temperature by the cooling medium inside the ring. Figure 44 shows a photograph of such a cooling ring.

The cooling ring is normally placed approximately 50 cm (i.e., about 1.6 ft) above the upper edge of the die, concentrically arranged with respect to the annular die. The inflation by the air stream inside the film tube makes the latter form a slight bubble beneath the cooling ring in the still-near-molten state as well as above the cooling ring in an

Figure 44 Application of a cooling ring in blown film casting. (Courtesy of Chemiefaser Lenzing AG.)

already solidified state. This arrangement has been proven to act especially well in the production of film tapes from polypropylene.

For further intensification of the cooling of the primary film tube, another procedure has been developed which can be applied in vertically downward extrusion. This cooling procedure uses a water film as a cooling medium. After the air cooling the inflated film tube is wetted with water, which carries the heat from the polymer film through vaporization. This procedure is known under the name "TQ-process."

Behind the nip rolls the cooled and flattened film tube is either cut and spread by a special device or it is immediately fed to the inlet rollers of the stretching unit or to a razor blade bar preceding the latter as a double layer. In the case of double layer processing, film tapes coming from the area of the folded edge show a higher tendency to fibrillate. This tendency can be removed by heating up the tapes (Reifenhäuser KG, 1974). A novel variation of the blown tubular film process, consisting of welding the two layers of the flattened film tube together under heat and pressure and stretching the doubled film, was suggested by Chemiefaser Lenzing AG (1975a).

The application of the blown tubular film process makes higher demands on the polymer properties than does the flat film extrusion method. The most important prerequisites for a successful tubular film extrusion is sufficient tenacity of the polymer film in its molten state and a high melt viscosity, respectively, thus avoiding super inflation of the film tube. Additionally, the crystallite melting point should be as low as possible to ensure that the frost line develops in the vicinity of the die slot. Therefore high-density polyethylene and isotactic polypropylene are very well suited for tubular film extrusion. The same is true for low-density polyethylene, though for heavier packaging film or for film tape production it is only interesting as a blending partner. Polyamides and polyethylene terephthalate, however, are not at all suitable for the blown tubular film extrusion.

A low polymer temperature at the die exit improves the stability of the film tube and has a favorable effect on operational stability. Molecular orientation takes place beneath the frost line. Because the axial velocity gradient and the viscosity of the film in this area are higher than immediately behind the exit of the die slot, the highest tensions build up there. Orientation takes place primarily in machine direction. Depending on the inflation ratio (i.e., the ratio of the tube diameter to the diameter of the annular die slot), the polymer molecules also orient in transverse direction. This cross orientation is often not removed completely during subsequent monoaxial stretching. As a result, film tapes from tubular blown film, as already pointed out earlier, have a markedly lower fibrillation tendency then those made from flat die extruded film. Accordingly, the tensile strength is somewhat lower and the handle is smoother. The latter property is important for fabrics from fibrillated film tapes because a smooth handle is generally desired by customers.

In Table 14 a comparison is made between tensile strength, elongation at break, and fibrillation tendency of polypropylene films, which have been produced by the flat film process and by the tubular film process. It can be recognized from the data that stretched tubular film, due to residual cross orientation, shows a tendency to lower strength and lower fibrillation, while elongation at break is substantially higher. The molecular orientation increases with the rate at which the melt film cools off. The cooling velocity is largely dependent on the film thickness, the melt temperature, and cooling air velocity.

The blown tubular film process offers a number of technical and economical advantages, especially at film gauges below 60 to 80 μm:

**Table 14** Differences Between Chill-Roll Casting and Blowing Techniques with Respect to Maximum Stretch, Achievable Tenacities, and Elongations, and Fibrillation Tendency of Polypropylene Films (mfi 1.7 g/10 min; Extrusion Temperature 240°/200°C respectively; Stretch Temperature 135°C)

| Processing method | Stretch ratio | Maximum stretch | Tenacity g/dtex | Elongation % | Relative fibrillation tendency |
|---|---|---|---|---|---|
| Chill-roll casting | 1:6 | | 3.4 | 33 | 2.8 |
| | 1:8 | | 5.3 | 22 | 5.0 |
| | 1:10 | | 6.1 | 16 | 7.0 |
| | | 1:10 | | | |
| Blown | 1:6 | | 2.7 | 54 | 2.2 |
| | 1:8 | | 3.8 | 27 | 3.8 |
| | 1:10 | | 5.8 | 19 | 6.7 |
| | | 1:11 | | | |

1. The die needs little space even for large film widths. This means that even films with a thickness of only 20 to 30 μm can be produced economically with a relatively high output.
2. There is no need for edge trimming during extrusion and the waste is small.
3. The adjustment of the slot width is easier than with flat dies. In case of the production of split or slit fibers from the film, a rotating motion of the die head or of the complete extruder is not necessary.
4. Investments for a blown tubular film extrusion line are markedly lower than the expenditure necessary for the aquisition of a flat film extrusion line.

Considering the technical advantages on the one side and the shortcomings of currently available film extrusion systems on the other, the decision for a certain system will be mainly determined by the desired thickness of the film or tape product. Film tapes for cordage goods and heavy duty containers have to be, in general relatively coarse and thick. They are therefore favorably produced by the flat die extrusion and water bath immersion quenching technique. The manufacturing of light fabrics for small bags and less demanding packaging purposes asks for thin and narrow tapes, which can be produced more economically from tubular blown films.

The thickness of the primary film is exceedingly important for the economic efficiency of the production process and properties of film tapes. The thinner the primary film the smaller the output obviously is. Due to the faster cooling the maximal obtainable stretch is also markedly lower and the achievable tenacity level is reduced. On the other hand, a quicker cooling rate in the production of thin primary film allows the use of higher draw-off speeds. Table 15 shows some of these interrelations based on material presented by Skoroszewski (1971). Further references dealing with these subjects are Gouw and Skoroszewski (1968), Balslev (1971), Fitton and Gray (1971), and Nott (1971).

Not only the film thickness but also the film draw-off speed is of great importance for stretchability and the attainable mechanical data. Table 16 illustrates that at a constant film thickness the maximum achievable stretch ratio decreases with increased inlet speed. Under the same conditions tensile strength decreases while elongation at break increases. The correlation between primary film thickness, draw-off speed, stretching behavior, and attainable textile mechanical properties finds its explanation in the different degrees of melt orientation. The thicker the primary film the more slowly it cools and the smaller, therefore, is the retained melt orientation and the higher the maximum achievable stretch ratio. On the other hand, a high draw-off speed results in an increased melt orientation in the die slot, whereby the stretchability of the resulting primary film is limited, as mentioned above.

**Table 15** Influence of Film Manufacturing Technique and Thickness on Polyethylene Tape Processing (mfi 0.4 g/10 min; Density 0.962 g/ml)

| Processing method | Film thickness μm | Maximum inlet speed for even stretch m/min | Maximum stretch ratio |
|---|---|---|---|
| Blowing | 50 | 20 | 8.5:1 |
| | 75 | 13 | 10:1 |
| | 100 | 8 | 11:1 |
| Chill-roll cast | 50 | 20 | 7:1 |
| | 75 | 13 | 8:1 |
| | 100 | 8 | 10:1 |
| Water quenching | 100 | 20 | 7:1 |

With respect to the stretching behavior, specific polymer characteristics also have to be considered. The maximum possible film thickness with high-density polyethylene is limited by its overall inferior stretchability. This becomes more critical if the heating of the film is not very well controlled. Naturally, thick tapes heat up more slowly than thin ones. To provide optimum conditions for a normal process performance in the stretching step, the preorientation of the undrawn tapes has to be kept as low as possible by drawing them off and cooling them slowly. The upper limit for the primary film thickness in the processing of high-density polyethylene under normal stretching conditions is in the range of 100 to 150 μm. When this limit is surpassed without reducing the stretching speed, drastic stretch ruptures occur and tensile strength declines. In the processing of polypropylene on the other hand, it is possible to extrude primary film with a thickness of 500 μm at 3 to 5 m/min and stretch it with a ratio of 1:8 to a tape gauge of 170 μm.

Certain chances to increase film thickness in the processing of high-density polyethylene can be obtained from a proper choice of the stretching equipment. In the stretching of high-density polyethylene film tapes in a hot air oven one has to deal with two opposing effects. On the one hand, it is beneficial to enter the stretch zone with a thick primary film since it has cooled more slowly and therefore shows only limited preorientation. On the other hand, stretchability is reduced by insufficient heating, which is more difficult for a thicker undrawn

**Table 16** Influence of Inlet Speed on Stretch Ratio of Polyethylene Tapes (mfi 0.4 g/10 min; Density 0.962 g/ml)

| Processing method | Film thickness μm | Inlet speed m/min | Maximum stretch ratio | Tenacity g/dtex | Elongation % |
|---|---|---|---|---|---|
| Blowing | 50 | 8 | 12:1 | 7.0 | 9.5 |
| | 50 | 13 | 10:1 | 6.0 | 11.5 |
| | 50 | 20 | 8.5:1 | 5.4 | 14.0 |
| Chill-roll cast | 50 | 13 | 9:1 | 6.0 | 22.0 |
| | 50 | 20 | 7:1 | 5.1 | 28.0 |

primary film. A solution to these problems is offered by the utilization of stretching units which transfer the necessary heat to the polymer film by direct contact, such as hot plates or heated roller assemblies. With such equipment, primary high-density polyethylene film up to thicknesses of 200 to 250 μm can be stretched with good results.

Besides the interrelations between film thickness, tube cooling, stretchability, and production capacity, the fibrillation tendency has always to be kept in mind. When deciding in favor of a certain film thickness it should not be overlooked that thin film tapes always have a greater fibrillation tendency than thick ones. This is true for both high-density polyethylene and polypropylene. However, at a given final film thickness and at the same stretch ratio, fibrillation appears more pronounced for polypropylene tapes. This is advantageous in the manufacture of ropes or twisted tape yarns for weaving. However, it is a definite disadvantage in the production of tape fabrics for bags or packaging purposes. These relations will be further discussed in a later chapter.

## III. SLITTING FILM INTO TAPES

The slitting of film into film tape can serve two purposes. The resulting film tapes can be used as such in many fields of application, such as ropes or binder twine, in fabrics or raschel knits, or as intermediate products to be processed into fibrillated split- or slit-film yarns or fibers. The packaging sector today is unthinkable without film tapes.

In principle, there are two ways to produce film tapes: (1) cutting of undrawn primary film using a bar equipped with razor-bladelike knives and stretching the resulting film ribbons lying side by side; (2) monoaxial stretching of the primary film over its whole width and slitting the film tapes afterwards from the stretched film. Both methods are applied in practice. They demand somewhat different technologies and lead to tapes with differing property characteristics. Though the stretching process will be discussed in detail later in Sec. IV, the two slitting methods will be described here in order to allow a comprehensive comparison.

### A. Cutting of the Undrawn Primary Film

The cutting or slitting of film into tapes is generally done with razor-bladelike cutting tools placed in a holding bar extending over the whole width of the film extrusion line. Examples of such film slitting devices are given in Figures 3 and 45. The film slitting requires the application of a certain cutting tension which has to be sufficient to press the film into the blades. The cutting tension is supplied by driven roller assemblies which are arranged before and behind the blade

Figure 45 Knife bar tape-cutting unit. (Courtesy of Reifenhäuser KG.)

equipped bar. With the help of a suitable speed difference between the two pairs of rollers, the cutting tension can be adjusted at will.

The design of the cutting bar and the position of the cutting tools on the bar has to be adjusted to the desired dimensions of the film tapes to be produced. Proper selection of the blade distance is important. The cutting width ($b_1$) of the primary film is a function of the stretching factor ($\gamma$) and the desired final tape width ($b_2$). The relation between these parameters is given by the formula

$$b_1 \text{ (mm)} = b_2 \text{ (mm)} \cdot \sqrt{\gamma}$$

Apart from the tape width, the tape titer (T) is also important. It is calculated from the final tape width and thickness (d) using the equation

$$T \text{ (dtex)} = b_2\text{(mm)} \cdot d(\mu\text{m}) \cdot 10 \cdot \delta \text{ (g/cm}^3\text{)}$$

At a given tape titer and a given tape width the film gauge ($d_2$) is adjusted through the proper selection of polymer output and draw-off speed.

The cutting width is determined by the dimensions of spacers between the individual cutting pieces; it is the sum of the width of the

**Table 17** Hardness of Inorganic Pigments

| Pigment type | Hardness (in Mohs' scale) |
|---|---|
| Carbon black | 1.0–2.0 |
| Cadmium sulfide | 3.0–3.5 |
| Chrome yellow | 2.5–3.0 |
| Iron oxide | 5.5–6.5 |
| Titanium dioxide | |
| anatase | 5.5–6.0 |
| rutile | 6.0–6.5 |
| Zinc sulfide | 3.0–3.5 |

spacer and the cutting blade width. With proper cutting, tape widths down to 1.3 mm can be realized (Hensen, 1978b). Film tapes for fabrics or raschel knits normally have widths between 1.5 and 6 mm at tape thicknesses of 20 to 50 μm. In extreme cases tape thicknesses of up to 10 mm are used. Normal tape titers range from 500 to 1100 dtex. Contrary to this, the heavy film tapes which are converted after fibrillation into ropes have widths between 20 to 40 mm and film gauges between 60 to 100 μm. The titer of such tapes lies in the range of 16,500 to 27,500 dtex.

For a clean and notch-free cut, sharp blades are absolutely necessary. Worn and blunt knives or blades cause dents on the tape edge which may lead to ruptures during the subsequent stretching process. Knife wear is dependent on the kind of film to be cut. Especially disadvantageous to knife wear is delustering of the film with titanium dioxide and mass dyeing with inorganic pigments. The harder the pigment or additive the shorter is the lifetime of the knives. In Table 17 the hardness of some frequently used pigments is listed according to their Mohs' hardness (Evans, 1971).

Naturally, the thickness of the primary film also has some influence on the slitting process. Thick films are harder to press into the knives because they have a greater cutting resistance. Favoring the cutting is a slight grooving of the film along the machine direction, which can be introduced by minute trapezoidal incision in one of the die lips. The peculiarities of profiled films will be discussed in detail later in this book. A double film layer from processing flattened tubular film does not cause any problems in cutting and stretching operations. The double film tapes running one on top of the other can be separated easily and without difficulties after the stretching step.

For the production of film tapes from freshly extruded primary film several systems have been developed.

### 1. *Bars Equipped with Blades*

The blade bar is the simplest and cheapest system. In Figures 3 and 45 it was demonstrated that the blade bar consists of razor or industrial blades, arranged parallel, separated by spacers of optional thickness, and mounted on the profiled ledge serving as tool holder. The blade bar is fixed into the film line in such a way that the cutting edge of the blades forms a well defined angle with the lower side of the film. A rectangular angle of incidence with the film and the blade edge is disadvantageous.

For the success of the cutting process the quality of the blades is of utmost importance. Hardened high-grade steel blades offer the cleanest cut and assure the longest tool life. A particular form of tool holder is described by Messerschmidt-Boelkow-Blohm GmbH (1978). It consists of a bar with a great number of slits at the smallest possible distances arranged vertically to the film web into which cutting blades can be placed according to actual needs. A drawback to the blade bars lies in the fact that the cutting edge is utilized only at one point. According to a patent by Windmöller & Hölscher (1978) this shortcoming can be avoided by a continuous swinging movement of the blade bar. In this way the film is hitting the blade edges constantly at different places along the whole length of their sharp faces. Rapid blunting of the blades is thus avoided and the service time considerably prolonged.

### 2. *Circular Knife Rollers*

Even in its optimized form, such as the swingable form suggested by Windmöller & Hölscher (1978), the blade bar always has the disadvantage of a pressing cut. The blades always have to be pressed into the film material and thus are prone to produce ruggedly cut edges. This shortcoming can be overcome by the application of circular cutting knives, such as shown schematically in Figure 46.

On a shaft rotating in the same direction as the film, circular knives are mounted and held at desired distances by corresponding spacers. Combined with the circular knife roller another rotating grooved roller is arranged in such a way that the circular blades of the cutting bar run in the grooves which constantly press the film into the rotating knife blades. The sheer cut achieved in this way furnishes cleaner tape edges (Domininghaus, 1969, p. 432). This is of great significance, especially in the cutting of films highly sensitive to fibrillation, such as foamed films. Naturally, the circular knife roller can be used without the counteracting grooved roller. One then has the advantage of an even wear of the sharp faces of the circular knives.

Normally the cutting tools for undrawn primary film are integral elements of film tape production machines. However, there also exists

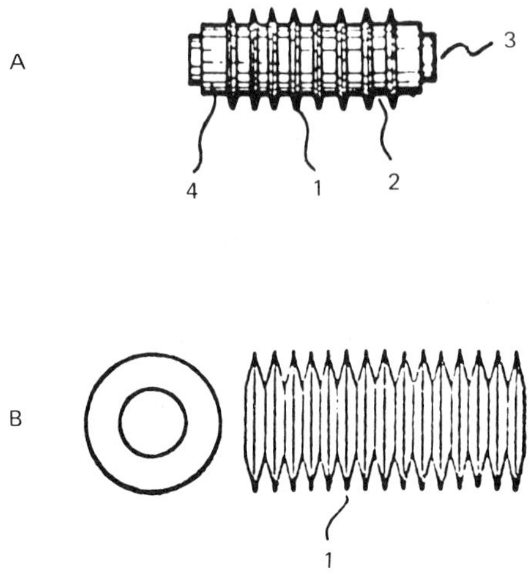


Figure 46 Schemes of rotary film tape (A) or film-fiber (B) cutters: 1, circular knives; 2, distance plates; 3, axis with screw threads for holding screws on both ends; 4, holding screws. (From Krässig, 1977.)

the possibility of combining these cutting devices constructively with textile machinery. Brändle (1977) suggests feeding film directly to a raschel knit machine equipped with tape cutting and drawing devices. The undrawn or partially predrawn film is thus cut right on the rashel into tapes which are hot stretched and fed continuously to the rashel operation and transformed into a knit fabric. An improved version of this cutting and stretching attachment to rashel knit machines is described by the same author in a later publication (Brändle, 1978).

Generally film tapes are produced and sold in the form of cylindrical cross-wound spools for further processing in textile mills. The first step in the textile processing consists in the assembling of tape ends to a beam which is later used on weaving or knitting looms. A great simplification of further processing is achieved when tape production and beaming is done in one continuous operation. Such a technique has been described by Chemiefaser Lenzing AG (1977). When film tapes are wound onto a beam the problem arises that, unlike spool doffing, in beam doffing all tape ends have to be stopped in case of tape rupture at the same time. Since the extruder supplies film continuously and can not be stopped during correcting tape rupture or during beam doffing, the film has to be stored during such actions. This is done with film storage consisting of a fixed row of rollers and a movable wheel sledge of rollers. This storage device is placed between the extruder and the tape cutting equipment. The roller storage device will accommodate the film during the holding time for the correction of tape ruptures or of beam doffing by the wheel sledge of rollers moving apart from the fixed row of rollers. Emptying the film storage is accomplished by a transient increase of the beam wind-up speed after the tape rupture correction or after the next beam has been doffed with all tape ends. Figure 47 shows a scheme of such a film storage and illustrates its function.

## B. Cutting of Monoaxially Stretched Wide Film

The slitting of already stretched flat film is in many cases a superior alternative to the cutting of undrawn film. In this case the freshly extruded primary film is first stretched as such in a single or double layer over heated rollers. Only after this uniaxial stretching is the film cut into film tapes. This may either be performed directly in a processing step immediately following the stretching operation with the help of a blade bar, or in a completely independent step.

It is of particular advantage to first produce a bobbin of monoaxially stretched film and connect it as a beam directly to the weaving or raschel knitting machine. With the help of a controlling release device the film is now cut on the weaving or knitting machine into film tapes. This technique has been developed by Chemiefaser Lanzing AG (1972a, 1972b, 1973; Hossak, 1971) and has become known as the "split-weaving" or the "Lenzing System." In applying this technique, the stretching of the primary film has to be done in such a way that the film gets the textile mechanical properties desired for the tapes. Furthermore, the neck-in in stretching has to be avoided as well as possible in order to avoid thickening of the film edges and an overly high titer of the outside tapes. The great advantage of this technique consists in bypassing the beaming or warping operations.

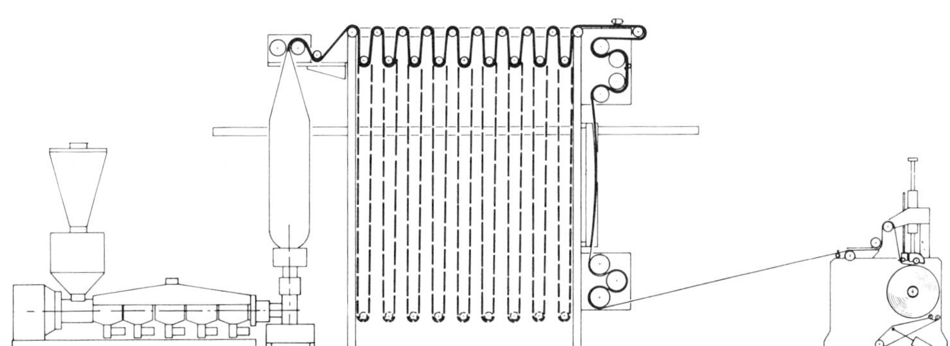

Figure 47 Scheme of a film storage in a direct-tape beaming line. (Courtesy of Chemiefaser Lenzing AG.)

The slitting of drawn film on the weaving or knitting loom is performed with the aid of a razor blade bar (see Figure 3 or 45). The lower limit of tape width is between 1 and 2 mm at a minimum film thickness of 15 to 20 μm. The cutting tension, which has to be applied to press the film into the blades, needs constructive means to control the release of the film to the cutting device under constant tension. The weaving or knitting loom must be equipped with a corresponding attachment.

The characteristics of film tapes produced in this way are obviously not identical with those having been obtained by in-line processing, where the cutting is done prior to the stretching operation (see Sec. III.A). The differences will be discussed in more detail in Sec. IV. In order to calculate the titer of tapes cut after the stretching operation, one has to take into account the missing neck-in phenomenon. The thickness of the stretched film is equal to the quotient of the thickness of the primary film and the stretch ratio. This has consequences in realizing the desired tape dimensions. Thick and narrow tapes are more difficult to produce with this method than by stretching single tapes cut from primary undrawn film.

Especially problematic in the utilization of tape cutting from already stretched film is the inevitable wear of the cutting blades. Rugged edges of the film tapes decrease their tensile strength. In a patent, Sekusui Jushi Company, Ltd. (1978) described a method to avoid this problem by rounding and smoothing the tape edges with heat after the cutting. Use of cutting the stretched film is made particularly in the manufacturing of metallized fibers. A polyester film with a thickness of 12 μm is covered with vaporized aluminum in a vacuum. The metallized surfaces are varnished and the film is then cut by means of a blade bar into flat tapes of 0.1 to 0.4 mm width (La Cellophane S.A., 1971; Lurex N.V., 1974).

## C. Cutting Resistance of Films

Depending on the polymeric material, eventual additives, and the processing method, the films show different cutting resistances against the cutting tools. It is often favorable to measure this resistance in order to select the suitable blade steel and the most suited knife shape. A method developed by Chemiefaser Lenzing AG for the measurement of the cutting resistance has been designed in close relation to the practical cutting in film tape manufacture. The razor blade used as cutting knife is mounted to a force recording device. At a cutting speed of 50 m/min (i.e., approx. 165 feet/min) the profile of the cutting force is recorded. The measured range of the cutting force for monoaxially stretched polyolefin film with thicknesses between 20 to 40 μm was found to be between 3 and 10 p*. With a digital integrator an average

*p = pond = 981 dyn ≅ 0.1 cN

of the cutting force over a longer film length can be obtained. It is a measurement of the long-term cutting energy and allows a better judgement than the transient, fluctuating instantaneous value. Normally the cutting resistance of monoaxially stretched tubular film in the cross direction is about 10% higher as in the machine direction.

Geitel (1976) defines the splitting resistance as splitting tension related to an anisotropy factor of one. The splitting tension can thus be calculated from the cutting tension with the help of the following formula:

$$W(s) = \frac{\Delta s \cdot R(q)}{R(1)}$$

where: W(s) = splitting tension, R(1) = tensile force in machine direction, R(q) = tensile force in cross direction, $\Delta s$ = cutting tension. The cutting tension is obtained by dividing the cutting force by the film thickness. The cutting force is measured according to Geitel in a cutting test with commercially available razor blades; the blade meet the film in a defined angle and penetrates it. The cutting force is influenced by the cutting angle and the depth of the cut.

Further aspects of this topic can be found in another publication by Michels et al. (1975). Geitel (1977) also examined the influence of characteristic features of blades of different origin on the cutting force. He studied, for example, the influence of blade thickness and of the side cutting edge angle on the cutting force. In a patent granted to Consolidated Bathwirt, Ltd. (1973), a method is described for the cutting of thermoplastic polymer film with the help of heated blades. By pressing the film with the tip of the heated blade against a supporting element, the cutting resistance is markedly lowered. Unfavorable is the fact that the hardness of the cutting blades decrease rapidly at higher temperatures.

## D. The Sequence of Cutting and Stretching and its Significance for Processing of Tape Products

The combination of different processing steps in the processing of fabrics and knits from film tapes offers numerous possibilities in meeting specific end-use requirements. Thus for every application the technically and economically most suitable system can be developed.

For the manufacture of flat or tubular fabrics and raschel knits there are, for example, three ways to produce the warp material:

1. The warp is produced by assembling the tape ends from spools which have been produced on film tape extrusion lines.
2. The warp material is produced on a film tape extrusion line and directly put on a beam.

3. Instead of a beam, one uses the film on the loom and cuts it into tapes needed as warp by a special tape cutting attachment.

For carpet-backing fabric production the traditional operation according to (1) and (2) is the best choice, whereby (a) the weft tapes are produced on a film tape extrusion line in the form of cross-wound spools, and (b) the warp tapes are made in the same way and assembled to beams on beaming units.

For packaging material, such as covering fabrics or canvases, the best advantage is offered by film cutting on the weaving or raschel machines with attachments for "split weaving". For some applications it is essential to coat or laminate the fabrics. This is done in the most economical way by melt extrusion coating with flat dies. Thus, coated flat or tubular fabrics are produced for the manufacture of sacks and flexible containers for heavy goods. Additional applications are in the lamination of tape fabrics with paper or aluminum foil for a variety of

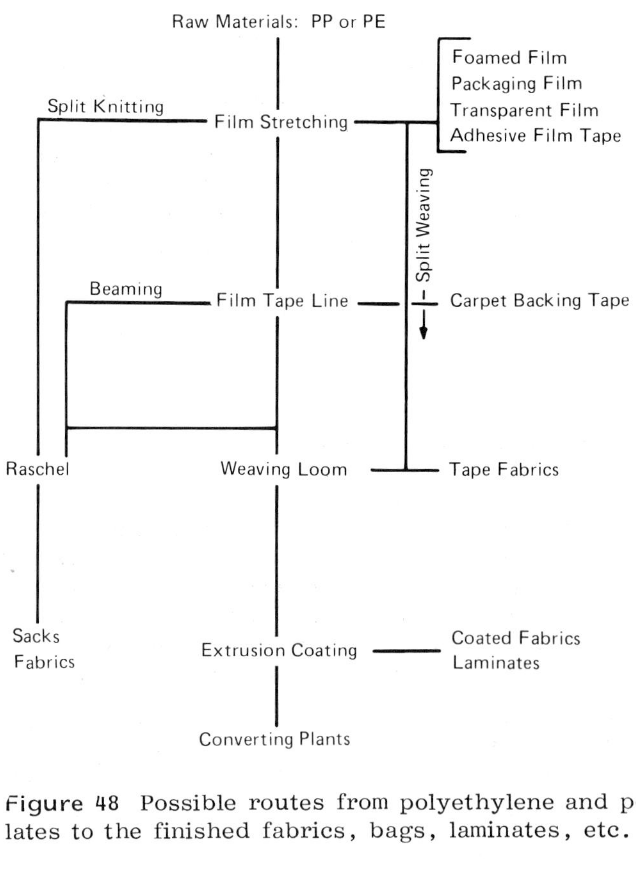


Figure 48 Possible routes from polyethylene and polypropylene granulates to the finished fabrics, bags, laminates, etc.

end-uses. Figure 48 gives a survey of all technologies used today in the manufacture of different webs from film tapes.

## IV. THE STRETCHING PROCESS

By the application of suitable stretching conditions it is possible to impart into films or film tapes most of the properties necessary for the applications. For the most important applications of film tapes, namely for sack fabrics, primary carpet backing, raschel fabrics, rope cords, binder twine, strappings, etc., the following quality characteristics can be influenced by a well directed regulation of the stretching procedure: tensile strength, elongation at break, elastic modulus, fibrillation tendency, titer consistancy, cross-section uniformity, plane laying, stiffness or flexibility, and gloss. Determining for the property profile, which is prescribed by the end use, are the structure characteristics of the undrawn film, the stretching conditions, and to some extent the stretching equipment used.

All the stretching operations used in practice have the following in common: The film which may be either wide or cut into ribbons is stretched either over a short stretching gap or in a stretching field between feeding and pull-off devices under heating to many times its original length. This stretching causes essentially a machine orientation of the polymer molecule chains or crystallites, respectively, under a simultaneous increase in the degree of order or crystallinity. These structural changes result in an elevation of the tensile strength and concurrently in a reduction in the elongation. The increase in strength and the reduction in elongation is more pronounced the higher the stretch ratio. The stretching temperature is also of great influence on the effect of the stretching operation.

The significance of the stretch ratio and the stretching temperature for the development of the crystalline texture of polypropylene films has been examined thoroughly by Samuels (1968). According to his studies the stretching begins with an orientation of the spherulites in the stretch direction. With increasing stretch ratio the spherulites begin to deform. The molecules in the lamellas unfold themselves and are increasingly oriented with their molecular axis in the stretch direction. As the spherulites disappear, the length of the crystallites increases (i.e., the crystalline phase grows at the expense of the amorphous or less ordered phase). The structure of the crystalline phase gets more perfect the higher the stretching temperature. This can be recognized from the structure data presented in Figure 49. The linear growth of the long periods (L) is largely dependent on the stretch ratio and the stretch temperature.

From a certain stretch ratio on, however, the long periods remain constant independent of a further increase in the stretch ratio. At

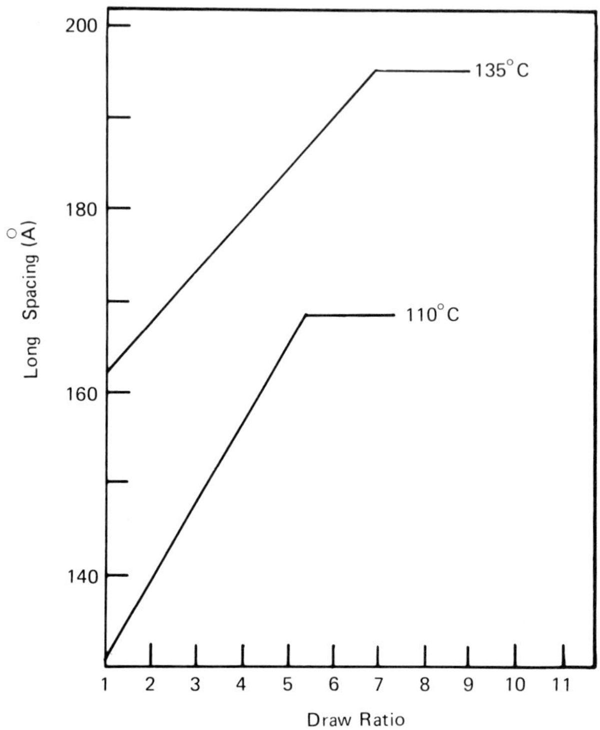


**Figure 49** Long spacings as a function of draw ratio for isotactic polypropylene films at drawing temperature of 110° and 135°C. (From Samuels, 1970.)

this point, according to Samuels (1968) and Peterlin (1972), begins the formation of microfibrils through lengthwise splitting of the lamellas and molecular unfolding. The microfibrils represent nearly rectangular blocks of crystallites or lamellas arranged along the fibrillar axis. They are oriented more or less into the direction of stretch. The molecules which form links between the individual microfibrils are less ordered but also oriented in the draw direction to a relatively high degree. Also the amorphous segments linking the crystallites or lamellas show a certain orientation. At very high stretch ratios one can observe that the film gets a fiberlike fine structure. Overstretching can finally lead to film rupture.

The formation of a fibrillar texture gives an understanding for the fibrillation tendency of films stretched to a very high degree. Bouriot et al. (1976) have pointed out that the fibrillation index rises with increasing stretch ratio and shows a sudden increase at the point where the crystallites or lamellas in the microfibrillar strings lose their identity period. At this point the density differences between the crystalline and the amorphous phases disappear. Polovikhina and Zverev (1969) reported on the improvement of structure orientation through stretching above the crystalline melting temperature of polypropylene. For the property characteristics of film tapes the stretching speed, the method of heating before and during the stretching process, as well as the technical features of the stretching machinery are also of significance, besides the stretch ratio and the stretch temperature. However, before starting the discussion on the individual processing details, the significance of the nature and the structural status of the primary film shall be described.

## A. Effect of Polymer Characteristics on Stretchability

The major condition for stretchability is that the polymer consist of linear unbranched macromolecules. This means that among the polyolefins the branched low-density polyethylene can be stretched only to a limited degree. As has been shown earlier in Sec. I.D.2, this does not apply to blends of low-density polyethylene with high-density polyethylene or with isotactic polypropylene, respectively. A certain loss of working potential, as expressed by the area under the stress-strain curve, can be observed for all blend ratios.

Comparing high-density polyethylene with polypropylene, it is possible to obtain higher stretch ratios and thus also higher tenacities at lower elongations at break in the case of film from polypropylene, as illustrated by Figures 50 and 51 (Skoroszewski, 1971; Malcomesz and Blechschmidt, 1977; Gouw and Skoroszewski, 1968; Evans, 1971; Moorwessel and Pilz, 1969). The more difficult stretching of high-density polyethylene may perhaps be explained by a certain degree of branching. On account of the dispersion forces which have to be overcome during stretching, one should expect that polypropylene is harder to stretch than high-density polyethylene (Michels and Franz, 1978). According to these authors the deformation energy of polypropylene is 1.5 kJ $cm^{-2}$, distinctly higher than that of high-density polyethylene, which is only 0.8 kJ $cm^{-2}$, as shown in Figure 52. Even higher deformation energies are necessary to stretch polyethylene terephthalate and polyamide films where the intermolecular forces are intensified by dipole attraction and hydrogen bonding.

An important draw parameter, which is determined by the nature of the polymer, is the drawing temperature. In the case of high-density polyethylene the drawing temperature is in a relatively narrow range between 110 and 130°C. For the stretching of polypropylene film or film tapes, stretching is possible over a significantly wider temperature range, namely from 120 to 170°C. High stretching temperatures are an essential requirement for high line speeds. The more

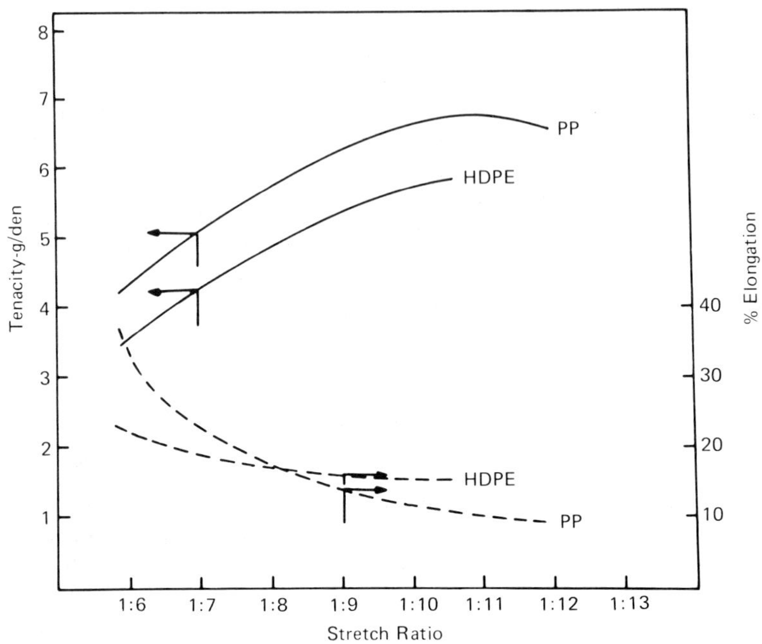


Figure 50 Effect of stretch ratio on tenacity and elongation of polypropylene and high-density polyethylene film tapes. (From Skoroszewski, 1971.)

sensitive stretching conditions for high-density polyethylene become evident through the difficulty in the stretching of thick primary films without disturbance. Increasing film thickness demands lowering of the line speed to avoid stretch ruptures and to obtain an equal film thickness. In the case of primary polypropylene films with thicknesses up to approximately 1,200 μm, stretch ratios of 1:11 and line speeds of 25 to 30 m/min (i.e., 80 to 100 ft/min) at the inlet side of the stretching equipment are possible without problems. The stretching of films from polymer blends, however, is more difficult. For more information, refer to the studies of Michels and Franz (1978).

## B. The Influence of the Structure of the Primary Film on Stretching

As already pointed out in Sec. II.A., one has to distinguish principally two kinds of primary films, namely flat extruded primary films solidified either by water bath or by chill-roll quenching, and tubular blown primary film solidified by air cooling. The most significant difference between these films lies in the degree of order or crystallinity and the degree of orientation. The higher preorientation of the chill-roll cooled, flat die extruded film in the machine direction becomes apparent in the lower maximum stretch ratio achievable at given inlet speeds. A tubular air-cooled film can be stretched to a noticeably higher degree due to its lower degree of preorientation and, to a certain degree, of cross orientation. This is illustrated for the case of high-density polyethylene films in Figure 53 (Skoroszewski, 1971).

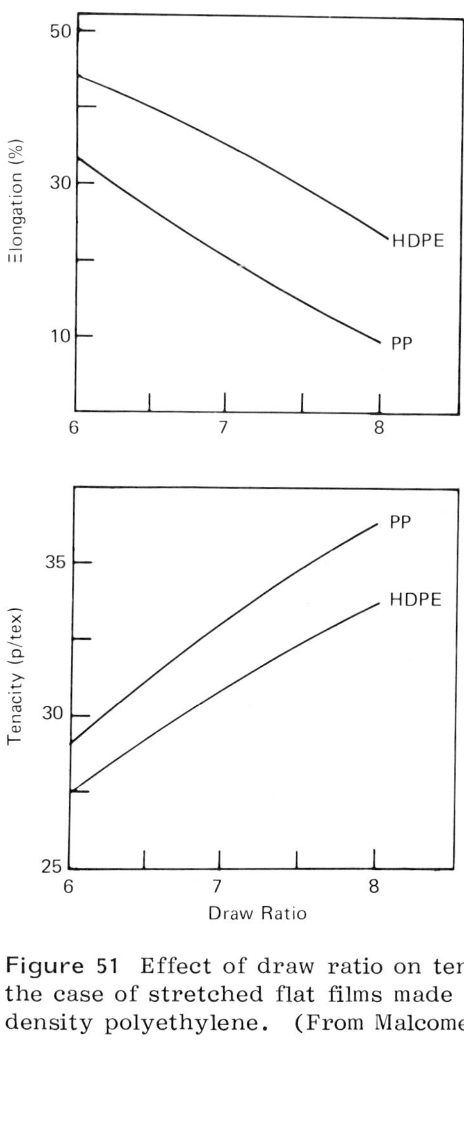


**Figure 51** Effect of draw ratio on tenacity and elongation at break in the case of stretched flat films made from polypropylene and high-density polyethylene. (From Malcomesz and Blechschmidt, 1977.)

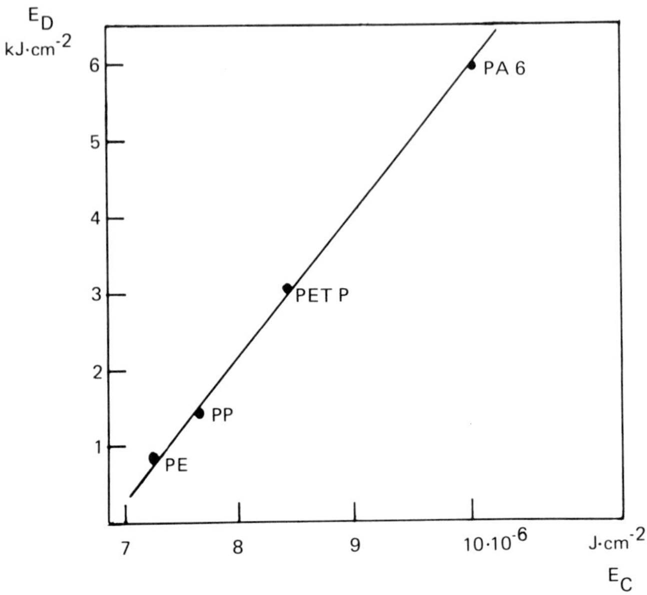


**Figure 52** Deformation energy ($E_D$) as a function of the cohesive energy ($E_C$) for different polymer films produced. (From Michels and Franz, 1978.)

With both film types, the maximum attainable stretch ratio decreases with increasing extrusion speed, since the melt orientation in the die slot rises with increased speed. The higher the velocity of the polymer melt passing the die slot the higher is the degree of preorientation and the lower the achievable maximum stretch ratio. A high stretch ratio, however, does not mean a high deformation energy or work consumption (described by the product of tenacity and elongation at break). Figure 54 shows that the work consumption of polypropylene films may well decrease with rising stretch ratio (Moorwessel and Pilz, 1969).

It is noteworthy that at the same stretch ratio the deformation energy of tapes from tubular film is always somewhat lower than that of tapes from flat die extruded chill-roll cooled film. This result can be explained by the cross orientation of tubular film which is not completely removed during monoaxial stretching. This reduces the breaking strength, but the elongation at break does not become accordingly higher. The lower fibrillation tendency of tubular blown film caused by a lower degree of orientation at a given stretch ratio can be seen from Figure 55 (Badische Anilin & Soda Fabrik AG, 1970b).

For polypropylene the fibrillation tendency of blown tubular film at the same stretch ratio is significantly lower than that of flat die extruded film. Since the fibrillation tendency of a film increases with the degree of monoaxial orientation, the results given in Figure 55 indicate that the orientation of a flat die extruded polypropylene film in machine direction is, under otherwise similar stretching conditions, higher than that of tubular blown film. The cross orientation of tubular blown film can be completely eliminated only at very high stretch ratios. However, it is possible to control the cross orientation of blown film by adjusting the inflation ratio during extrusion. This can be varied easily from a ratio of 1:1.2 up to a ratio of 1:1.6.

Tubular blown film also has one additional feature which must be considered. As the inflated film bubble is normally only air cooled from the outside, the primary film has a slight asymmetric—laminar—structure (bicomponent structure) with regard to degree of orientation and crystallinity in comparison to water-quenched flat die extruded film (Schuur and Van de Vegt, 1975). This may result in unevenness and a curling tendency of film tapes produced from blown film.

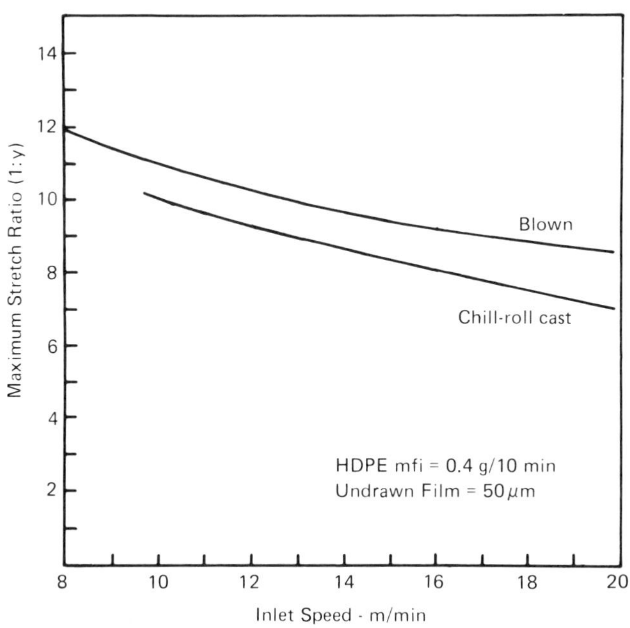


**Figure 53** Influence of inlet film speed and film production technique on stretchability of high-density polyethylene film. (From Skoroszewski, 1971.)

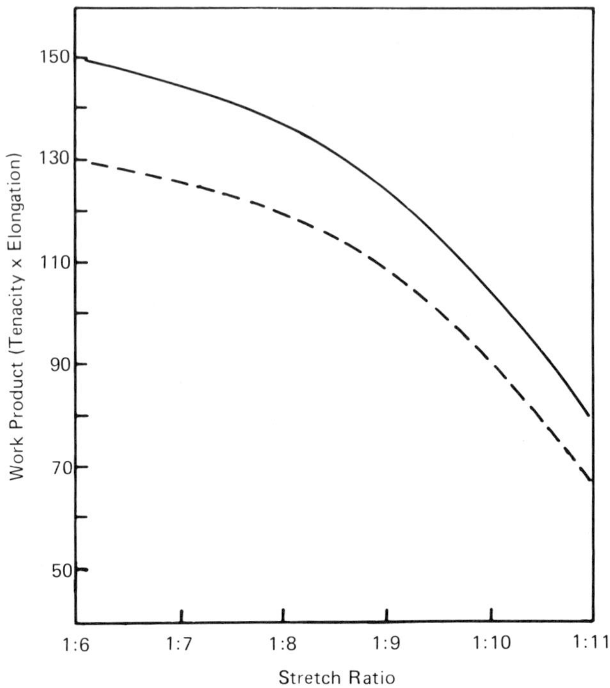


Figure 54 Effect of stretch ratio on the work product of weaving tape made from polypropylene: —— , flat die film; - - -, blown film. (From Moorwessel and Pilz, 1969.)

Such tapes have a tendency to curl up in the transverse direction, which interferes with further processing and can be of disadvantage in certain end-use applications.

In conclusion it can be said that the cooling-off speed and the thickness of the primary film strongly influences stretchability. However, since these two parameters are always acting in close connection with the stretching speed, they will be discussed later.

## C. Film Stretching and Tape Stretching

In Sec. III we addressed the two techniques for the production of film tapes, namely (a) the stretching of primary film before slitting into tapes, and (b) the stretching of film tapes after cutting the primary film.

The first technique comprises the stretching of the primary film in its entire width, while the second consists in the stretching of a flat array of narrow film tapes. The two procedures result in tapes of widely differing characteristics. Polypropylene film tapes produced from films stretched prior to tape cutting have, for example, a much lower fibrillation tendency and also a somewhat greater tendency for tearing in the cross direction compared with film tapes made by stretching from narrow film ribbons cut from primary film prior to the stretching. The reason for this difference is to be found in the peculiarities of the stretching of film in its entire width and of the stretching of narrow film ribbons. In the latter case the tape edges are thickened by the "neck-in", giving higher resistance against tear forces. Narrow tapes, furthermore, respond to a given stretch force easier than wide films; this explains the somewhat higher fibrillation tendency.

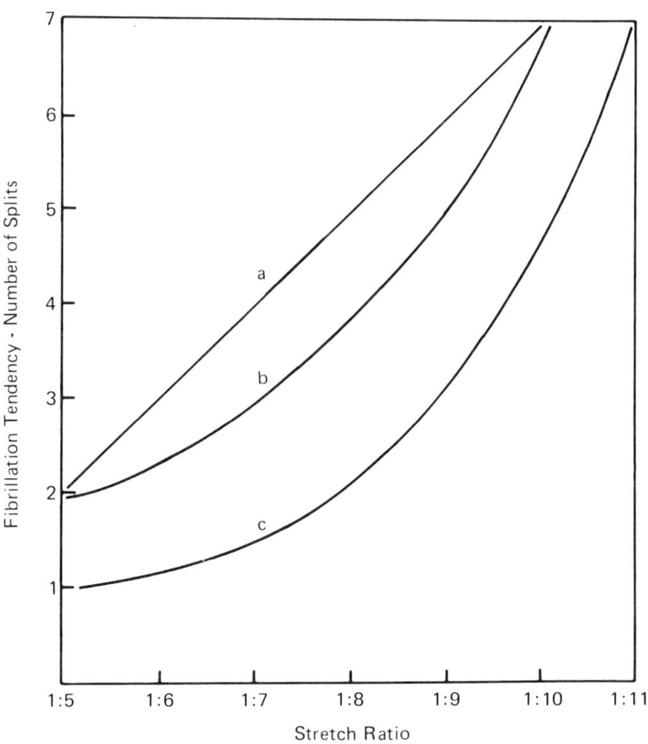


**Figure 55** Effect of stretch ratio on fibrillation tendency of polypropylene weaving tape (100 μm film): a, flat die film; b, blown film; c, blown film made from a blend of polypropylene with 10% low-density polyethylene. (From Badische Anilin & Soda Fabrik, 1970b.)

If one stretches the film in its entire width, a certain degree of cross orientation persists, even at high stretch ratios. The stretching of narrow tapes, however, results in a more ideal monoaxial orientation of the macromolecular assemblies as each individual tape goes through its individual neck-in at the yield point. The width reduction enables an undisturbed rearrangement of the crystallite assemblies into the stretching direction. Particularly pronounced is the difference between the properties of film tapes cut from already stretched film and film tapes cut from unstretched film followed by subsequent stretching in the case of polypropylene tapes. Polypropylene tapes cut from already stretched film show a distinct tendency for tear ruptures in the transverse direction. This characteristic property proves very unfavorable in the application of monoaxially stretched polypropylene film for split weaving or split knitting. The monoaxially stretched high-density polyethylene film behaves much more favorably in this respect. Since the fibrillation tendency of monoaxially stretched high-density polyethylene film is also much less expressed than that of polypropylene, film tape made by cutting from already stretched polyethylene films finds wide application for the manufacturing of raschel-knit fabrics. On the other hand, such stretched polyethylene films show sufficient fibrillation in needle roller processing of fibrillated film yarn or film fibers.

The facts and observations presented above show that film tapes which are to be processed into split fibers should be produced preferably from tapes stretched subsequent to film cutting. The same applies to the manufacturing of tape yarns for carpet backing fabrics. In the tufting operation an enhanced fibrillation tendency of the film tapes in the carpet backing fabric is desired in order to keep the unavoidable drop in fabric tenacity caused by the needle perforation as low as possible. Figure 56 shows very clearly the different behavior of film tapes produced either by stretching in a narrow roller gap without width reduction or in a hot air stretching oven over a longer distance with width reduction (the first procedure resembles the properties of tapes cut from stretched film).

In the stretching of film in its entire width the stretching field should be very short in order to prevent width reduction and extensive thickening of the film edges, making edge trimming necessary. For this reason roller stretching equipment allowing short gap stretching should be used. The heated metal rollers should be equipped with rubber-coated pressure rollers and other devices reducing transverse film shrinkage and keeping the film width constant. In the stretching of film tapes cut from unstretched primary film in a flat array, the drawing is favorably performed in longer stretching fields, giving each individual tape the opportunity to stretch evenly. Such conditions are ideally found in hot air ovens or, with some limitations, on hot plates. However, satisfactory stretching can also be achieved on heated roller stretching equipment.

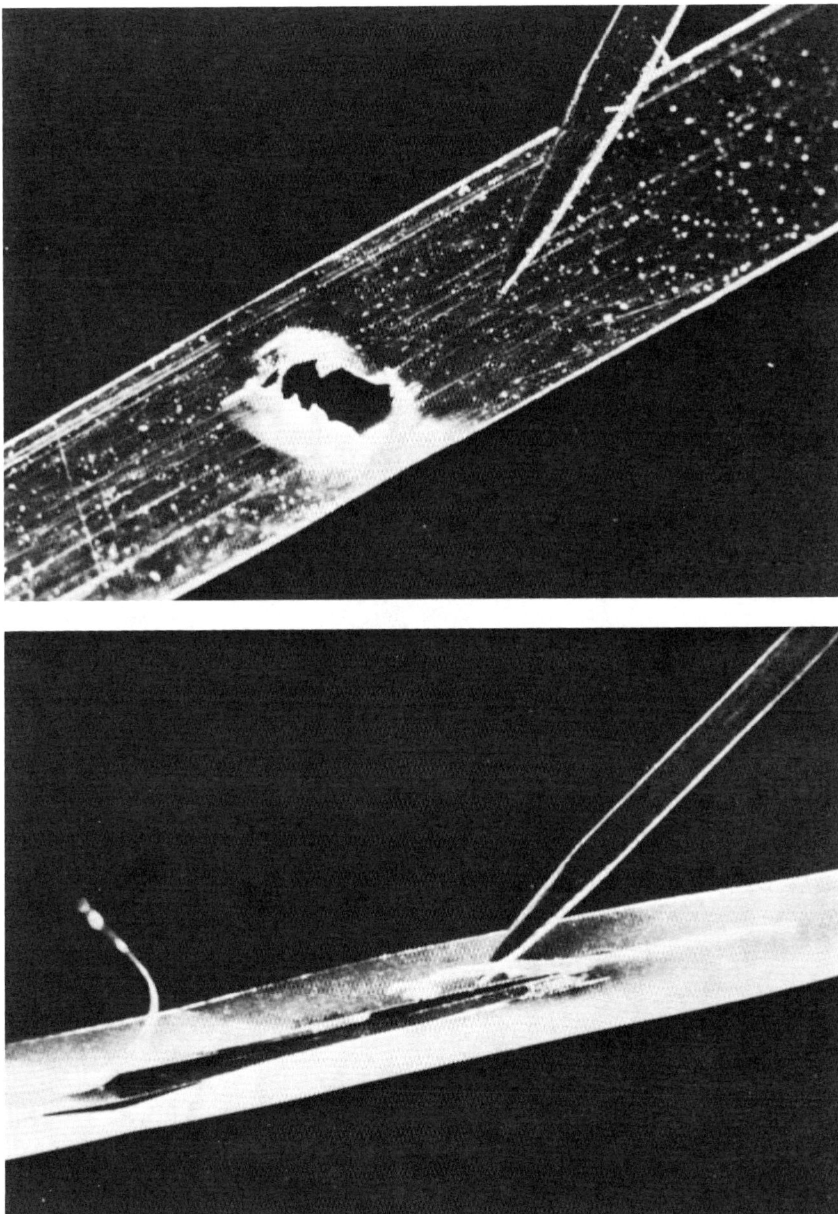

Figure 56 The effect of needle punching on polypropylene tapes. Top, tapes stretched in a narrow tangential gap; bottom, tapes stretched in a hot air oven. (From Schuur and Van de Vegt, 1975.)

## D. The Significance of Stretch Ratio for Film or Film Tape Properties

Tenacity, elongation at break, modulus, heat shrinkage, and many other properties of films and film products are dependent on and can be adjusted by the proper selection of the draw ratio. While tensile strength increases with rising stretch ratio, the elongation at break and the deformation energy decreases with rising stretch ratio, as demonstrated in Figure 57 (Malcomesz and Blechschmidt, 1977). This aspect is of importance with regard to the manufacture of film tapes used in the processing of bags or sacks. For such tapes a maximum work area is desired in order to realize the impact strength necessary to sustain high drops.

As mentioned before, increasing stretch ratio decreases elongation at break. More important for the end use, however, is the proportioned elastic elongation. It increases with rising stretch ratio. As a result, the form stability of articles made from highly stretched tapes under tension is improved. Figure 58 gives some result of the relation between elongation at break and elastic elongation at various ratios of stretch. The loop strength also increases proportional to the stretch ratio.

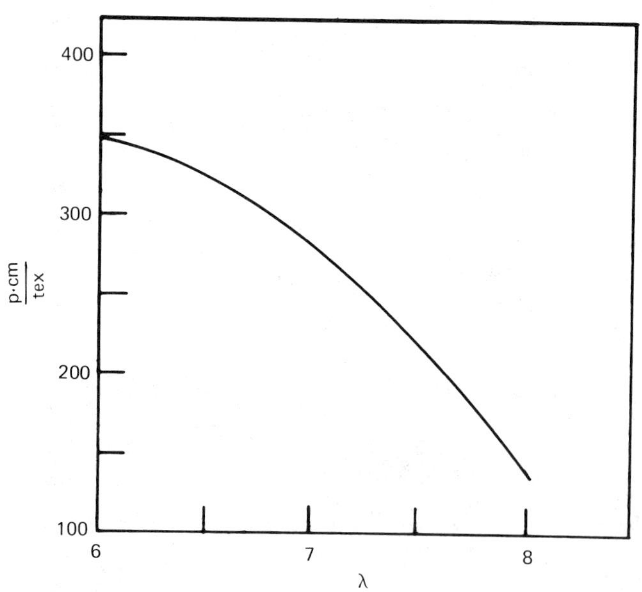


**Figure 57** Effect of stretch ratio λ on the work product of stretched high-density polyethylene films. (From Malcomesz and Blechschmidt, 1977.)

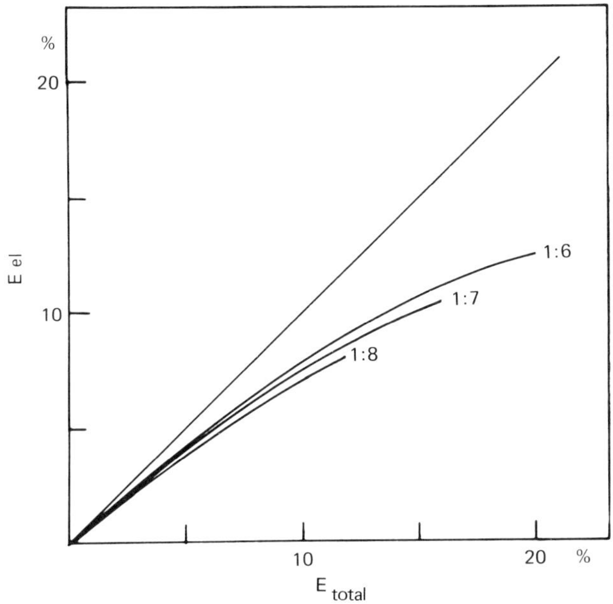


Figure 58 Elastic elongation ($E_{el}$) and total elongation ($E_{total}$) of stretched high-density polyethylene films at a load of 1.6 kp as a function of the stretch ratio. (From Malcomesz and Blechschmidt, 1977.)

This effect of stretching is illustrated by corresponding observations made by Malcomesz and Blechschmidt (1977) and is shown in Figure 59. The changes of tensile modulus and elongation at break as a function of the stretch ratio are demonstrated in Figures 60 and 61.

The effect of the stretch ratio on tenacity and elongation is shown in Figure 62; there is no linear relationship. Both properties asymptotically approach a final value with increasing stretch. The course of the interrelation is to a great deal dependent on the stretching temperature, as will be discussed in Sec. IV.E. From a certain point on, the tensile strength declines with increasing stretch ratio, as shown in Figure 63. This is very probably due to overstretching and molecular ruptures in the structural system of the film or film tape. An increasing fibrillation tendency may also contribute to the observed drop in strength. At a high tendency to fibrillate, a lengthwise splitting may occur under the forces applied in testing and individual breaks of thin fibrillated strands will take place before the actual break load is reached.

As already mentioned, a high stretch ratio applied in the production of polypropylene films or film tapes is causing poor thermal

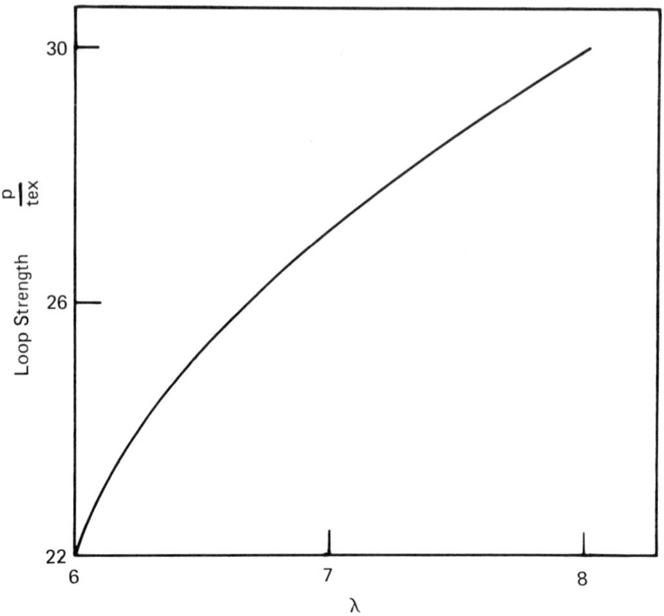


**Figure 59** Effect of stretch ratio λ on the loop strength of high-density polyethylene film. (From Malcomesz and Blechschmidt, 1977.)

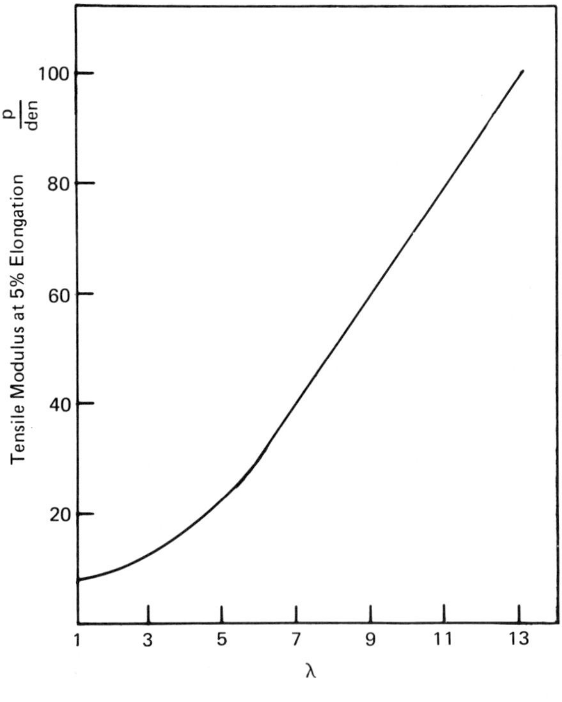


**Figure 60** Effect of stretch ratio λ on the tensile modulus of polypropylene film tape (stretching temperature 140°C). (Courtesy of ICI.)

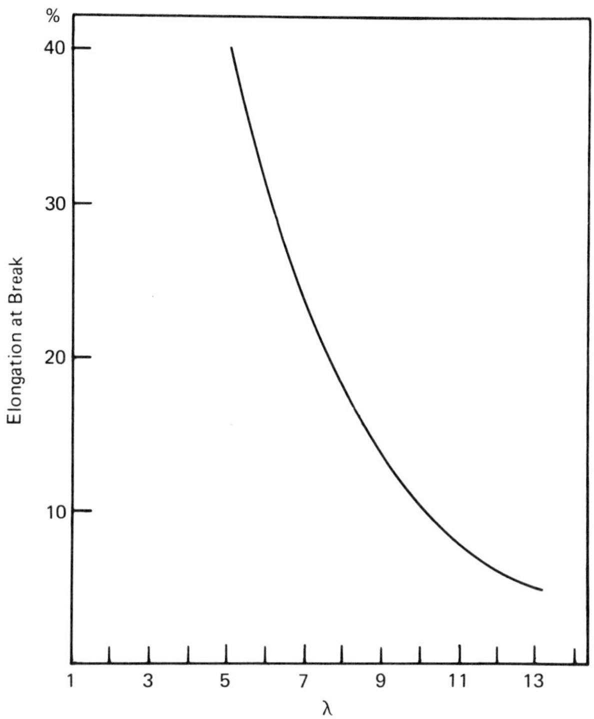


**Figure 61** Effect of stretch ratio λ on the elongation at break of polypropylene film tape (stretching temperature 140°C). (Courtesy of ICI.)

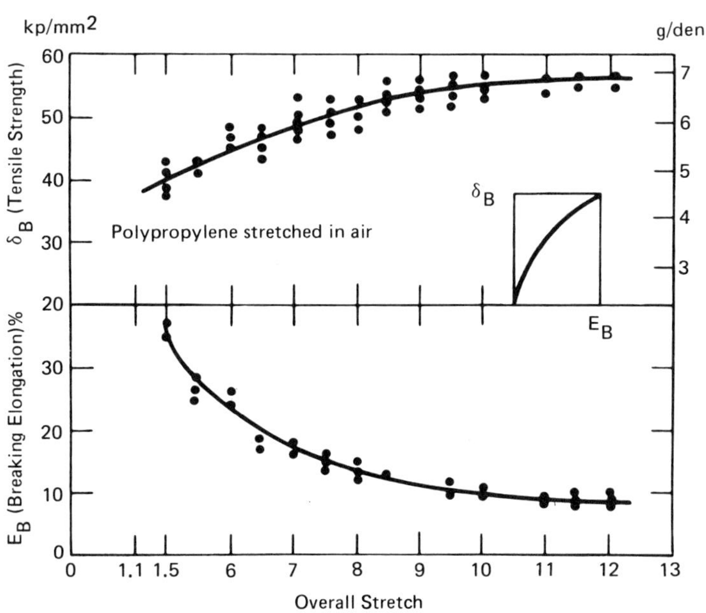


**Figure 62** Tensile strength and elongation at break of polypropylene tape yarns as a function of the stretch ratio. (From Hossack, 1971.)

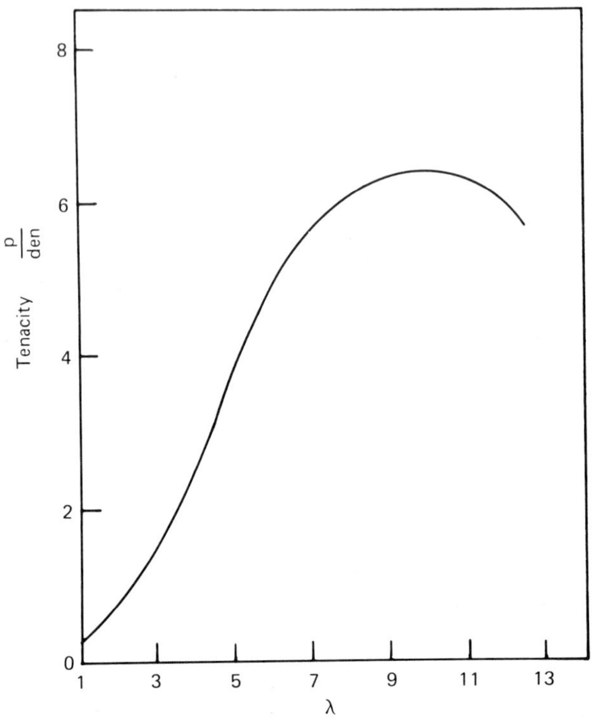


Figure 63 Effect of stretch ratio λ on the tenacity of polypropylene film tapes (1000–1800 denier). (Courtesy of ICI.)

shrinkage behavior. Shrinkage behavior is measured on test samples under constant tension in hot air or hot oil. The effect of an increasing stretch ratio on the thermal shrinkage of polypropylene weaving tape is shown by Figure 64 (Moorwessel and Pilz, 1969).

While one does not observe dimensional changes at shrinkage temperatures up to 110°C, shrinkage temperatures higher than 120°C cause drastic changes as soon as the stretch ratio exceeds 1:7. It can be seen from Figure 63 that at stretch ratios between 1:11 and 1:12 there exists a maximum of thermal shrinkage, measured at 120° or 130°C. This range of stretch ratios and their effect on thermal shrinkage behavior has been examined thoroughly by Cooper (1975). He states that stretching of polypropylene film which has been extruded at high temperatures and cooled slowly results in a significant reduction of thermal shrinkage, even at high stretch ratios of more than 1:10. This effect is especially distinct at relatively low stretch temperatures. The shrinkage maximum flattens out with increasing stretch temperature and disappears completely at temperatures above 160°C, as can be seen from Figure 65.

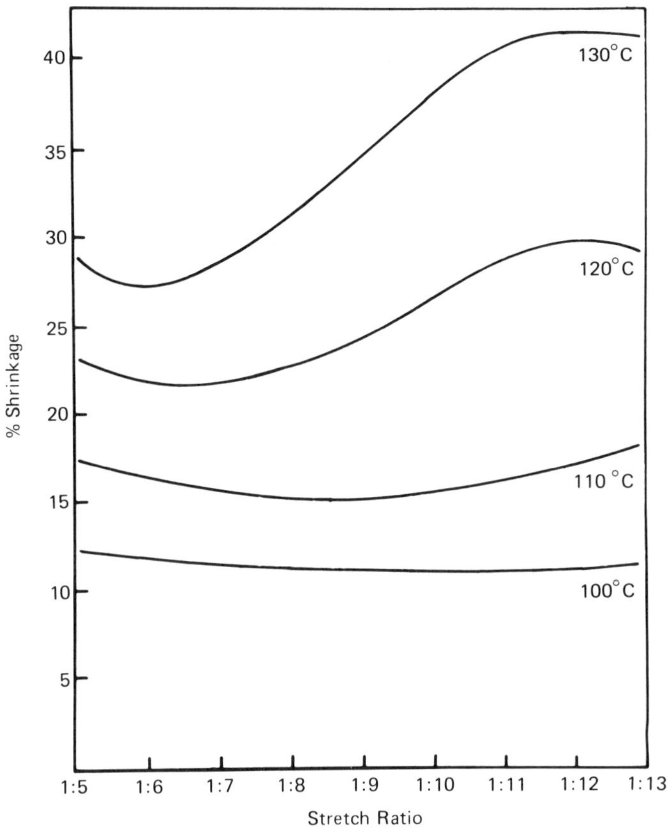


**Figure 64** Effect of stretch ratio on the heat shrinkage of polypropylene weaving tape at varying shrink temperatures (blown film; undrawn film thickness 100 μm; no heat setting treatment.) (From Moorwessel and Pilz, 1969.)

For practical operations this means that good dimensional stability for polypropylene film or film tapes at elevated temperatures can be achieved, even for high stretch ratios, by applying high stretch temperatures. High stretch temperatures in the proximity of the crystalline melting point are, however, always somewhat problematic and can cause ruptures during the stretching operation. Nevertheless, the production of shrink-proof polypropylene tapes demands high stretch temperature, especially when high stretch ratios are used.

In addition, Figure 65 shows how tensile strength and elongation at break develop independent of the stretching temperature in the range where overstretching occurs. Gansfield et al. (1976) describe

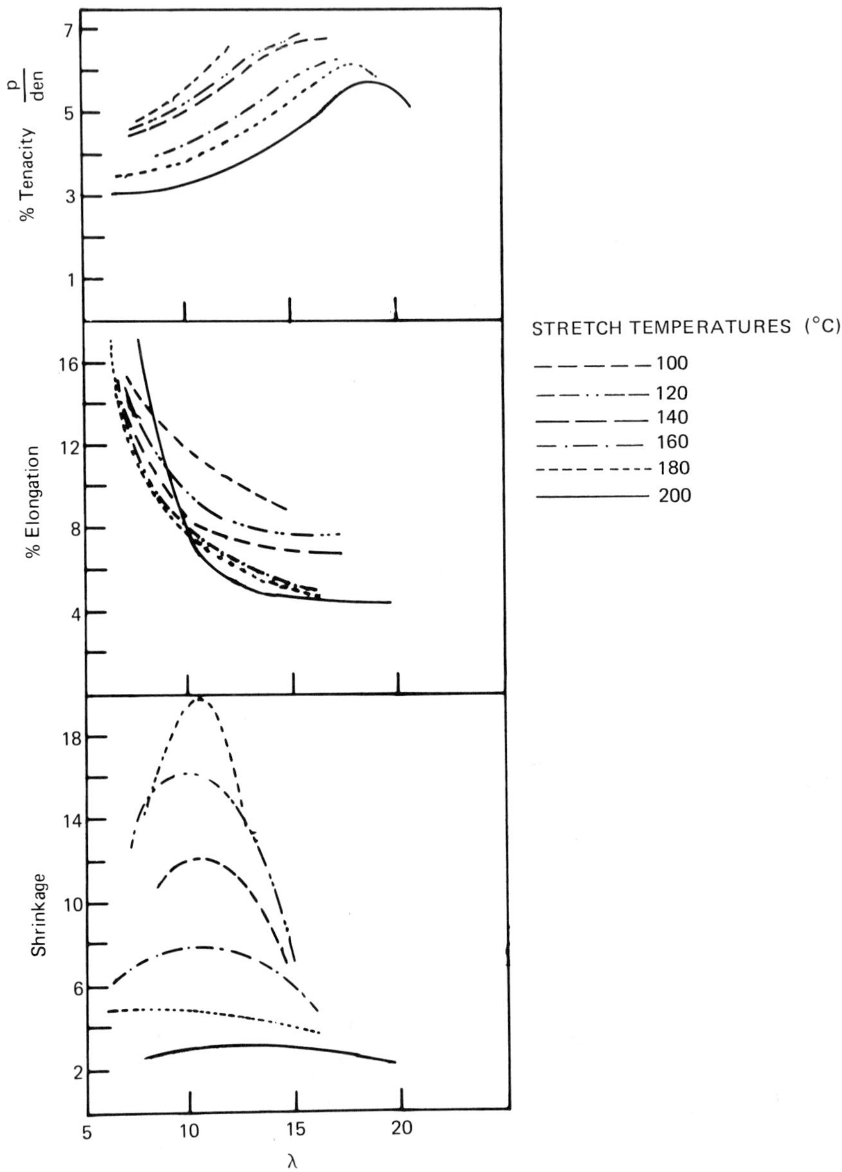


Figure 65 Effect of stretch ratio λ and stretching temperature on tenacity, elongation at break, and heat shrinkage at 130°C of polypropylene split fiber yarn. (Courtesy of ICI.)

the stretching of polypropylene film to extremely high moduli. Under optimum conditions a modulus of $1.9 \times 10^{10}$ $Nm^{-2}$ was obtained. As mentioned earlier, the range of high stretching ratios has technical relevance also with regard to fibrillable film tapes. The higher the achieved molecular orientation the easier is the subsequent fibrillation into split or slit fibers (see Figure 55). The stretching of films or film tapes to extreme high stretch ratios, however, is often connected with difficulties on production scale because under such conditions tape ruptures are more likely to occur.

For a smooth stretching process the following requirements should therefore be fulfilled:

1. The polymer melt should be well filtered and free of inhomogeneities.
2. Application of slow air cooling should be made in order to enable high stretch ratios.
3. The primary film should be free of folds and creases in order to minimize the tendency for stretch ruptures.
4. The film surface should be free of indents which means that chill-roll surfaces have to be absolutely smooth.

## E. The Significance of Stretch Temperature for the Stretching Process and for Tape Properties

The molecular orientation of high polymers under tension is an exothermic process. Nevertheless, the stretching has to be performed with external heat application in order to give the macromolecules sufficient mobility. The stretching temperature (i.e., the temperature of the heating elements or of the ambient heating medium), should be as near as possible to the softening point of the polymer to be processed. For this reason the optimum stretch temperature of high-density polyethylene lies between 110 and 125°C and for isotactic polypropylene between 140 and 175°C. For polypropylene a much wider range of stretching temperatures is thus possible. High stretching temperatures are especially important in combination with high line speeds in order to compensate for the shorter residence time in the stretching aggregate.

The stretching temperature has a noticeable effect on a number of important film or film tape characteristics, such as tenacity, elongation at break, thermal shrinkage, and fibrillation tendency. Figure 66 shows the dependence of the tensile strength of polypropylene film from the stretch ratio applied and from the stretching temperature. It can be seen from the results presented in the graph that maximum tenacity is achieved at stretch temperatures of 130 to 150°C. The decrease in tenacity of polypropylene films above the optimum stretch temperature is also shown by Figure 67. It can be seen from this graph

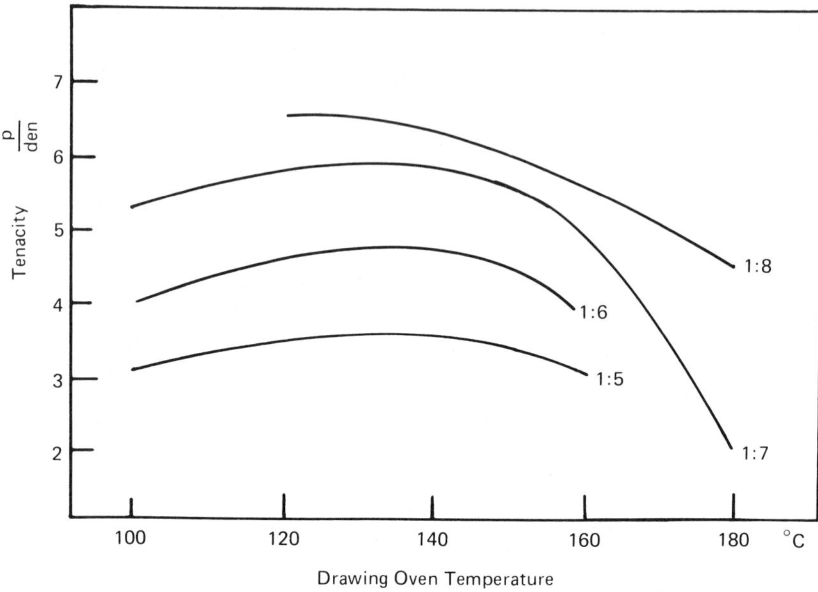


**Figure 66** Effect of oven temperature on tenacity of polypropylene tape yarn at different stretch ratios. (Courtesy of ICI.)

that a high extrusion temperature (up to 300°C) facilitates the stretching process but limits the maximum achievable tensile strength (see also Sec. II.A.).

On the other hand, it is very difficult to stretch a film at low temperature especially when it was also extruded at low temperature. The good stretchability at high temperatures is an immediate result of the thermal plasticity of the given polymer. Under these conditions the polymer exists in a plastic state so that the monoaxial deformation can be performed without creating excessive internal tensions. A stretched film free of internal tension shows an improved dimensional stability and also a reduced relaxation tendency. The effect of the stretch temperature on thermal shrinkage is illustrated in Figure 68 (Moorwessel and Pilz, 1969). Further data on this subject may be found in publications by Gouw and Skoroszewski (1968) and by Weber (1967).

The fibrillation tendency of monoaxially drawn films is markedly reduced by elevated stretching temperatures (Moorwessel and Pilz, 1969). Polovikhina et al. (1978) stated that subsequent heating of a stretched film shows the reverse effect. When polypropylene film stretched at a stretch ratio of 1:10 and at a stretch temperature of 100°C is subsequently heated to 120°C, a 50% reduction in transverse tenacity was observed, indicating an increased fibrillation tendency. This effect could be traced to an increase in the degree of orientation and of crystallinity through the heat setting under tension. Heat setting with simultaneous relaxation, however, leads to a reduction of internal stresses and reduces the fibrillation tendency.

As mentioned above, overstretching phenomena can be observed in the stretching of polyolefin films. When such films are stretched at low temperatures and with a high stretch ratio, a particular phenomanon appears. Numerous microscopic voids or fissures are formed in the film and give it a silvery look. Polypropylene films especially show this phenomenon when they are stretched at temperatures below 100°C. The tenacity is somewhat reduced by these voids. Films of this kind may find application as ornamenting tapes (Schuur and Van de Vegt, 1975). It should be mentioned, however, that films containing pores, so-called "foamed films," are generally produced by different means, namely through the addition of gas generating agents.

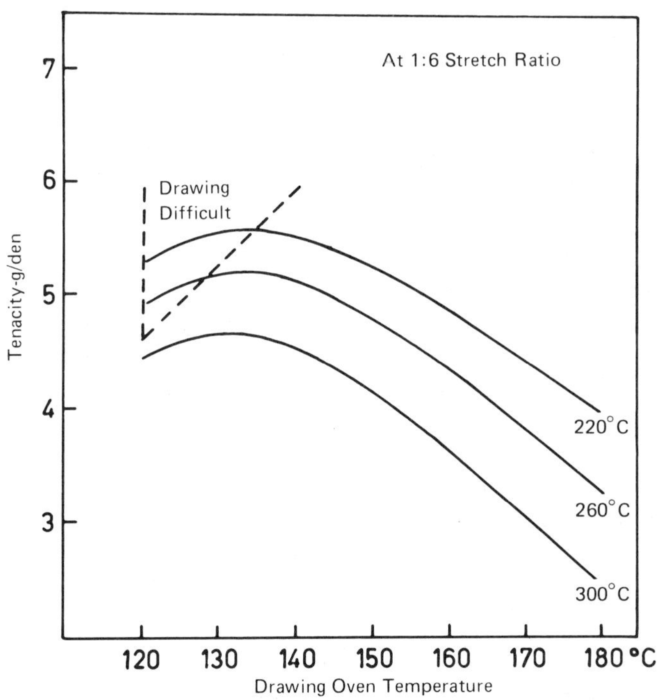


Figure 67 Effect of drawing oven temperature and of extrusion temperature at a given 1:6 draw ratio on the tenacity of polypropylene film. (From Evans, 1971.)

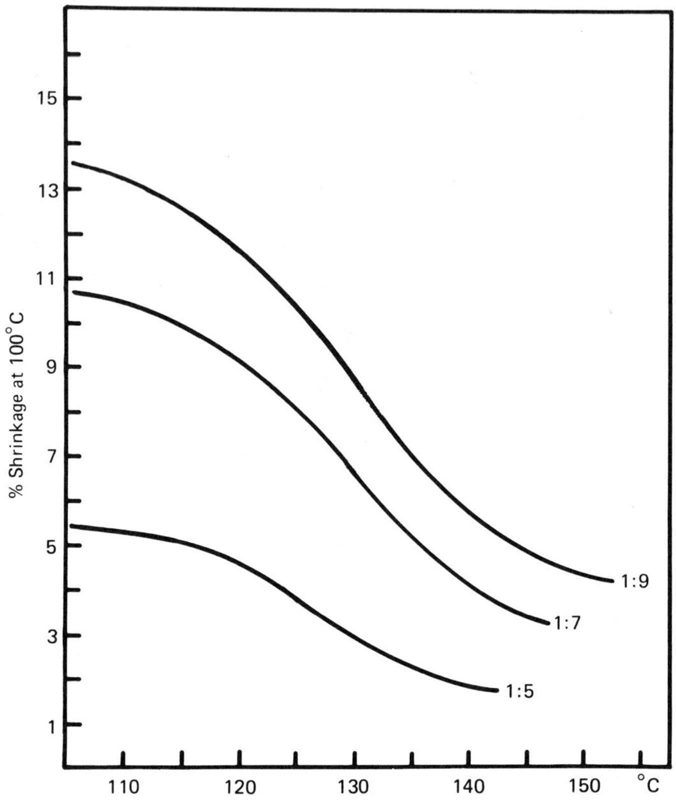


**Figure 68** Effect of stretching temperature on the drawn film heat shrinkage at 100°C (after heat setting at 125° under 15% length release). From Moorwessel and Pilz, 1969.)

In the drawing of wide flat film the stretching temperature should be—according to Toyobu Co. (1974)—for polypropylene film, not more than 20°C below or above the crystalline melting point in order to obtain the highest possible orientation parameters (measured as the difference of the refractive indices in length and in transverse direction in the range of 1.3 to 2.7 × $10^{-2}$). Tapes which are cut from film drawn in this way have maximum tenacity in machine direction at a high level of transverse tenacity. According to Mitsubishi Rayon Co. (1974), a two-step stretching treatment of polypropylene films at 175°C leads to films which in cutting result in tapes having very good mechanical properties. A method applying stretch temperatures which gradually increase in line direction is described by Ohta (1977).

## F. The Effect of Film Thickness and Inlet Speed on the Stretching Process

The correlations between film thickness and film speed on the one side and the operational safety and textile mechanical properties of the final product on the other side is relatively complicated. The statements made in Sec. II.C. should be recalled at this point. Some of the aspects of these correlations have already been discussed there. The significance of the properties of the primary film was specifically illustrated in this earlier discussion. The thicker the unstretched primary film the higher must be the stretch ratio in order to achieve a high level of tensile strength at low elongation at break. Only at high stretch ratios can the low degree of preorientation in slowly cooled thick primary films be balanced. Figure 69 (Moorwessel and Pilz, 1969)

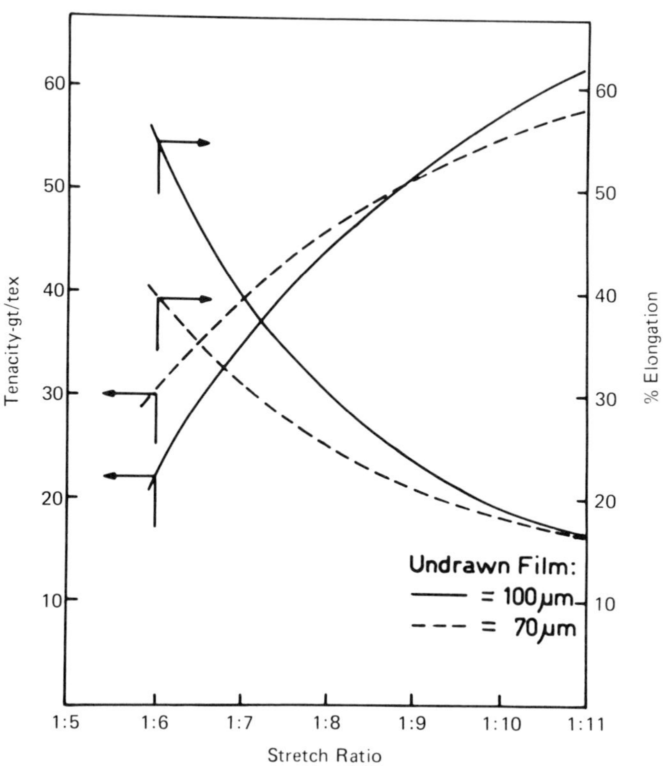


**Figure 69** Influence of film thickness and stretch ratio on the development of tenacity and elongation in polypropylene weaving tape. (From Moorwessel and Pilz, 1969.)

shows this correlation for polypropylene films. The same relationship has also been observed for high-density polyethylene. A slow cooling of the film always necessitates the application of higher stretch ratios.

Between high-density polyethylene and isotactic polypropylene, however, exist a marked difference with regard to the operational regularity of the stretching process in relation to the evenness of film thickness. In the case of polypropylene, stretch ratio and stretch temperature are not limited by the thickness of the primary film. In the range of 40 to 1,200 μm the stretch ratio can be varied from 1:8 to 1:11 and higher without reduction of the line speed. Only at very high molecular weight (i.e., extreme low values of the melt-flow index) does the inlet speed of the primary film into the stretching operation have to be reduced in order to achieve optimum tenacity levels. With high-density polyethylene, however, a greater primary film thickness demands a reduction of the inlet speed to avoid excessive stretch ruptures due to nonuniform film heating.

As Figure 53 has already shown, the maximum possible stretch ratio diminishes with increasing film inlet speed for a given film thickness of the unstretched primary film (Skoroszewski, 1971). A high film speed creates high melt orientation in the die slot, which reduces the stretchability. This gives rise to some overstretching and thus to a reduction in the achievable optimum tenacity at high stretch ratios (Evans, 1971).

In Figure 36 the danger of overstretching at high inlet speeds (i.e., extrusion speeds) was illustrated. The data in the same figure, however, also indicate that high tenacity values can be obtained with high line speeds when the stretch ratio is properly choosen. As already pointed out, it is possible to suppress the melt orientation by increasing the extrusion temperature at the die and by widening the die slot. A thick film and a high line speed make thorough and quick heating of the film or the film tapes entering the stretching procedure essential. The danger, however, that at high inlet temperature the film or individual tapes could melt—causing rupture—must be taken into account.

Thus, the three factors: film thickness, inlet speed, and stretch ratio, are of utmost importance for the efficiency of film or film tape production, not only with respect to line capacity but also with optimum end-product properties. Figures 70 and 71 illustrate to which degree the production capacity of a manufacturing unit for polypropylene tapes can be raised through an increase in film thickness, winding speed, and stretch ratio.

It should be mentioned that the thickness of the film tapes is important for the UV light stability of fabrics produced therefrom. When a film tape of 40 μm thickness loses 50% of its initial tenacity during 5 months exposure to UV radiation, the half-life period of 80-μm-thick film tapes from the same polymer is approximately 9 months. The

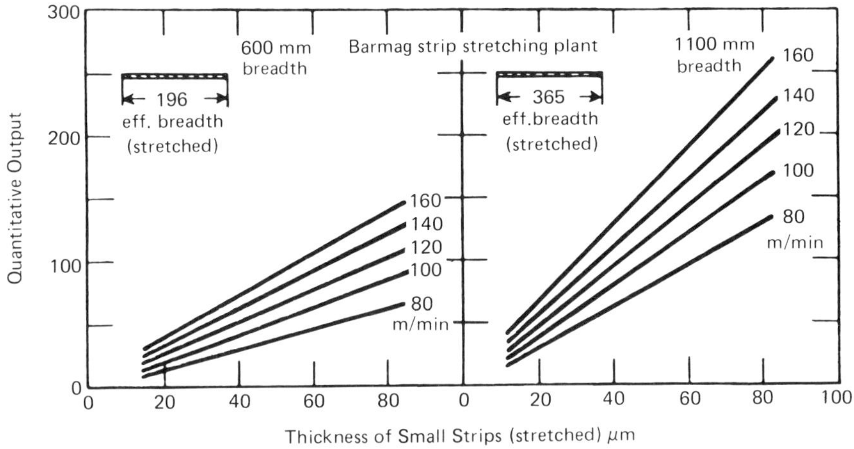


Figure 70 Output of a polypropylene tape yarn production line as a function of film thickness, film breadth, and winding speed at a stretch ratio of 1:7.5. (From Hossack, 1971.)

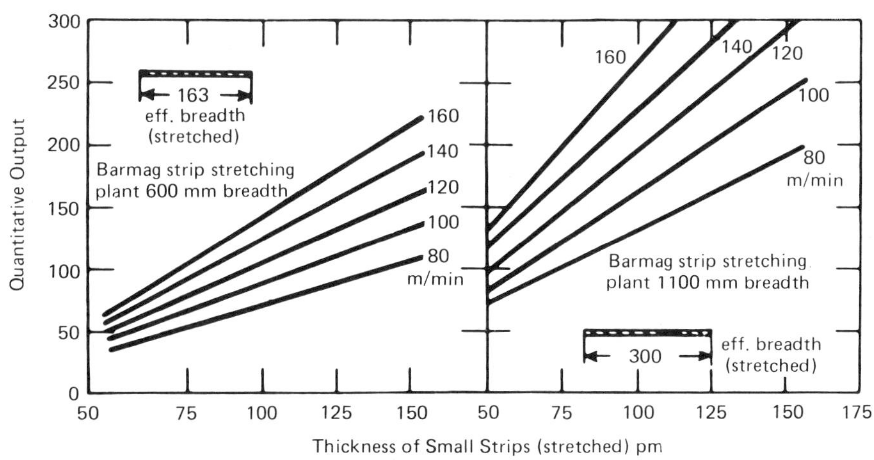


Figure 71 Output of a polypropylene tape yarn production line as a function of film thickness, film breadth, and winding speed at a stretch ratio of 1:11. (From Hossack, 1971.)

inner layer in the film tapes of larger thickness are somewhat protected by the outer layers (Hossack, 1971).

## G. Stretching Technologies

For the monoaxial stretching of polyolefin films, several machine systems were developed. A stretching unit consists fundamentally of the following elements. The feeding unit acts as holding device and supplies the holding force, while the traction unit running at a higher speed corresponding to the stretch ratio supplies the stretching force. The necessary stretching force is transferred to the film through friction and the tension force or by means of pressure rollers. The tension $S_1$ between the feeding and the stretching unit is dependent on the tension $S_2$ behind the stretching unit, on the friction coefficient $\mu$ between the film and the stretch roller surfaces, and on the loop angle $\alpha$ of the film contact with the stretch rollers. The following relation exists between these units:

$$S_1 = S_2 \cdot e^{\mu} \cdot \alpha$$

This means that the higher the friction coefficient and the higher the loop angle of the film with the stretch rollers, the lower the tension behind the traction unit to prevent the occurrence of backlash slippage has to be (Hensen and Braun, 1978b).

The distance between the feeding unit and the traction unit is variable. It can be only a short gap, as in a roller stretching machine, or several meters, as in heat stretching ovens or in hot plate stretching equipment. The heating of the film can be accomplished on the heated rollers of the feeding unit or between the feeding and the traction unit with the aid of hot air or a hot plate. Besides the stretching in one step, double-step stretching using two stretching ranges arranged one behind the other is also applied in practice. Other elements of the stretching units for film tapes should be mentioned, namely suction devices for broken film tapes and winding units for edge trimmings.

### 1. *Units for the Stretching of Film Tapes*

Fundamentally, all stretching systems for film tapes can be classified into two groups:

1. Tape stretching units combined with slitting the primary film in front of the stretching field.
2. Stretching of wide primary film combined with cutting the stretched film behind the stretching field.

All units for the stretching of film tapes have, as a common feature, a 2.5- to 4-m long stretching field. The holding force on the inlet side of this field is achieved by entwining the film on one-side supported rollers or galettes of the 3-, 5-, or 7-roller inlet holding unit ("trios," "quintets," or "septets"). Since trios or quintets the friction alone is in most cases not sufficient for slippage-free force transfer, the tapes have to be pressed to one of the three or five rollers with a rubber coated pressure roll. This calls for enforced bearings for the corresponding trio or quintet rollers. Frequency regulated roller or galette drives allow for a constant stretch ratio, equalizing variations in the line speed.

The rollers or galettes of the feeding and traction units can be heated with oil. In the case of the feeding unit, heated rollers bring the film up to the stretching temperature more quickly. Heating of the traction rollers assists the heat setting of the film tapes, as will be described in detail in Sec. V.C. In the actual stretch field between the feeding and traction unit, the tapes have to be heated thoroughly. This can be accomplished either in a hot air duct or through passage over a hot plate. The principle of the heat stretching ovens or of hot plate stretching units is demonstrated in Figures 72 and 73.

Depending on stretch temperature, film speed, and stretch ratio, a more or less sudden occurrence of the neck-in leading to the drop in tape titer takes place. Due to the form of the tape extension in the moment of the occurrence of the stretch, one speaks of the "necking zone". The place at which the neck-in appears should be in the first third of the stretch field in order to make thickenings in the tapes disappear in the further course of the stretch. The necking zone should be as short as possible. This requires a very uniform temperature distribution over the stretch field. In many cases, therefore, infrared heated ovens with air circulation are in use. For an equal temperature distribution in the stretch oven an air velocity of more than 20 m/sec is needed. An equal distribution of the hot air stream across the width of the stretch oven is also of decisive importance.

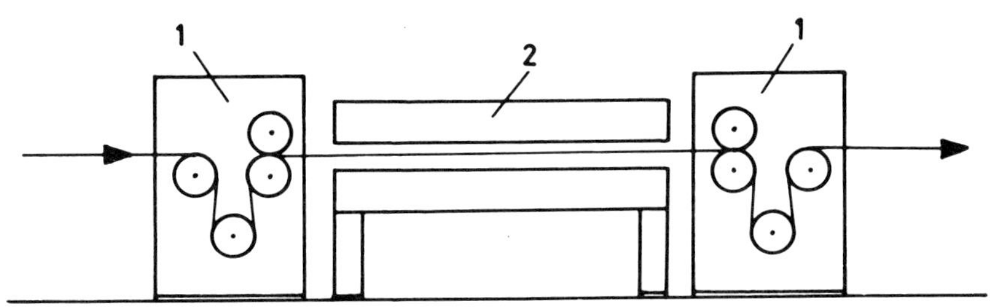


Figure 72 Scheme of a heat-stretching oven: 1, trios (for the incoming undrawn film at lower inlet speed on the left-hand side; for the outgoing stretched film at higher takeoff speed on the right-hand side); 2, stretching oven. (From Krässig, 1977.)

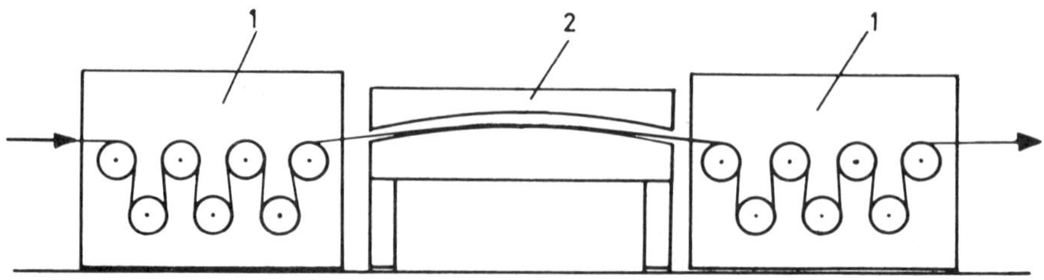


Figure 73 Scheme of a hot plate stretching apparatus: 1, septettes (for the incoming undrawn film at lower inlet speed on the left-hand side; for the outgoing stretched film at higher takeoff speed on the right-hand side); 2, hot plate stretching unit. (From Krässig, 1977.)

Besides this system, double circuit air circulating ovens with air heating through electric heating rods are also in use. At maximum air temperatures of 300°C and air velocities of 10 to 20 m/sec, the maximum temperature deviation in the conduit should be no more than ±1%. The advantages of these systems lie in the very uniform heating of the tape flock from all sides. Disadvantageous however, is the relatively high energy consumption and the difficulty in taking up and hauling broken tape ends. For film tapes which are later being fibrillated, hot air stretching is an advantage because the necking is not hindered by any frictional resistance. Thus, there is no residual transverse orientation.

With respect to the handling of tapes during the stretch operation, stretching on hot plates offers certain advantages in comparison with the stretching in hot air ovens. With such systems the heat is transferred by direct contact with the polished hard chrome plated steel surface of the stretch plate. The heating of the stretch plate is performed with hot oil circulation. Somewhat problematic, however is the uniform temperature distribution across the whole stretch plate. At overly high temperatures the film tapes will soften and will tend to stick to the metal surface. At excessively low stretch temperatures the probability for the occurrence of stretch ruptures increases and often the desired textile mechanical properties can not be achieved.

Causes for stretch ruptures can be either residual water from the water cooling bath, uneven melt-flow index of the polymer used, decomposed polymer residues from the die, rough spots on the chill-roll surface, or thickness variations of the film. Stretching performance can be improved by the application of a mixing zone at the head of the extruder screw and by a partition of the hot plate into two or three heating zones. A temperature profile with increasing plate temperature in machine direction supplies the best conditions for a smooth stretching process. The film or film tapes stick less to the hot plate if its molecular orientation is higher. A quickly cooled primary film tends normally to stick more to the hot stretching plate than a slowly cooled film.

The stretching of tubular blown film on a hot stretching plate can be performed in any one of the following three ways:

1. The tube is slit open behind the draw-off nip roll pair and opened to its whole width. The single layered spread film is lead to the stretching field. In the case of film tape manufacture the spread film is cut into ribbons on a razor blade beam in front of the inlet to the stretching unit and passed into the stretch field.

2. The film tube is stretched as a double layer. In the case of film tape processing the cutting can be performed on the flattened double layer film tube. In the stretching field the upper layer of film ribbons is stretched on the upper side of the specially designed hot plate, while the lower layer of film ribbons is stretched on the bottom side of the plate. The width reduction occurring in the stretch allows joining of both tape arrays and guides them with the help of comb guides to the stretch outlet unit without enlarging the working width. In this case the folded edges of the film tube have to be discarded as trimmings.

3. The film tube is both cut and stretched in double layer. Though in the stretching step the upper layer has no direct contact with the hot stretching plate, this has, in most cases, no adverse effect on the course of the stretching operation. This processing method is the most simple, and from the line capacity point of view, the most economical way. However, it has the disadvantage that heat setting is not possible. When the tapes are heated in double layer to high temperatures in order to be relaxed, the two tape layers often stick together.

The length of the stretching oven or of the hot plate is important for the line speed. The higher the intended line speed, the longer the stretching oven or hot plate should be. Also, in a long stretch field a better possibility exists that thickenings in the film or in film tapes are stretched out and equalized, which in turn improves the winding behavior. After the stretching and the often necessary heat setting the film or the film tapes are wound onto beams or cylindrical cross-wound spools. The winding motor speed is controlled by a tension sensor which continually scans the film or tape tension. The film beams contain up to several thousand meters of film length. The tape spools normally weigh between 2 and 4 kg.

The further processing of tape spools is done on a spool creel. Besides the time-consuming work to charge the creel, this type of processing has the additional disadvantage that the tapes get a draw-off twist due to the overhead unwinding, resulting in bended spots in the tapes, giving the woven tape fabric an uneven look. As mentioned in Sec. III.A., there are two ways to avoid these shortcomings: (1) the tapes are wound directly in the processing onto a beam, or (2) applying the "split-weaving" technique, where a stretched film is cut by an appropriate cutting attachment to the loom into warp and processed into a fabric in a continuous operation. A more recent suggestion advocates the use of only partially stretched film being cut and stretched by an attachment on the loom allowing the performance of both functions.

### 2. *Systems for the Stretching of Films*

The basic principles of stretching a film in its entire width have been treated in Sec. IV.C. The major design features of the short-gap stretching units are determined by the requirement to avoid width reductions in the stretching zone. The schematic drawing in Figure 74 illustrates the design principle of such a hot roller short-gap stretching unit.

On the upper three to five heated rolls the incoming film is preheated to the stretching temperature and then stretched to the desired stretch ratio between the gap of the following two stretch rolls. The gap in which the stretch is performed is normally 1.8–30 mm (i.e.,

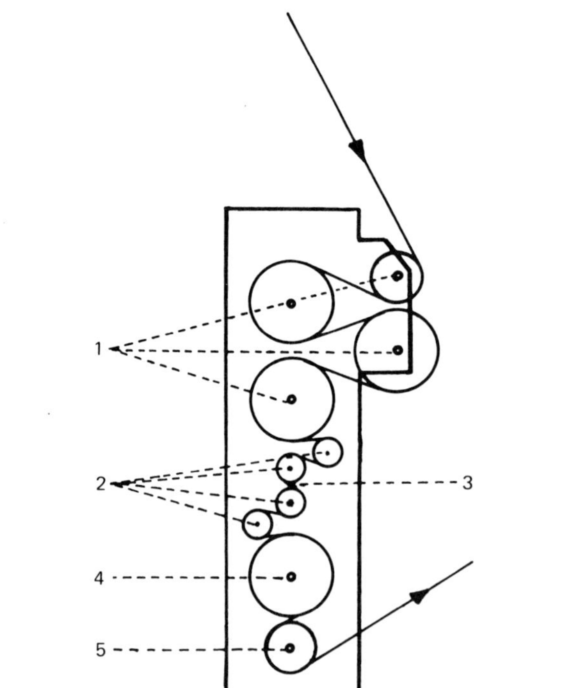


**Figure 74** Scheme of a hot roller short-gap stretching unit: 1, preheating rolls; 2, short-gap stretching rolls; 3, stretching gap; 4, heat-setting roll; 5, cooling roll. (From Krässig, 1977.)

0.07–1.2 in.) wide. The rollers of the two stretch duos are a heated steel roller and a rubber-coated pressure roll. The pressure load is provided by eccentric disks. When a low stretch ratio with a small pulling force is applied, it is possible to operate without applying holding pressure. The intimate contact between the film and the surface of the stretch rolls gives a good adhesion, thus largely avoiding the reduction in film width.

The width of the stretched film ($b_2$) can be calculated with the following formula from the width of the primary film ($b_1$):

$$b_2 = b_1 - \frac{8 \times b_1}{100}$$

Adhesion of the film to the stretch roll can be further improved by the application of an electrostatic field. After the stretching the film passes a heat-setting roller and a chill roll. With a detection device the film is inspected for possible holes. Behind the stretching unit the edge trimmings are cut off and either wound on bobbins or compacted and cut into pieces for recycling back into the extruder inlet funnel. All rollers in such a short gap-stretching machine are driven individually and heated with oil. Besides machines with only one stretch field there are also those with two stretching zones arranged one after the other. Through double stretching, films can be produced with even better mechanical properties, such as improved tenacities.

The following processing aspects are essential for achieving good film quality: (1) crease-free film at the stretch inlet, (2) even temperature distribution over the whole width of the stretch rolls, and (3) avoiding slippage of the film in the stretch zones. With short-gap stretching units, normally tubular films are processed. They are slit open before entering the stretching step and therefore fed into the stretch field in a single layer. However, double-layer stretching is also possible with roller short-gap stretching equipment. This system can also be used for welding both film layers together into one film.

Through the enforced width preservation the monoaxial stretching does not result in a complete length orientation. Films stretched with short-gap rollers, therefore, show a markedly reduced fibrillation tendency and are thus preferably used in the production of woven or raschel tape fabrics. The processing of monoaxially stretched high-density polyethylene films on raschel machines has therefore found much industrial application in the manufacture of a wide variety of bags. However, restrictions exist for uniaxially stretched polypropylene film with respect to its application in this area. Polypropylene film is normally much stiffer and shows too low a tear resistence in cross direction, which results in tape ruptures during weaving.

While on a short-gap roller stretching unit, the film is heated on individually motor driven heating rolls and stretched in a short gap between the stretch pairs, in a simple roller stretching unit the stretching is done on several idling heated rolls. These heated rolls are arranged in such stretching units between two pairs of motor driven rolls running with different speeds in accordance with the intended stretch ratio. In this case the stretching is achieved over a longer distance in the passage over several heated rolls and under a distinct width reduction. The zone in which the stretching occurs extends over a certain path, according to the roller temperature, the polymer, and the stretch ratio. With the "sandwich" method (i.e., the stretching of double layered tubular film) on such simpler roller stretching units it is possible to laminate the two film layers together using the heat generated by the polymer in the course of the drawing.

The stretching of double layered tubular film enables the use of circular dies with a larger width than that of the line width when single layer stretching would be used. By this means a high extrusion capacity can be achieved (Balk, 1978). Since on these simple roller stretching units some width reduction can not be avoided, films produced this way show a somewhat greater fibrillation tendency than those produced on short-gap stretching units.

Tape fabrics produced from monoaxially oriented film applying the split-weaving technique are noted for their very good and even coverage and their smooth look, since the warp tapes do not show any twists normally occurring in the beaming of tapes. This is a good advantage especially in packaging fabrics. On the other hand, the low fibrillation tendency forbids application in the manufacture of carpet backing for tufted carpets, since the lack of length fibrillation causes too high a tenacity loss during the needle tufting operation.

According to a patent by Kalle AG (1968), the insufficient fibrillation tendency of films stretched on short-gap roller stretch units can be increased by accomplishing the monoaxial stretching in two steps. In the first step the primary film is prestretched in short-gap stretching. In the second step the prestretched film is cut into tapes which are than after-stretched in a long distance stretch field. The subdivision of the stretching procedure into two steps results in very good tenacity values, in machine as well as in transverse direction. In conclusion it should be mentioned that the method of split weaving is very suitable for the production of a wide variety of fabrics for all kinds of end uses.

## V. THE RELAXATION AND HEAT SETTING

Every freshly stretched thermoplastic polymer film has a more or less distinct tendency toward length shrinkage. Immediately after the drawing the length shrinkage begins, even at room temperature. This shrinkage tends to disperse internal stresses in the molecular segments of the amorphous phase. Among other factors, the amount of internal stress depends on the stretching tension applied in the stretching of the film or tapes. Measurements of the stretching tension during the drawing of film tapes show clearly that high draw tension results in tapes having a high heat shrinkage.

The length of shrinkage of freshly stretched tapes and films is generally termed "relaxation"; shrinkage increases as the applied temperature increases. This phenomenon is called heat shrinkage and can be observed with all film or film tapes manufactured from thermoplastic polymers. It is defined as the percentage length reduction observed in a certain period of time at a given preload when exposed to a constant elevated temperature. Heat relaxation is commonly ascribed to the memory of the polymer to its structural state prior to the orienting drawing. Due to the higher mobility of the chain molecules at elevated temperature, the molecular segments in regions of lower order tend to achieve a conformational state of a lower energy level, closer to the thermodynamic equilibrium. A certain disorientation of the crystallites and, in consequence, a shortening of the film or film tapes in machine direction is the unavoidable result. At high shrinkage temperatures in the vicinity of the softening point, far-reaching changes in the mechanical properties take place.

Peterlin and Baltá-Calleja (1969) have examined the changes in crystalline texture occurring during the relaxation of stretched polypropylene film very thoroughly. With increasing temperature and tempering time a continuous growth of the crystallites takes place. At the same time the intensity of the long-period x-ray scattering increases. This proves that during the heat relaxation the crystallites get more perfect through the healing of order defects, while at the same time the molecular segments and the molecules in the amorphous phase disorganize and entangle. The increase of the long periods is indicative of the length growth of the crystallites or lamellas (see also Baltá-Calleja and Peterlin, 1971).

In certain applications of monoaxially stretched film tapes or split fibers the dimensional stability under the influence of elevated temperatures is of exceeding importance, especially for carpet backings and for extrusion-coated tape fabrics. (Ivett, 1968; Anderson, 1974; Cooper, 1975). In these applications the heat shrinkage measured at 132°C (i.e., 270°F) of 0.5 to 1.5% is being demanded (Scheiner, 1977). Many tape fabric specifications today ask for a heat shrinkage of less than 1%. Also very important in this connection is a narrow distribution of the heat shrinkage behavior among the single warp tapes in tape fabric manufacture. When the difference in heat shrinkage values of tapes going into the same fabric gets too large, unevenness and buckling will result.

For the production of unstabilized tapes or split fibers it is advisable to relieve the tension in the tapes or fibers after their exit from the stretching field by allowing a defined relaxation before leading them to the winding equipment. Otherwise, spools are obtained which become very hard through subsequent relaxation in the tape package. In all heat-setting treatments one has to consider that the parameters are not only adjusted so as to achieve a certain low residual heat shrinkage level; the tenacity, elongation, and fibrillation tendency also have to be kept in mind in accordance with the requirements of the intended end use. Heat setting also effects these properties, therefore heat-setting conditions have to be selected in such a way as to achieve a desired tenacity and elongation level at the required residual thermal shrinkage.

## A. The Effect of Production Conditions of Monoaxially Stretched Films or Tapes on Heat Shrinkage

Heat-setting and relaxation conditions always have to be brought in line with the shrinkage force and maximum length shrinkage of the stretched film or film tapes. Therefore, production of the primary film has to be done under conditions resulting in a material having basically a low shrinkage tendency. Various possibilities exist for achieving this goal. They will be discussed in the following chapters.

### *1. Extrusion Conditions and Heat Shrinkage*

Here also the choice of polymer is of decisive influence. Generally a high melt-flow index results in a low hot-air shrinkage. Furthermore, the shrinkage tendency of a film or film tapes is determined by the cooling rate. Figure 75 shows the correlation between the thermal shrinkage of stretched tapes and the surface temperature of the chill roll in the case of flat film extrusion of polypropylene.

It is clearly demonstrated that heat shrinkage decreases with increasing chill-roll temperature. The reason for this behavior lies in the fact that the coarse spherulitic crystal structure resulting from slow cooling leads to less internal stresses during drawing than the smectic and strongly interlinked structure formed on fast cooling. This also explains the experience that slowly cooled primary films have a higher stretch limit than rapidly cooled films (Samuels, 1968). Generally, a well crystallized primary film leads to final products of better dimensional stability. In primary film of lower crystallinity molecules in less ordered regions tend to a larger degree to adopt conformations of lower energy at increasing temperatures.

Heat shrinkage is also a function of the extrusion speed, as illustrated by Figure 76. The heat shrinkage of unstabilized film increases sharply with increasing extrusion speed. This effect is especially pronounced at relatively high stretch ratios (Gouw and Skoroszewski, 1968). The reason for this is the higher melt orientation of the macromolecules at the higher extrusion output. With an increased degree of molecular orientation the thermal shrinkage generally increases.

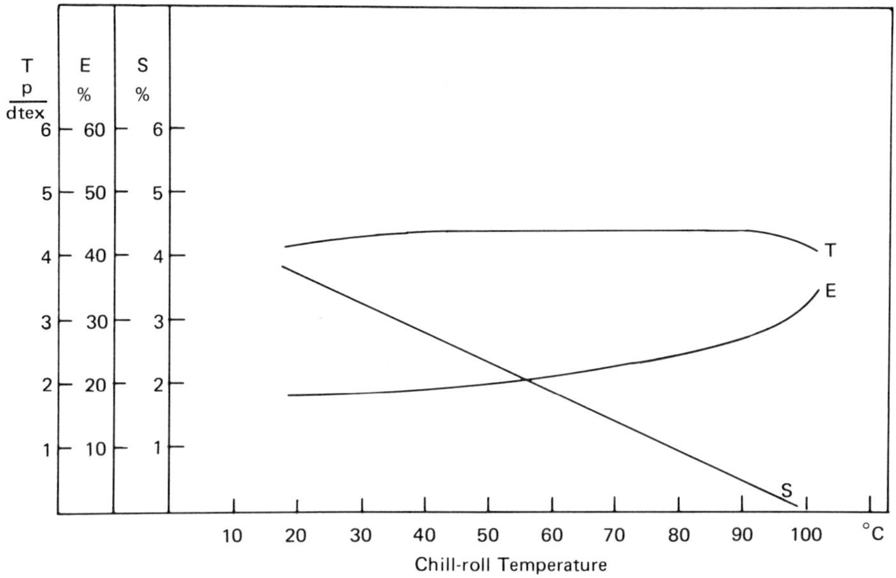


**Figure 75** Effect of chill-roll surface temperature on tenacity (T), elongation at break (E), and shrinkage at 130°C (S) of polypropylene film tapes. (From J. Lenz, unpublished work.)

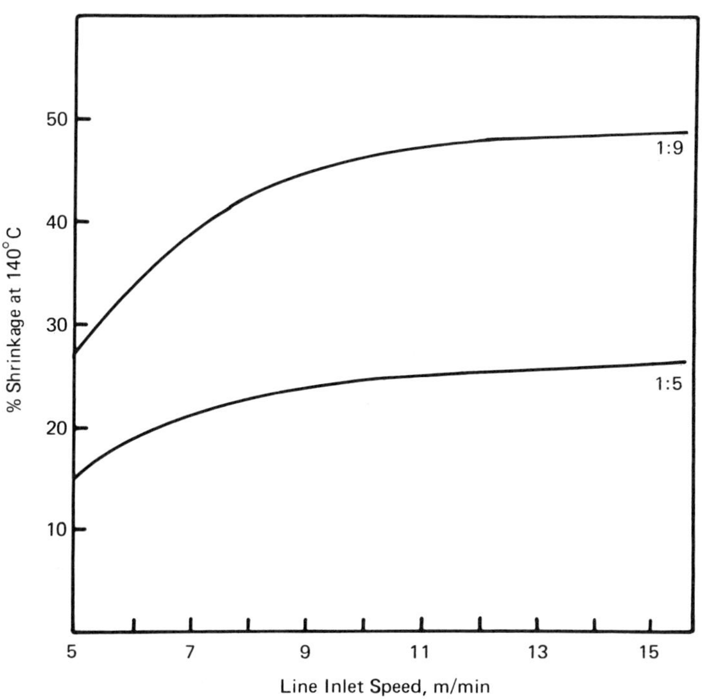


**Figure 76** Influence of stretch ratio and inlet speed on heat shrinkage at 140°C. (From Gouw and Skoroszewski, 1968.)

### 2. Stretching Conditions and Heat Shrinkage

Considerably greater than the influence of the extrusion conditions is the effect of stretching parameters on the heat shrinkage tendency of thermoplastic films and film tapes. The detailed statements made in Sec. IV.D. and the data presented in Figure 64 should be recalled here. At shrinkage temperatures up to 100°C an increase of heat shrinkage with stretch ratio cannot yet be observed. A sharp increase, however, occurs as soon as this temperature level is surpassed. It was previously mentioned that at stretch ratios higher than 1:12 the shrinkage tendency diminishes again. This fact can be explained with the previously mentioned two-step model of plastic deformation developed by Peterlin (1972). This model assumes the formation of well oriented microfibrils connected by intrafibrillar molecular bridges, "tie molecules." Increasing stretch ratio leads to an improvement of the fibrillar structure and to a reduction of the number of extended tie molecules between the fibrils.

The practical significance of high stretch ratios shall be illustrated with an example of the manufacture of polyethylene tapes having low heat shrinkage of less than 0.5% measured at 130°C. Figure 77 shows the dependence of tenacity, elongation at break, and residual heat shrinkage on the stretch ratio under operating conditions, including continuous heat setting after the stretch field. The data given in Figure 77 make it obvious that for the manufacture of "zero-shrinkage" tapes there are two sets of processing conditions which are of interest:

1. Stretch ratios of 1:5 and 1:6; under these conditions tapes are obtained with approximately 3.5 cN/dtex and about 40% elongation at break.
2. Stretch ratios of 1:13 and 1:14; at these stretch ratios tape tenacity amounts to approximately 5.5 cN/dtex and the elongation at break to about 20%.

At high stretch ratios there is, of course, the danger of tape rupture in the stretch field. It is therefore necessary to carefully avoid inhomogeneities or surface damages in the unstretched primary film. High draw ratios generally demand for a desired final film thickness a correspondingly thick primary film, which results in slow cooling and thus good crystallization. Furthermore, in such cases very even heating in the stretching field is necessary. It is understandable that it is much more difficult to achieve a low level of shrinkage in combination with high tensile strength and low elongation at break, than at low levels of tenacity and somewhat higher elongation. Barmag-Barmer Maschinenfabrik AG (1973) described a modified stretching procedure for the production of tapes with low heat shrinkage, applying double stretching with limited draw tension in each of the two stretching zones. This process design is supposed to allow a better reduction of the internal stresses in the stretched film tapes.

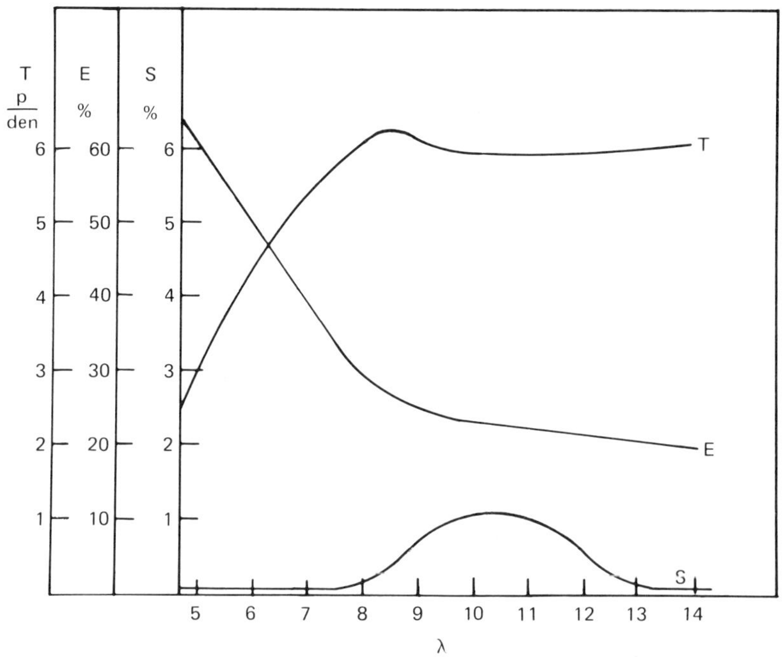


Figure 77 Effect of stretch ratio on tenacity (T), elongation at break (E), and heat shrinkage at 130°C (S) of polypropylene tapes produced under zero-shrinkage heat-setting conditions. (From J. Lenz, Lenzing Plastic Information, 1976.)

An additional way to reduce the shrinkage tendency of stretched films or film tapes is given by the variation of the stretching temperature. Generally it can be observed that the shrinkage tendency decreases with increasing stretching temperature, as illustrated by the data given in Figure 78. Additional information on the effect of the stretching temperature on the heat shrinkage was previously given in Figure 68.

This behavior may be explained by the thermoplasticity of polyolefins. With increasing stretch temperature, the necessary stretching tension decreases. As a result, low internal stresses are formed in the stretched film. A high stretching temperature results in larger and better developed crystallites. Consequently the number of extended tie molecules tending towards an equilibrium conformation between the crystallites is decreased. This means in practice that a second means toward film tapes with high tenacity, low elongation at break, and low heat shrinkage exists in stretching at temperatures in the vicinity of the softening or melting point.

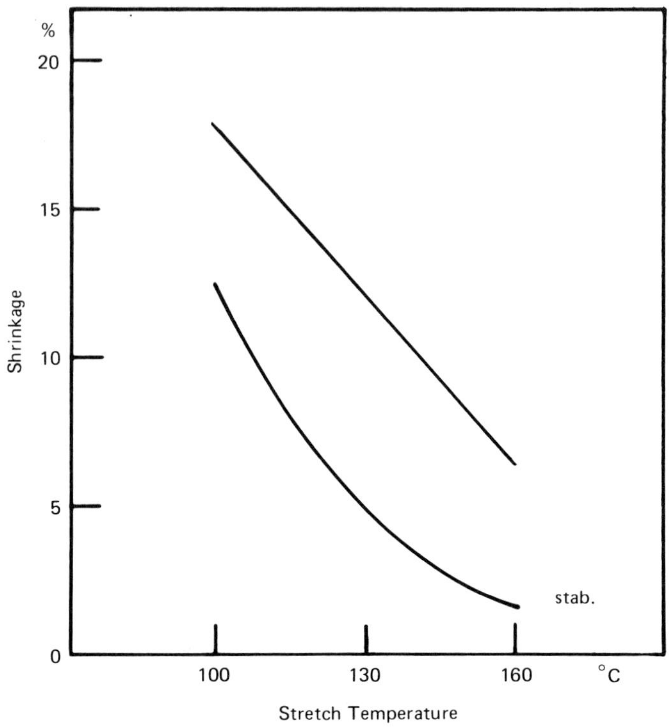


Figure 78 Effect of stretch temperature at a 1:7 stretch ratio on the heat shrinkage (130°C/20 sec) of a 1000-denier polypropylene tape yarn stabilized at a heat-setting temperature 10% higher than the stretch temperature and of unstabilized tape yarn. (Courtesy of ICI.)

## B. The Fundamentals of Heat Setting

Heat setting of thermoplastic polymer film or film tapes is generally accomplished following the stretching treatment in a continuous or discontinuous way. It is common to all heat-setting techniques that the relaxation shrinkage is initiated by exposure to elevated temperatures and by allowing length shrinkage to a defined extent. A measure for this relaxation shrinkage is the relaxation or shrinkage ratio expressed as a percentage of the original length dimensions. The higher the shrinkage temperature and shrinkage ratio are adjusted, the lower is the residual heat shrinkage of the final product. In adjusting the level of the residual heat shrinkage by a high shrinkage ratio in the heat setting, it has to be kept in mind that the shrinkage ratio also strongly influences the tensile strength and elongation at break, as illustrated by Figure 79.

Polypropylene tapes which had been stretched to a 1:9 stretch ratio can be brought to a residual heat shrinkage—measured at 180°C—by heat setting with a shrinkage of 25% at a setting temperature of 160°C. Similar experience has been reported by Gouw and Skoroszewski (1968).

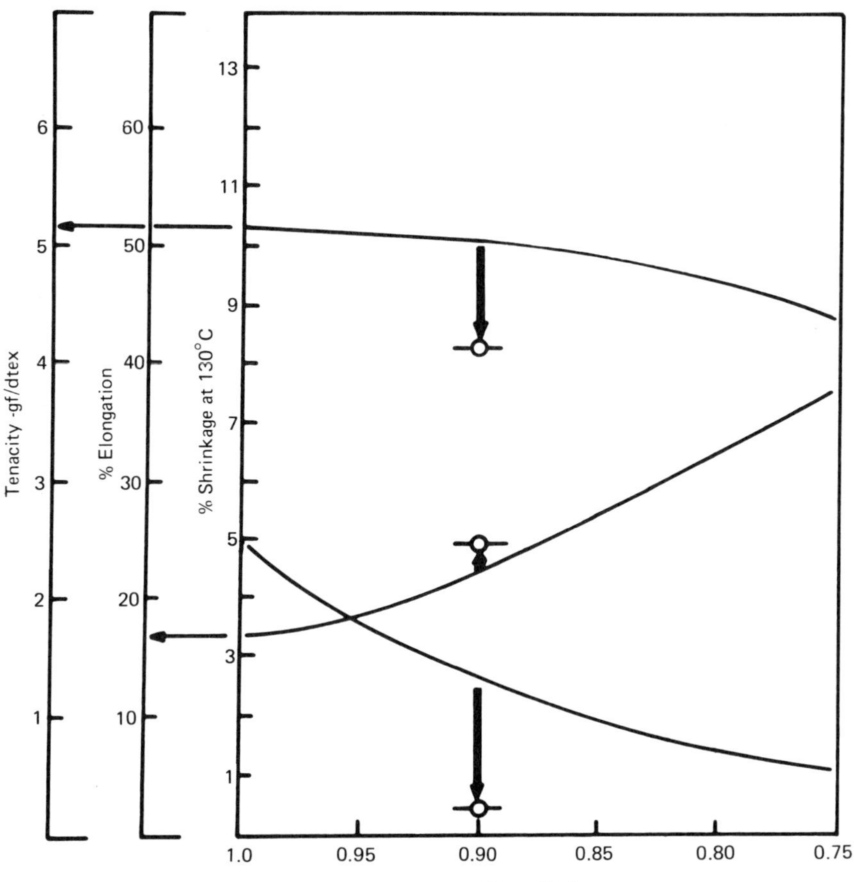


Figure 79 Effect of creep ratio (length release) in heat setting on tenacity, elongation at break, and heat shrinkage at 130°C of polypropylene weaving tape (flat die extruded film; 90 μm primary film thickness; stretch ratio 1:9; stretch temperature 105°C at hot plate; annealing temperature 160°C on hot rollers; arrows indicate property changes when extruded with hot chill roll and treated at higher stretch temperature).

However, treatments at such high shrinkage ratios increase the elongation at break to such an extent that the resulting tapes are unsuited for certain end uses. In Figure 80 complete information of the interrelations between heat-setting conditions, such as temperature and shrinkage, and the resulting mechanical properties of highly stretched polypropylene tapes, is presented. (See Imperial Chemical Industries, Technical Service Note, p. 116b.)

As can be seen from the data given in this figure, the proper selection and coordination of the heat-setting temperature and shrinkage is very important in order to achieve the desired final properties. The higher the heat-setting temperature the higher must the shrinkage be adjusted in order to achieve a desired low residual heat shrinkage in the annealed tapes. In the presented case, "zero shrinkage" can be achieved only with a heat-setting temperature of 200°C and shrinkage of 15%.

In practice the heat-setting temperature and relaxation ratio have to be adjusted in such a way that the film or tape tension in the shrinkage area is kept as low as possible. Unfortunately, operating in the heat-setting treatment with low tape tension is difficult due to unsteady running of the tapes and to sagging tape threads. Fluttering and slack tapes often cause trouble in annealing treatments. Slack tapes tend to stick to heated rollers, to touch the overheated insides of hot air heat-setting ovens, or to stick together when they touch each other. It is therefore advisable, for a safe heat-setting operation, to adjust the residual heat shrinkage with a relaxation ratio as low as possible. A relaxation ratio of 1:0.95 is generally sufficient to remove most of the internal stresses. The annealing temperature should be so high that the film or the tapes have enough tension for smooth film or thread motion. For this it is often necessary to be content with approximately 2% residual heat shrinkage.

Figure 81 shows how tenacity, elongation at break, and residual heat shrinkage change with the shrink temperature when a shrink ratio of 1:0.95 is allowed. (See Imperial Chemical Industries, Technical Service Note, p. 116b.) The position of the minimum residual heat shrinkage in Figure 81 allows deduction of the optimum heat-setting temperature at 5% relaxation. Above 170°C the shrink force of the material surpasses the shrink ratio adjusted in the machine, and tape tension as well as residual heat shrinkage increase again. Below the optimum setting temperature the heat supply is not sufficient to give the polymer molecules enough mobility for removal of all internal stresses. Therefore, the residual heat shrinkage still remains high. With decreasing film or tape tension and increasing annealing temperature, tensile strength decreases somewhat while the elongation at break is markedly higher.

From the above, it is obvious that the relaxation ratio has to be well adjusted to the shrink force which is determined by the stretch

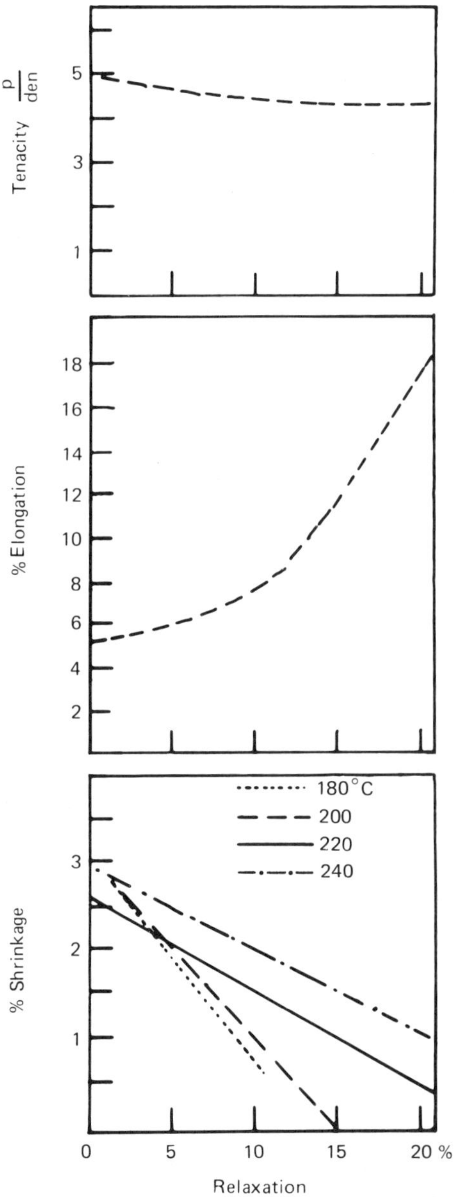


**Figure 80** Effect of relaxation (length release) in heat setting at annealing temperatures between 180 and 240°C on tenacity, elongation at break, and heat shrinkage at 130°C of polypropylene split-fiber yarn. (Courtesy of ICI.)

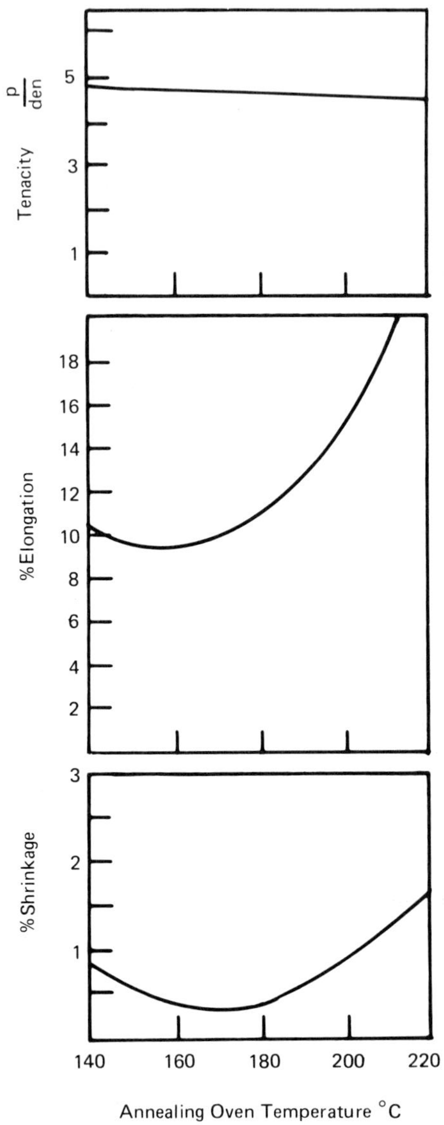


**Figure 81** Effect of annealing temperature at a relaxation (length release) of 5% on tenacity, elongation at break, and heat shrinkage at 130°C of polypropylene split-fiber yarn. (Courtesy of ICI.)

ratio and the shrinkage temperature. The production engineer has to decide in each case which compromise he has to make between a desired low level of residual heat shrinkage of his final product, the best possible mechanical properties of the latter, and an acceptable process performance. After having passed the annealing treatment, the film or the film tapes have to be wound on a film beam or on spools. During winding the film or thread tension should be as low as possible. In the case of tape winding the tension should be just sufficient to prevent the tape yarn from sliding off the front surfaces of the cross-wound spools. At excessively high wind-up tensions the residual heat shrinkage of the film or tape may well increase again in storage.

## C. The Techniques of Heat Setting

The various known techniques of heat setting can be subdivided into continuous and discontinuous procedures which, however, may also be combined. The continuous procedure is performed either by annealing on heated rollers or by heat setting in a hot air zone. One way of doing this is utilizing heated stretch outlet roller units and performing the relaxation between these and an unheated multiple roller unit (i.e., a septette). In this case the array of tapes is brought to the desired shrinking temperature utilizing the stretch temperature. Subsequently the tapes are contracted in the relaxation zone between the two units, allowing the desired shrinkage ratio. The advantages of this system lie in space saving and in the possibility for eliminating the shrinkage in several steps by the utilization of additional heated rollers which are operated at different speeds.

According to Hensen and Braun (1978a) and to Balk (1978), in film tape plants for the production of carpet backing fabrics, a hot air oven for heat setting is not absolutely necessary. The Lenzing system also operates without a hot air oven. A well established production scheme using heat-setting rollers for the annealing treatment is shown in Figures 82 and 83.

Applying this annealing method, an even temperature distribution with a maximum deviation of ±1°C over the whole operation width of the heat-setting rollers has to be achieved. If the temperature distribution is uneven with temperatures too high in some locations the film tapes will tend to stick to the roller surface, leading to undesired wrappings. Obviously, the speeds of the stretching and stabilizing rollers must be accurately controllable.

For the annealing, hot air-circulating ovens are frequently applied, as mentioned in Sec. IV.G.1. In these ovens the hot air moves normally in two circuits synchronous with the tapes. The temperature

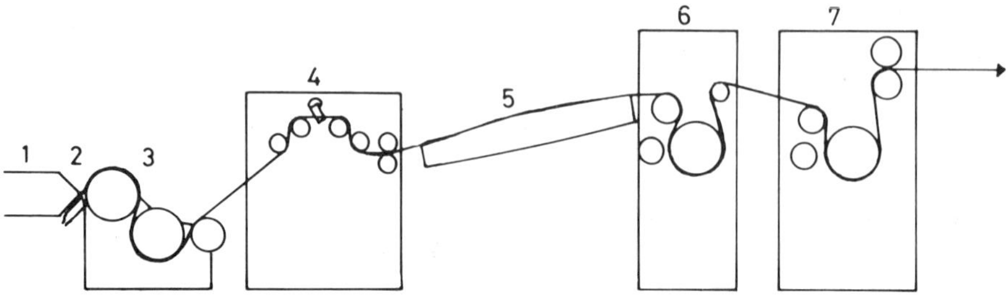


Figure 82 Film tape line for the production of tapes for carpet backing with zero shrinkage. 1, die head; 2, air suction device; 3, chill rolls; 4, slitting knives; 5, heated plate for long distance stretching; 6, stretching rolls supplying the stretching force; 7, heat-setting roll.

Figure 83 Production line for polypropylene tape yarns with heat setting resulting in a product having approximately 1% heat shrinkage at 130°C (heat-setting roller be seen on the right-hand side). (Courtesy of Chemiefaser Lenzing AG.)

accuracy has to be around ±1°C. Hot air ovens are supplied by the machine manufacturers with operating widths from 300 mm up to 1450 mm (i.e., 1 ft to approx. 5 ft). Under conditions of appropriate adjustment of the stretching and annealing conditions it is possible to produce film tapes with residual shrinkage of less than 2% (measured at 130°C) with still favorable levels of tensile strength and elongation at break.

In practice today, however, a combination of hot air stretching and heat setting on heated rollers is most often used because such working conditions allow the best control of shrinkage uniformity (Scheiner, 1977). Film tapes with a guaranteed residual heat shrinkage of only 0.5% can be produced by annealing heat-setted tapes on cross-wound spools with saturated steam in an autoclave. For this purpose autoclaves are used which find wide application in the textile and fiber industry for heat setting twisted continuous filament polyamide or polyester yarns in order to achieve twist stabilization. Also, storing film tapes on spools for several hours in hot air chambers reduces tape shrinkage to practically zero. In discontinuous heat-setting treatments one has, on the other hand, always to face the problem that the internal tape layers in the vicinity of the spool core contract less than the tapes in outer layers. The inner layers are hindered in their shrinkage and, in addition, the heat enters only slowly into the interior of the spools. For these reasons the annealing on the spools in autoclaves or in hot air chambers should be applied only to continuously heat-setted tapes prior to the batch annealing.

## D. The Measurement of Residual Heat Shrinkage

The heat shrinkage of film tapes can be determined either in oil or in hot air.

### 1. *The Determination of Oil Shrinkage*

The British Standards Institution and the European Association for Textile Polyolefines (EATP) propose in their standard testing recommendations BS 4611/11 and EATP Nr. 1/10, respectively, the following method for the determination of heat shrinkage.

The film tape has to be immersed for 20 sec into a temperature controlled hot oil bath. The tape has to be tensioned by a small preload related to its titer. Subsequent to the exposure in the hot oil, the treated tape will be allowed to cool for 10 min in a relaxed state. The length change, related to the original length (approx. 200 mm) is reported as hot oil shrinkage in percent for a given temperature. The temperature, time, and distance measurements have to be kept within narrow tolerances as ±0.5°C, ±1 sec, and ±0.5 mm, respectively.

Compared with measuring the heat shrinkage in hot air, the hot oil shrinkage has the advantage of being independent of the temperature of the surrounding air. The results obtained with the hot oil shrinkage method are therefore more exact. Contrary to this, the advantage of the hot air shrinkage determination rests in the easier and quicker performance and is therefore better for the quality control in film tape production.

### 2. *The Determination of Hot-Air Shrinkage*

The hot-air shrinkage apparatus consists of two heated plates arranged horizontally, one above the other. The film tape is clamped in an appropriate holding device outside the plates and tensioned with a small preload related to its titer. After adjusting the measuring scale to the length mark on the tape, the holding device carrying the film tape is pushed between the heated plates for a 10-min exposure. The temperature of the hot plates has to be adjusted to the selected heat treatment temperature with an accuracy of ±1 to 2°C. During the heat-shrinkage test the apparatus has to be protected from any draft from the surrounding air. Additionally the tape should not come into contact with the thermometer. At the end of the exposure, when no further length change occurs, the length reduction is measured. From this reading the percentage shrinkage in relation to the original length can be calculated.

## VI. THE SPLITTING PROCESS

The most important step in the production of fibrous products from films is the splitting or slicing process. According to the applied technique, this processing step may be performed before, during, or after the stretching. For the controlled separation of film in individual continuous filaments or capillaries it is often advisable to slit it before the stretching procedure. This may be done, for example, in analogy to film slicing into tapes by means of special cutting tools. The separation into more or less continuously separated filaments can also be achieved by lengthwise grooving of the film surface during extrusion with the aid of profiled dies. Naturally, the controlled slicing or splitting of smooth or grooved films into individual filaments of relative regular titer during or after the stretching step is also possible. In all these cases multifilament yarns are obtained, closely related to those produced by melt spinning through spinnerets. The main difference, however, lies in the fact that filament titer of fibrous products produced by controlled film splitting or slicing techniques are normally coarser. Filament titers lower than 8–10 dtex (7–9 denier) are hard to achieve with film slitting or splitting techniques.

Yarns of a fundamentally different character are obtained in uncontrolled mechanical or chemomechanical splitting or fibrillation of monoaxially stretched films. In the application of these splitting or fibrillation techniques the film is not separated into continuously separated, parallel single filaments. These techniques result in products in which the film is split into separated filament segments of widely varying length, forming a more or less regular network. These yarns are neither multifilament yarns nor staple fibers in the conventional sense. Greater fibrillation forces in the film splitting produce a more dominating character of yarn spun from staple fibers. However, from film-to-fiber yarns made by uncontrolled mechanical fibrillation it is difficult to classify them either with continuous filament yarns or with staple fiber-type yarns. There does not exist a sharp borderline in the character of such film-to-fiber products. There exist many intermediate forms which will be treated later in more detail.

All techniques for uncontrolled mechanical or chemomechanical splitting or fibrillation into film yarns have as a prerequisite the high splittability or fibrillation tendency of highly stretched films. A high splitting tendency for easy fibrillation under mechanical action is given when, with increased stretch, the anisotropy of strength (i.e., the ratio between longitudinal and transverse strength) is high. Under such conditions it is possible to produce split-film yarns with a more pronounced textile character.

The splitting of highly stretched film can be achieved in an uncontrolled more or less statistical way under the influence of various kinds of shearing forces such as twisting, milling, brushing, air jet treatments, etc. For a more controlled splitting or slicing the use of needle rollers, knife bars, or knife rollers is recommended. In the following, we will first deal with all those techniques of controlled film slicing or splitting, leading to endlessly separated multifilament yarns. Subsequently, those procedures will be discussed which result in split-film yarns using more or less uncontrolled splitting techniques on highly stretched smooth films. Finally, film slicing or splitting techniques will be presented through which texturized split-film yarns are obtained. It is self-evident that all film splitting or slicing processes leading to more or less continuously separated filaments can also be used for the manufacture of staple fiber products when the produced yarn strands are collected into larger tows and are cut into staple with conventional cutting machines.

## A. Procedures for the Production of Multifilament Split-Film Yarns

For the production of multifilament yarn products the film has to be separated into individually separate threads, similar to those obtained in conventional spinneret spinning. Ideally, the individual filaments should have circular or polygonal cross sections and should be as fine as possible in order to be applied in real textile applications. This goal can be attained by certain film splitting techniques. By lengthwise profiling the film, either during extrusion through the use of specially profiled dies or by film embossing directly after the film solidification, the splitting process can be predetermined and directed. The actual splitting of the film into more or less individually separated continuous filaments is accomplished during stretching or in the course of subsequently performed separation treatments. Besides this, there also exist splitting or slitting techniques which do not need specially pretreated films, but are based on the blending of the polymer with incompatible polymeric additives or the use of fine cutting devices.

### 1. *Profiling of Films with Special Dies*

The profiling of film in the form of a lengthwise grooving is aimed at easier splitting along the inserted grooves. The film, therefore, better splittability the larger the ratio between thick and thin areas of the grooved film. During stretching, profiled films already start to split as a result of the necking. If the thickness difference is not sufficient for a complete splitting into continuously separated filaments, the separation will have to be completed by applying additional mechanical stresses.

Profiling of the film can be carried out before, during, or after the extrusion of the melt through the die. According to Showa Yuka K.K. (1975a, 1975b), the melt stream leaving the extruder screw is divided in the adapter between the filter screen block and the die slot in many individual streams (up to 200). The melt is thereafter extruded through a normal annular die under reunification of the separated streams resulting in a lengthwise grooved tubular film. In stretching such a film to a stretch ratio of 1:8, it splits into a number of single filaments equal to the number of streams formed before the die slot. In such a way, continuous filament yarns with a defined number of threads and with a defined titer can be produced. This processing principle has also been applied to polyethyleneterephthalate (Imperial Chemical Industries, Ltd., 1971d). Here also the melt stream is divided by slots in the adapter, temporarily, into several partial streams which reunite just before the flat die. From the die emerges a polyester film which fibrillates very easily when rubbed transversely.

Relatively fine titers can be produced in film splitting through grooving of the film by means of a profiled die. English Courtaulds, Ltd. applied in 1962 for a patent for the first time based on this principle (Courtaulds, Ltd., 1968). Somewhat later the German company Barmag-Barmer Maschinenfabrik AG (1971b) was granted a patent for a closely related process known as the "Barfilex" process. According to the technique described by Barmag, the polymer melt is extruded from a wide and distinctly profiled flat die in the form of a lengthwise grooved film or narrow longitudinally profiled film tapes. In the latter case the profiled film strips have a width of 20 to 25 mm and in their original unstretched state a thickness of approximately 1 to 2 mm. Before solidification the array of profiled film tapes is drawn off the die at a ratio of 1:5 and up to 1:20. This is done by widely variable draw-off rollers. After solidification and cooling, the strongly profiled tapes are oriented by stretching on conventional stretching units with stretch ratios ranging from 1:10 up to 1:12. At this high stretch ratio the profiled tapes split up into individual filaments without any further mechanical treatment. The cause of the splitting is the known width reduction in the neck-in point. In this process the solidification of the profiled film or film strips can be accomplished either by chill-roll cooling or by water-bath immersion.

The cross-sectional shape of the single filaments is determined by the design of the die lips responsible for the film profiling. One of the die lips shows a conventional even effective surface, while the other lip has V- or U-shaped parallel grooves in the melt-flow direction. Through the splitting of the grooved film at the thin places the resulting filaments obtain a rough surface at the splitting edges. This makes the capillaries bulking well, thus they can be readily twisted and wound up and are arranged slip-proof in the rope or fabric structure.

A further feature of this process is the possibility that the profiled strips are connected by thin membranes, giving support for the array of profiled strips in their way, from the die to the draw-off rollers and to the stretching unit. These supporting membranes split during the stretching. The flow sheet for the Barfilex process is given in Figure 84. According to construction features of the die, the Barfilex

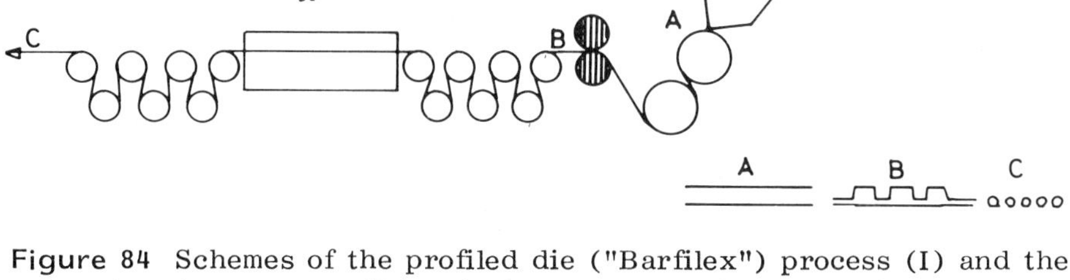


Figure 84 Schemes of the profiled die ("Barfilex") process (I) and the roll-embossing process (II). (From Hensen and Braun, 1978b).

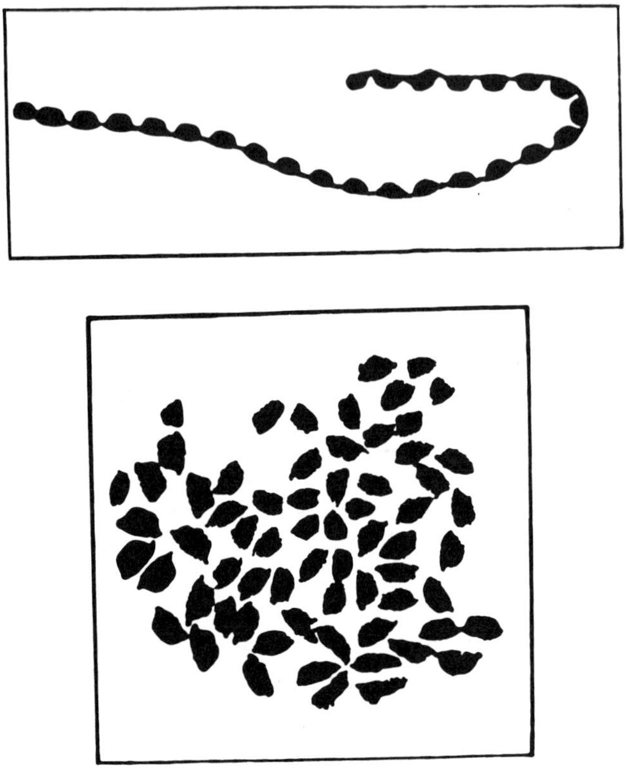

Figure 85 Cross section of the profiled film (top) and of split fibers (bottom) produced with the Barfilex process. (From Harms, Krässig, and Sasshofer, 1971).

process allows production of mono- or multifilament yarns which can be used in the manufacture of ropes, bags, floor coverings, etc. The process also allows the manufacture of texturized multifilament yarns and of staple fibers by cutting multifilament tows with conventional staple fiber cutting machines.

The individual filaments resulting from the Barfilex process normally show a triangular or trapezoidal cross section and titers between 20 to 25 dtex, as illustrated by Figure 85. A similar process is also described by Ube Industries, Ltd. (1973) in which polypropylene is extruded through a 200-μm-wide die slot with a sawtooth-like profile of one die face. By stretching the resulting profiled film with a 1:7 stretch ratio, it splits into continuously separated filaments of approximately 20 dtex.

In 1969, Hercules, Inc. (1969) was granted a patent protecting the inventors rights for a process of splitting a grooved film into endlessly separated continuous filaments. In this process the monoaxially drawn film grooved during extrusion or subsequently by embossing, is led under tension over a bar carrying sawtooth-like ridges over its entire width. The distance between two tooth points is equal to the distance of the longitudinal grooves in the film. The teeth of the splitting bar separate the film along the thin indentations in the film, forming continuously separated filaments. The inventors claim that this method of film slitting leads to fibers with latent crimp properties. The principle of this method and the splitting tool are illustrated in Figure 86. In Sec. VI.A.4. a system developed by Chemiefaser Lenzing A.G. (1970) will be described, which uses a similar film separation principle on smooth unprofiled film (see Figure 89).

According to Shell Internationale Research Maatschappij N.V. (1970b), fibrillable profiled film can also be produced with a die which is not a slot with one profiled face but a row of holes lying closely side by side over the whole width of the die. The melt stream emerging from the die holes join to a continuously lengthwise grooved film due to the Barus effect. The profiled film obtained in this way is cooled and solidified in a water bath and splits into single continuous filaments during stretching. Profiled film tapes themselves also find applications. According to Kuranray Co. (1972), profiled flat tapes produced applying the above-described principles show bastlike character and can be used in the manufacture of mats, wall covers, etc.

*2. Profiling of Smooth Primary Films by Roller Embossing*

In the roller embossing process a profile is impressed into the smooth film during, or immediately after solidification with a pair of rollers, one of which carries the pattern of the profile. The grooves of the profile cause the splitting of the film in subsequent stretching. Since

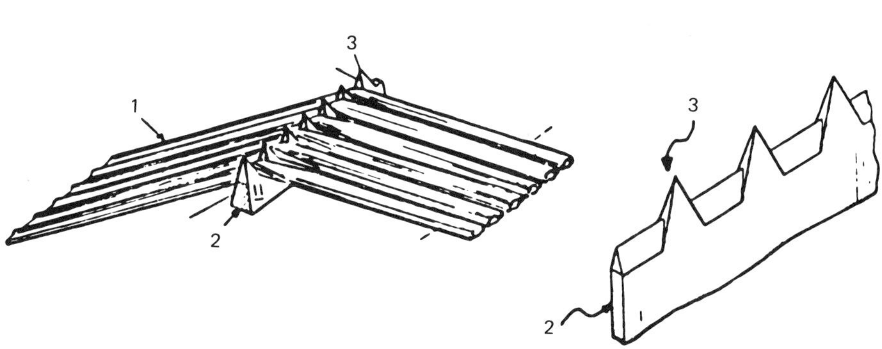


Figure 86 Scheme of the film-slicing tool used in a film-to-fiber separation process developed by Hercules, Inc. 1, striated film; 2, separation block; 3, sharp ridges.

the profiling is performed during or after the film is drawn down from the die, no thinning out of the profile pattern occurs, reducing the distance of the longitudinal grooves. The minimum achievable filament titers are therefore coarser than those in die profiling, as used in the Barfilex process.

In Figure 84, already shown earlier, the placing of the profiled roller duo between the chill roll and stretch inlet unit is shown. One of the two rollers carries the V- or U-shaped grooves, while the other smooth roller acts as backing roll, pressing the film into the grooves of the embossing roller. The profile of longitudinal grooves thus introduced into the still plastic film causes the complete separation of the film into single filaments during the stretching and, if necessary, in subsequent fibrillation steps such as twisting (Shell, Den Haag, 1977).

In the profiling operation the film is deformed at temperatures just below the crystalline melting point. A variant of this process is the "melt embossing" technique. Here the melt is extruded from a flat die into the slot between a smooth and grooved roller duo. In this embossing roller duo the film is cooled near the solidification point and thereby profiled in the desired manner. The final cooling is accomplished on a subsequently arranged chill roll (Dow et al., 1972).

Also in a process described by Tweedale (1974), the profiling of the film is performed in a semisolidified state of the primary film. The embossing roller carrying the profile pattern presses through the still-plastic primary film against the smooth backing roller, thus introducing into the film a distinct profile. By splitting the film in subsequent stretching and fibrillation treatments, filaments are formed whose degree of fineness is determined by the number of grooves per unit width dimension and by film thickness. Figure 87 shows a cross-section of a film profiled in this way.

This method comprises the possibility to accomplish the cooling and profiling of the primary film in one operation. According to a proposal made by Lambeg Industrial Research Association (1975), the polymer melt is extruded onto a chill roll carrying numerous radial grooves. To assure that the film is grooved properly, maintaining its shape, a rubber-lined backing roller presses the solidifying film against the profiled chill roll. This type of process can also be used in the manufacture of profiled film tapes. In 1971 a patent was granted to Shell Oil Co. (1971) for a process in which a finished solidified film is led at elevated temperature through two rollers of which at least one is embossed. Here also the profiled film separates into individual continuous filaments in subsequent stretching.

A profiling unit consisting of edged comb rollers for the production of fibrillated film is described by Shell Oil Co. (1975). An improved version of the device was publicised somewhat later by Shell Internationale Research Maatschappij N.V. (1979). The extruded polymer melt is led vertically down into a gap which is formed by a ribbed chill

Figure 87 Cross section of a roller-embossed film used for split-film yarn production. (From Tweedale, 1974.)

roll and a smooth support roll. A stretching of the profiled film in a ratio of 1:8 in a hot air oven splits the film into single filaments with a titer of 33 dtex. Both edges of the ribbed roller circumference are equipped with a shoulder and a groove which prevent a sidewise movement of the film during running through the embossing roller duo.

For the sake of completeness, it should be mentioned that it is also possible to emboss the not completely solidified or heated primary film with a special profile pattern in such a way that on stretching it splits into a networklike structured scrim. According to an invention of Smith & Nephew Plastics, Ltd. (1979), a melt extruded film consisting of a blend of a polyolefin with some added polystyrene is pressed in a gap between two rollers, one of which is profiled while the other one is cooled enough to solidify the embossed film. According to the kind of profile pattern in the embossing roller, the profiled film splits on stretching into a more or less fine network. The roll embossing process is also suitable for the manufacture of profiled film tapes with improved tensile strength (Aspin and Wall, 1979).

All methods of filament yarn production from profiled film result in yarns with high tensile strength. This is due to the relatively small mechanical stresses necessary to split embossed highly stretched film into separated filaments. This kind of film splitting does not introduce into the obtained filaments any mechanical damages. It will be shown later that fibrillation techniques performed on smooth film tend to result in a reduction of tensile strength, which is greater the finer the desired filament titer.

### 3. *Splitting of Films from Polymer Blends*

The use of polymer blends is another means for film modification favoring splitting into single filaments. Polymer blends suitable for this purpose consist of incompatible and immiscible components forming separate phases in the film. Through the influence of shear forces the blended film splits easily into fibers at the phase boundaries. The splitting may also be performed by film stretching or by removal of one of the film components. Rasmussen (1976a) describes an extruder with which two polymers are extruded in form of partial melt streams. These melt streams join on leaving the die, forming a film consisting of lamellas of the two polymers lying side by side. This film is subsequently treated mechanically or chemically to separate the lamellas or remove the soluble interlamellas-forming polymer, thus separating the filaments formed by the second polymer. The same principle can also be applied in a modified manner in such a way that the lamellas formed by the two polymers used are not completely separated in the fibrillation treatment (Rasmussen, 1976b), leading to network structures.

Another version of this processing principle is claimed by Mitsubishi Rayon Co. (1971c). Two incompatible polymers (e.g., polypropylene and polyamide-6,6) are led into the extruder die head in the form of several melt streams which are further subdivided by a layer of small steel balls. The melt, consisting of numerous fine streams, is extruded from the die in the form of 50-mm-wide and 0.1-mm-thick tapes. In the stretching of blended tapes to a ratio of 1:8 they split into numerous filaments with varying cross sections.

A further development of this principle is the extrusion of a blend of nylon-6 and isotactic polypropylene through a profiled die. When stretched-film tapes made from this blend are twisted to 80 t/m, a yarn results having a tenacity of up to 13 cN/dtex (12 g/den). As far as we know, these methods of chemomechanical film splitting did not find application in the industrial practice until now.

### 4. *The Controlled Slitting of Smooth Films into Defined Filaments*

In the foregoing chapters, processes were discussed in which film splitting is promoted by either special pretreatments of the extruded film or by applying polymer blends. In both cases the basic principle of enhancing the film splitting has been the introduction of weak spots into the film. These measures cause regular film splitting during the stretching operation as well as under the effect of mechanical forces, such as twisting, rubbing, bending, etc.

In addition to these more or less controlled splitting methods there exist a number of processes for film-to-fiber separation, starting from smooth unmodified film. In most cases greater separation forces or cutting techniques have to be applied, enhancing the danger of filament damages. The first process for film-to-fiber conversion starting from monoaxially stretched smooth polymer films, resulting in more or less continuously separated single filaments, has been described by Dow Chemical Co. (1956). The cutting tool used in the Dow process consisted of a roller duo, similar to that already mentioned in the discussion of the roller embossing processes. The upper roll of the duo is provided with V-shaped grooves with sharp ridges at a distance of 1 mm (1/25 in.), the lower roll is made from elastic material. Figure 88 (A) shows a scheme of this cutting duo and the principle of its action.

Eleven years later, Mirsky (1967) described a method in which the splitting device is a bar equipped with circular blades which are separated from one another by distance spacers of 2 mm thickness. The blade bar can be used stationary, rotating, or oscillating. According to Mirsky a backing roller is not supposed to be necessary [see Figure 88 (B)]. As one can see, these methods of film-to-fiber separation are somewhat related to the methods of slitting monoaxially stretched film into film tapes which were described in Sec. III.B. The difference with those methods lies in the intention to achieve a much finer filament titer.

A novel principle for film-to-fiber separation, aiming for the same goal, has been developed by Chemiefaser Lenzing AG (1970). According to this method the primary film which has been processed by extrusion of the polymer melt through a flat die and by cooling and solidification on two chill rolls is drawn under stretching over a heated bar equipped with sawtooth-like cutting tools arranged transversely with respect to the processing direction. On the sharp ridges of the cutting tools the film is slit into fine filaments (down to approx. 100 μm width) under prestretching to 1:2 up to 1:3 stretch ratio. The array of continuous filaments can be further stretched up to a total stretch ratio of 1:9 (for polypropylene). With cutting tools having 150 μm distance between the sawtooth edges, single-filament titers of 20 to 25 dtex have been obtained.

A special characteristic of this process is the use of a heated cutting tool. In doing so the following effects are achieved:

1. The mechanical forces necessary to achieve film slitting are reduced.
2. The cut side edges of the filaments are molten and thereby smoothed, and possible damage coming from cutting defects is eliminated.

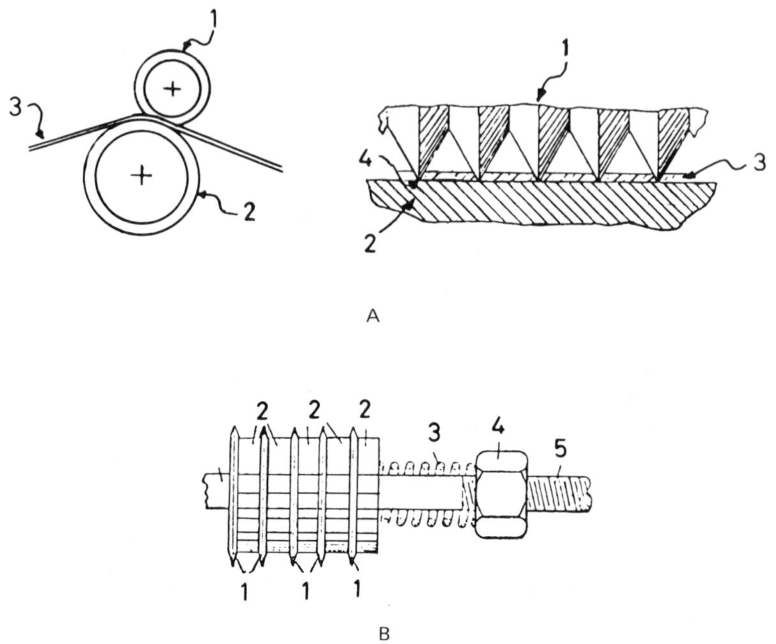


Figure 88 Schemes of circular cutting tools used in controlled film-to-fiber separation. (A) Roller cutting assembly: 1, roll carrying parallel narrowly spaced circular ridges; 2, resilient backing roll; 3, film or film tape; 4, sharp ridges. (From Dow Chemical Co., 1956.) (B) Circular film cutter: 1, circular cutter plates; 2, distance pieces (spacers); 3, compression spring; 4, threaded nut; 5, mounting bar with screw thread. (From Mirsky, 1967.)

3. Through the asymmetric heating of the filaments in the moment of their formation, a latent tendency for crimp development on heat exposure is being introduced (see also Sec. VII.).
4. The frictional resistance of the primary film being pressed into and cut at the sawtooth edges of the cutting blades causes, together with the heating of the latter, the localization of the stretching zone immediately onto the fibrillation zone. Therefore the freshly formed filaments taper off after their formation in relation to the speed ratio between the chill-roll speed and the speed of the feed rool of the stretching unit, thus preventing any sticking together. This can be assisted by an air knife mounted over the slitting bar which helps in the rapid cooling of the freshly formed single filaments.

Figure 89 shows schematically the design of the cutting bar and of the sawblade-like cutting tools used in the Chemiefaser Lenzing process.

Figure 90 shows the photograph of a production line. The nearly ideal filament uniformity achieved with this process was demonstrated earlier in Figure 11 (Chap. 1, Sec. III). A similar slitting device has been suggested by Shell Oil Co. (1971). The claimed device differs from the slitting tool described above since it rotates during the slitting action. In this connection reference should also be made to the procedure for the production of metallized tapes which was discussed in Sec. III.B.

## B. Methods for the Production of Fibrous Split Yarns

The application of film technology for the production of multifilament yarns remains, in the end always in competition with the traditional melt-spinning technique. Quality and efficiency are decisive in determining which of the two methods is to be given preference. Generally, a film tape or film fiber unit can be more efficient at a smaller production capacity than a sophisticated melt-spinning unit. This holds true especially in the production of yarns with higher single-filament titer and higher total yarn titer. Below single-filament titers of 10 dtex, conventional melt spinning is more efficient and gives markedly better quality products than film-splitting or film-slitting techniques. For single-filament titer above 15 dtex, the reverse is the case, especially with respect to cost efficiency. The investment costs for film-to-fiber

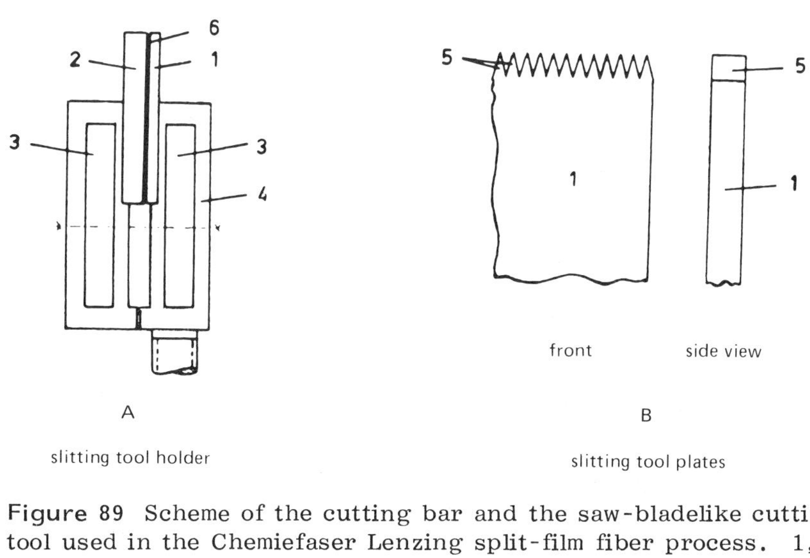


**Figure 89** Scheme of the cutting bar and the saw-bladelike cutting tool used in the Chemiefaser Lenzing split-film fiber process. 1, slitting tool plate; 2, film preheating plate; 3, heating elements or hot liquid; 4, metal housing and mount; 5, sawtooth ridges (top distances 60–500 μm); 6, insulation. (Courtesy of Chemiefaser Lenzing AG.)

Figure 90 Split-fiber yarn production line. (Courtesy of Chemiefaser Lenzing AG.)

lines are supposed to be approximately 25% less than for a corresponding conventional melt-spinning unit.

Both technologies show distinct analogies. The normal film processing technology has to be more or less modified in order to be able to produce multifilament yarns with single capillaries of defined and regular titer and length uniformity. The better the regularity, the more sophisticated the film-forming technique has to be. Proper die flow designs and the use of metering pumps are essential for high uniformity. Nevertheless, there are limits to the application of film-to-fiber technology with respect to titer uniformity and to cross-section characteristics. Thus yarns made by film separation are suitable mainly for technical applications and for certain textile uses where high uniformity and regularity are not required. To produce products for these applications, however, the use of production units for film yarn may in the end be more economical than conventional melt-spinning facilities.

The production conditions for the manufacture of split- or slit-film products from highly monoaxially stretched smooth film are relatively simple. They are closely related to film tape manufacturing. The installation of a suitable fibrillation unit into a normal line for the production of stretched film tapes either between the stretching and the heat-setting or (sometimes better) after the heat-setting, gives the potential of producing film-to-fiber products. Apart from the technical requirements connected with the installation of the fibrillator, and from the adjustment of the production parameters to achieve desired fiber product properties, normal film processing know-how can be applied. The fibrillation of a smooth stretched film, however, demands a generally higher stretch ratio in order to lower the transverse film strength, sufficiently enhancing fibrillation tendency at reasonably low mechanical splitting forces.

Depending on whether the forces inducing the film separation act on the film transverse to the stretching direction at random (e.g., by twisting, rubbing, or bending), or utilize a special splitting tool separating or splitting the film in machine direction at certain places, one distinguishes between an uncontrolled (or random) and a controlled film splitting procedure.

The uncontrolled or statistical mechanical or chemomechanical fibrillation procedures use various kinds of shearing forces acting transversely or at a certain angle to the stretching direction on highly drawn film or film tapes. These forces split the film or the film tapes, which can also be made from polymer blends, into fibrous structures. Often the resulting fibrous structures are still interconnected in a network-like fashion and can not be defined with respect to fiber length or titer. The application of these fibrillation methods is limited mainly to polyolefins or blends of these with other incompatible polymers.

The controlled fibrillation is in most cases based on the use of needle rollers or other tools able to introduce holes or incisions into primary or stretched film or film tapes. A geometrically defined pattern of holes or cuts is introduced into the film product, giving it network-like characteristics which can be altered by the fibrillator design and production conditions, such as subsequent drawing in very wide limits. Split-film yarns produced in this way have acquired some importance in the manufacture of ropery and fabrics. Needle-roller fibrillation is basically applicable to all monoaxially stretched films. Films from polymers with low splitting tendency, however, can cause problems. Blending with incompatible additives can be of help in such cases.

### 1. *Fundamentals of Fibrillation of Monoaxially Stretched Films*

The fibrillability or splittability of a film defines its tendency to split up under the influence of a given transversely acting force. The fibrillability is greater the smaller the transverse tenacity of the film is. The relation of tensile strength in the length direction to tensile strength in the transverse direction (e.g., the strength anisotropy) gives a relatively good indication of film splittability. Another test giving a good assessment of film splittability is tear strength in the stretch direction. At the rupture of a film tape sample, high fibrillation tendency can be recognized by subsequent breaking of split off sections. This effect is visible in a stepwise decline of the breaking tension in the recorded stress-strain curve. The number of stress decline steps can be used as an index for the film fibrillability. Counting split ends of broken film tape of a given width is a suitable method in practical work. Naturally, such a measurement shows a relatively wide spread. Better reproducible splittability values are obtained when the rupture is performed in an instrument designed for the measurement of impact strength. In the impact rupture more split ends are produced, showing improved reproducibility in repeat tests on the same material.

Geitel (1976) has developed a more sophisticated testing method based on the measurement of the dent breaking force. In this test tension is measured which is needed to press a notching pin of 1 mm thickness with a rounded point through a film sample. Reference has to be made to statements made in Sec. III.C. The measurement of splitting resistance, defined as the splitting tension reduced to an anisotropy factor equal to one, is thus suitable for the fibrillation property of a film in film-to-fiber fibrillation as well as in film tape cutting with a blade bar (Dawczynski et al., 1978). Finally, the measurement of the tongue-tearing resistance in the length direction is also a good indication of fibrillability.

Besides the measurement of resistance, information on the degree of fibrillation achieved under given conditions is also of interest (Geitel, 1976). A good measure of the degree of fibrillation should be the achieved fiber fineness. However, in network-like structures, which often result in film fibrillation, it is impossible to define fiber titer. There are no distinct fibers in the traditional sense. The microscopic determination of the length and thickness of separated fibrous segments seems to be of practical value in these cases. Measurements of the fineness of film fibers in split-film yarns and in film fiber nonwovens have been described by Kampf (1976). Characteristics for the definition of split-film fiber products can be regarded as: the width of fibers or fiber segments, their diameter and cross section, the separation length of fibrous segments, the number of cross points, and the hairiness. Methods of automatic image analysis are being used to characterize split-film yarn products with respect to those product properties. The evaluation of analytic results is performed with statistical methods (Reuter and Kampf, 1978).

Another very indicative measure of the fibrillation tendency of film tapes is the loss of tenacity upon twisting. The loss of tenacity of film tapes in twisting is smaller if the tape is fibrillated in the twisting procedure. Increasing fibrillability results in an improved tensile strength utilization (Brittain 1968; Broatch, 1968; Klust, 1968; Nichols, 1971, 1972). Table 18 shows that film tapes from polypropylene suffer

Table 18 Strength Twine Conversion Efficiency of Polypropylene Fibrillating Tape and of Polyamide and Polyester Nonfibrillating Tapes

| Polymer type | % Conversion efficiencies |
|---|---|
| Polypropylene | 90–105 |
| Polyamide | 30–60 |
| Polyester | 30–60 |

a substantially smaller loss of tenacity in twisting due to their fibrillation than those from polyamide or polyester showing no fibrillation. Measurements of the splitting resistance of films from polyethylene, polypropylene, polyacrylonitrile, polyester, and polyamide supporting the observations described in Table 18 were reported by Michels et al. (1975).

Good splittability of monoaxially drawn polyolefin films can be explained by the relatively small dispersion interactions between adjacent molecules. For polyamides and polyesters, however, much greater molecular attraction forces in the form of hydrogen bonding and dipole interactions exist. These intermolecular forces in polyamide and polyester films or film tapes increase the transverse strength and therefore decrease the ability to split. Among the polyolefins, polypropylene films show a markedly higher splitting tendency at a similar level of molecular orientation than polyethylene films. Apparently the methyl side groups along the polypropylene chain reduce the van der Waal's forces between neighboring polymer molecules. With respect to fibrillability the following order exists among the most important thermoplastic polymers: polypropylene > high-density polyethylene > polyamide > polyester.

For the sake of completeness it should be mentioned that in the last decade the fibrillation of polytetrafluorethylene films has also gained importance (Chemiefaser Lenzing AG, 1975b; TBA Industrial Products, Ltd., 1976). As polytetrafluorethylene is no thermoplastic polymer, the films have to be processed either by mechanical peeling off from sintered cylindrical blocks or by "ram extrusion." In the latter case the polymer powder is converted with paraffin oil into a thick paste which is pressed through a flat die with a piston extrusion device. By evaporation of the oil and after stretching, an easily fibrillatable film is obtained. From this film polytetrafluorethylene staple fibers can be made through needle-roller fibrillation or slicing with cutting tools.

Fibrillability of highly oriented polyolefin films has been studied thoroughly by Polovikhina et al. (1978). Figure 91 shows the relation

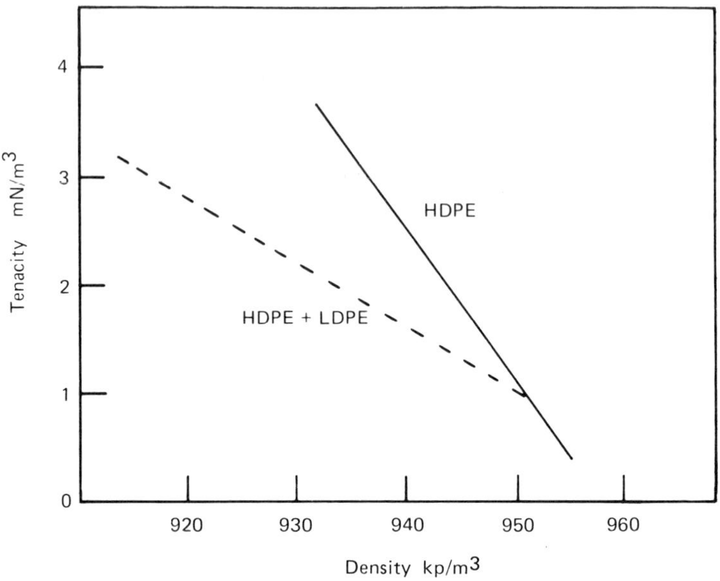


**Figure 91** The effect of density on the tenacity of polyethylene film in its transverse direction. (From Polovikhina et al., 1978.)

found by them between the film tenacity in transverse direction and the film density of high-density polyethylene films and of film made from a blend of high-density and low-density polyethylene. According to these results the fibrillability of films from the nonpolar high-density polyethylene is mainly determined by the degree of crystallinity which increases with the stretch ratio applied and also substantially in subsequent tempering treatment. Through the addition of low-density polyethylene the degree of crystallinity is suppressed, resulting in an increased split resistance. This observation corresponds to the behavior of blends from polypropylene and low-density polyethylene discussed in Sec. I.D.2.

The influence of processing conditions on the fibrillability of monoaxially drawn smooth films was discussed earlier. The most important factors influencing fibrillability are:

1. Polymer blends of incompatible components generally have a greater fibrillation tendency than the pure polymers.
2. Flat films are in most cases easier to fibrillate than tubular films since the latter show a certain transverse orientation coming from the inflation with air.
3. Fibrillability is increased by high draw-down ratios between the die and the drawing-off unit.

4. Fibrillability increases with increasing stretch ratio. Elevated stretching temperature, however, has rather the reverse effect.
5. Regarding the dependence of fibrillation tendency of highly stretched films on subsequent heat treatments, contradictory observations exist. According to Polovikhina et al. (1975) and Martinova et al. (1979), the splitting tendency of polypropylene film increases with rising annealing temperatures due to an increase in crystallinity. Hajamasy and Alberti (1976), however, claim that elevated annealing temperatures decrease the splittability. These authors explain their observations by the disappearance of fissures and micropores between the fibrillar crystalline areas through annealing. Hence these starting points for fibrillation are eliminated. Bouriot et al. (1976) also find a lower splitting tendency for annealed films from polypropylene and explain this by the partial conversion of microfibrillar elements into lamellas serving as adhesion points between the remaining lengthwise oriented fibrils.
6. Film tapes which were stretched after tape cutting possess generally a higher splitting tendency than those which were cut from already stretched film. The stretching of narrow tapes results in a higher degree of orientation.
7. Fibrillability can be influenced by additives. This will be treated in detail in Sec. VIII.C.

It should further be mentioned that according to Nawlicka (1977) film fibrillation causes a certain reduction in molecular weight.

### 2. *Techniques for the Controlled Fibrillation of Monoaxially Stretched Film*

In contrast to uncontrolled film-to-fiber separation techniques, controlled fibrillation has the advantage of controlling more exactly the characteristics and properties, such as fineness and strength, of the fibrous product by proper adjustment of the various process parameters. Because of the relatively good uniformity of the resulting products, this fibrillation technique has gained wide acceptance for the production of split-film yarns.

Either an array of film tapes or a widely spread film is fed to a fibrillation tool. In film yarn manufacturing it is advisable to cut the film into tapes before or after the stretching operation prior to the fibrillation step in order to exactly predetermine the titer of the resulting yarn and to facilitate the yarn winding. In staple fiber production it is also possible to fibrillate the film without precutting into tapes. The resulting fibrillated web is cut after the fibrillation into the desired staple length. Subsequent carding results in a complete separation into single fibers (Schuur and Van de Vegt, 1975). The fibrillation unit is adjusted in its design to the stretch mounts and arranged either behind the stretch unit or the heat-setting unit. With the aid

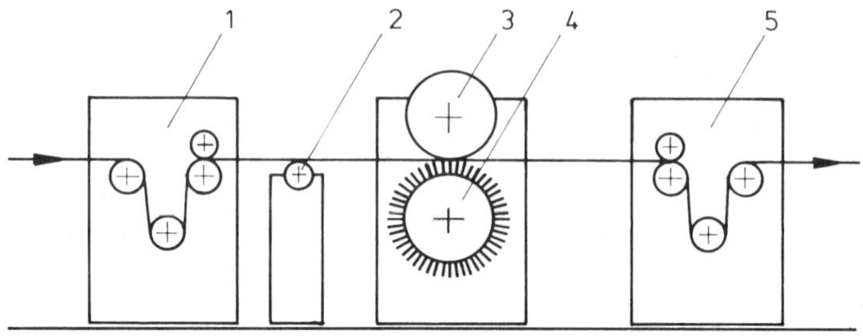


Figure 92 Scheme of a needle roller fibrillation unit: 1, inlet trio; 2, lubrication roll; 3, rubber-coated pressure roll; 4, pinned fibrillation roll; 5, outlet trio.

of a separately driven roller trio, the film tension during the fibrillation step is regulated (Balk, 1978). Fibrillated film tapes are subsequently wound onto cardboard or plastic tubes in the form of cross-wound cylindrical spools. Fibrillated webs which are to be processed further into staple fibers are laid off into spinning cans. The fibrillation unit generally consists of a roller armed with specially arranged needles or blades rotating, in most cases, in the film processing direction. Figure 92 schematically shows the action of such a fibrillation unit.

In operation the film is pressed under a variable angle of incidence by the film tension against the needles of the fibrillation roller. To promote penetration of the needles or blades into the film, a rubber-coated press roller mounted above the fibrillation roller forces the film into the needles or blades. The ratio of the inlet speed of the film to the rotational speed of the needle or blade roller is termed "fibrillation ratio." The character and properties of the fibrillated film product are dependent (a) on the number and arrangement of the needles or blades on the fibrillation roller, (b) on the fibrillation ratio, (c) on the diameter of the needle or blade roller, (d) on the angle of incidence and contact between the fibrillation roller and the film, and (e) on the film tension. Naturally, also the film characteristics, such as film thickness and degree of orientation achieved in stretching, are of importance for the outcome of the fibrillation. The fundamental rule of fibrillation implies that a small contact angle leads to networks with small mesh width and coarser fiber titers. Fibrillation rollers with larger diameters used at high contact angles result in long meshes formed by finer fiber segments. Such a texture is shown by Figure 93. The mesh width (L) in cm is a function of the contact length (a) in cm and the fibrillation ratio $V_f/V_r$ ($V_f$ = film speed; $V_r$ = rotational speed of the fibrillation roller at its outer surface in meters per minutes), and can be calculated from the following formula:

$$L = a\ (1 - V_f/V_r)$$

The ratio formed from the total titer (weight per unit length) of the film ($T_{total\ film}$) and from the average titer of single-fiber segments ($T_{single\ fiber}$) is termed "degree of fibrillation" (DF). This value can be calculated using the formula:

$$\frac{T_{total\ film}}{T_{single\ fiber}} = DF = a(V_r/V_f - 1) \cdot N \cdot m$$

in which N is the needle or blade density on the fibrillation roller given as the number of needles or blades per cm along the width of the roller, and m is the number of needle or blade rows per cm of roller circumference. Fisher (1969), however, applies N to the number of needle or blade rows over the total fibrillator circumference. Further literature on the same subject can be found in publications by McMeekin (1968), Badrian and Choufoer (1971a), and Choufoer (1975). The theoretically achievable titer of single-fiber segments ($T_{single\ fiber}$) can be estimated with the following formula:

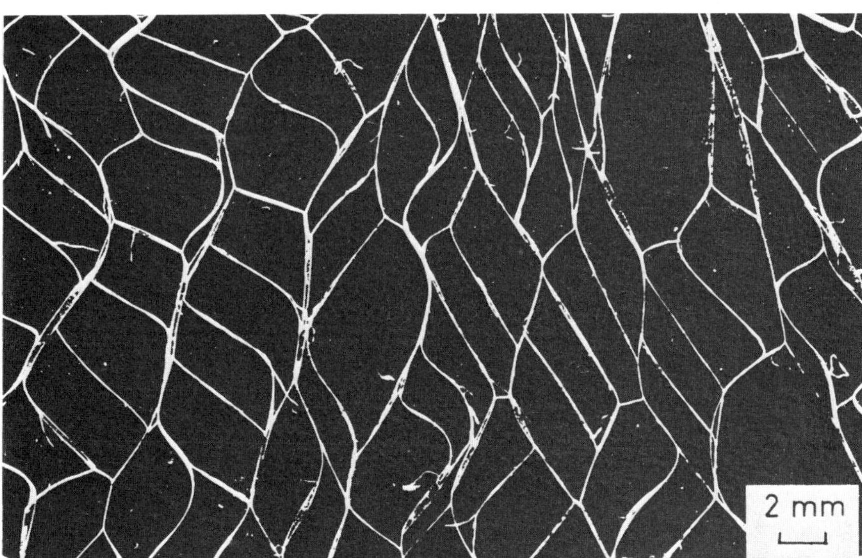


Figure 93 Fibrillated product obtained by passing uniaxially oriented film over a pinned roll fibrillator. (See Schuur and Van de Vegt, 1975.)

**Table 19** Examples of Needle-Roll Film Fibrillation Conditions with Respect to the Resulting Split-Film Fiber Fineness

| | Shell-LIRA process | |
|---|---|---|
| | MK-I | MK-II |
| *Tape characteristics* | | |
| Tape width (before stretch) | 2 cm | 2 cm |
| Tape thickness (before stretch) | 100 μm | 100 μm |
| Stretch ratio | 1:8 | 1:8 |
| Tape width (after stretch) | 0.7 cm | 0.7 cm |
| Tape thickness (after stretch) | 35 μm | 35 μm |
| Tape titer (after stretch) | 2200 dtex | 2200 dtex |
| Tape titer per unit film width | 3150 dtex | 3150 dtex |
| *Processing conditions* | | |
| Line speed ($V_f$) | 120 m/min | 120 m/min |
| Fibrillator roll speed ($V_r$) | 360 m/min | 360 m/min |
| Fibrillation ratio (FR) | 3.0 | 3.0 |
| Fibrillator diameter | 12.5 cm | 20.0 cm |
| Contact length | 2.5 cm | 15.0 cm |
| N needles per cm of needle bar | 13 | 13 |
| M bars per cm of roll circumference | 1.1 | 1.4 |
| *Resulting film-fiber titer* | | |
| For MK-I process: | | |
| $\frac{3150}{2.5 \cdot (3-1) \cdot 13 \cdot 1.1} =$ | 44 dtex | |
| For MK-II process: | | |
| $\frac{3150}{15.0 \cdot (3-1) \cdot 13 \cdot 1.4} =$ | | 6 dtex |

*Source:* Badrian and Choufoer, 1971a.

$$T_{\text{Single fiber}} = \frac{T_{\text{total film}}}{a\ (V_r/V_f - 1) \cdot N \cdot m}$$

However, this relation is strictly valid only if the ratio $V_r/V_f$ is not too far away from 1.

The achievable degree of fibrillation, however, differs rather widely from film to film. It depends to a large extent on the splitting tendency of the polymer film and is strongly related to the chemical and physical characteristics of the polymer forming the film. The use of the formulas mentioned above allows one to roughly estimate the degree of fibrillation and fineness of fiber segments to be expected from the thickness of the starting film, based on construction of the fibrillation roller and from operating conditions.

Especially important for the result of fibrillation is the fibrillation ratio, the fibrillator dimensions, and the contact length as shown by the two examples given in Table 19 (Badrian and Choufoer, 1971a). In reality the calculated values normally are somewhat higher than the actually achieved fiber segment titers (e.g., when the calculation predicts a segment titer of 40 dtex, actual titers of approx. 30 dtex are obtained). This is due to an additional self-splitting of the film in the course of mechanical stresses during fibrillation. In most cases this self-splitting can not be avoided. Additives, such as pigments or light stabilizers act as nuclei, activating additional splitting. Also, films from polymer blends tend to self-fibrillation, a fact which facilitates the manufacturing of film yarns and film yarn products with a high degree of fibrillation, showing certain advantages in their end-use, as already mentioned on several occasions.

When the needles of the fibrillator roll do not penetrate the film equally deep, coarse single-segment titers are the result. It is very important to maintain equal pressure of the film on the fibrillator roller over its whole width. The desirable degree of fibrillation depends to a large extent on the end use of the fibrillated film yarn or film product. Fibrillated film yarns for wovens, for the manufacture of special cordage, and for other articles with flexible or good textile hand ought to be manufactured with a high degree of fibrillation. Also, the manufacture of staple fibers by way of film fibrillation calls for a high degree of fibrillation. The manufacturers of high-density polyethylene and isotactic polypropylene have developed a number of polymer grades giving films with different fibrillation tendency (Badische Anilin & Soda Fabrik AG, 1970a, 1970b, 1970c, 1970d; Mitsubishi Rayon Co., Ltd., 1970a, 1970b; Phillips Petroleum Co., 1970a). The geometry of the network texture of fibrillated film products can be controlled by the arrangement of needles or blades on the fibrillator. Regular meshes result from needles or blades arranged in staggered lines. According to the distance from one needle or blade to the next, the pattern can be varied. Regular networks are advantageous because of the greater tensile strength compared with fibrillated film products with irregular network texture. Furthermore, the distance between the needle or blade rows is of importance. Distances between the needle or blade bars on the fibrillator which are too small will cause the "fakir effect." In this case too many needles or blades touch the film at the same time and they can not penetrate but "carry" it. By increasing

Figure 94 Photograph of a pinned mechanical fibrillator roll with straight rows of pins. (From Schuur and Van de Vegt, 1975.)

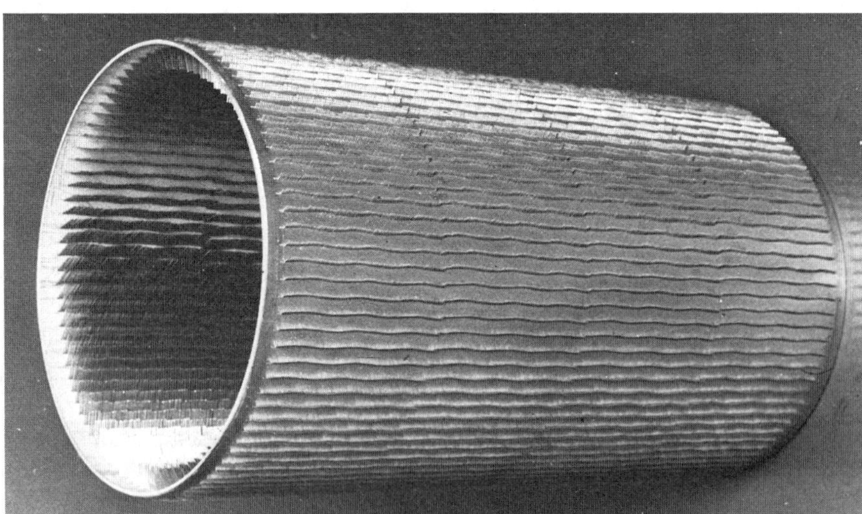

Figure 95 Photograph of a pinned mechanical fibrillation tube (without end flanges) showing sinuous wave form of pinning developed to minimize the "fakir" effect. [Courtesy of Wm. R. Stewart & Sons (Hacklemakers), Ltd., Scotland.]

Figure 96 Mechanical fibrillation machine with the fibrillation tube withdrawn. [Courtesy of Wm. R. Stewart & Sons (Hacklemakers), Ltd., Scotland.]

the pressure in order to eliminate this undesired effect, numerous pins will only seize already existing slits and not introduce new ones.

Figure 94 shows a fibrillator with very fine needles placed at narrow distances, while Figure 95 shows a fibrillator tube with "sinuous wave" placement of the pins, counteracting the fakir effect. Figures 96 and 97 show fibrillator units with one or two fibrillator rollers. The existence of two fibrillation rollers in the same unit offers the possibility of bringing the second fibrillator roll pneumatically into operating position when wrapping of loose ends occurs on the first fibrillation roll, thus allowing the processing to continue without interruption (Feuerböther, 1976). Whatever type of needle or blade fibrillator happens to be used, the resulting film-to-fiber product will always have a relatively wide spread in fiber fineness and rectangular cross sections, as shown in Figure 98 (see also Figures 6 and 7).

Fabrics made from fibrillated film yarns normally have a somewhat coarse touch and a filmlike luster. Therefore, great efforts have been made to impart in fibrillated film yarns a softer and more textile handle.

Figure 97 Fibrillator unit with two pinned rolls. (From Feuerböther, 1976.)

The best results toward this goal have been achieved by needle-roller fibrillation of longitudinally striated film (discussed in Sec. VI.A.1). The thin indentations can be penetrated easier by the needles or sharp edges of the blades, thus facilitating separation of the fibrils and making the processing conditions more controllable.

Disadvantageous in needle- or blade-roller fibrillation are the relatively high losses in tenacity and elongation. The finer the network the more fiber ends result and the greater is the loss in work potential, as defined by the product of tenacity and elongation at break. The decreasing tenacity of fibrillated film yarns can partly be compensated, however, by subsequent twisting. Figure 99 shows the correlation between fibrillation ratio and tensile strength, elongation, and hot-air shrinkage of fibrillated film fiber products. In spite of certain limitations in titer uniformity and to other textile characteristics, such as textile handle and filmlike luster, needle- or blade-roll fibrillation is probably today the most important film-to-fiber processing method and is therefore under permanent further development.

Shell Internationale Research Maatschappij N.V. (1974) described a fibrillator device consisting of two pinrolls arranged one upon the other. The film runs through a slot between the two rolls with an inlet speed somewhat below the rotational speed of the fibrillator rolls measured at their circumference. Stewart & Sons (1976) suggests the adjustment of a fixed angle between the needle or blade-roll axis and the film processing direction. By a divergence from the normally used right angle by some 1 to 3°, the fineness of fibrillar segments of the

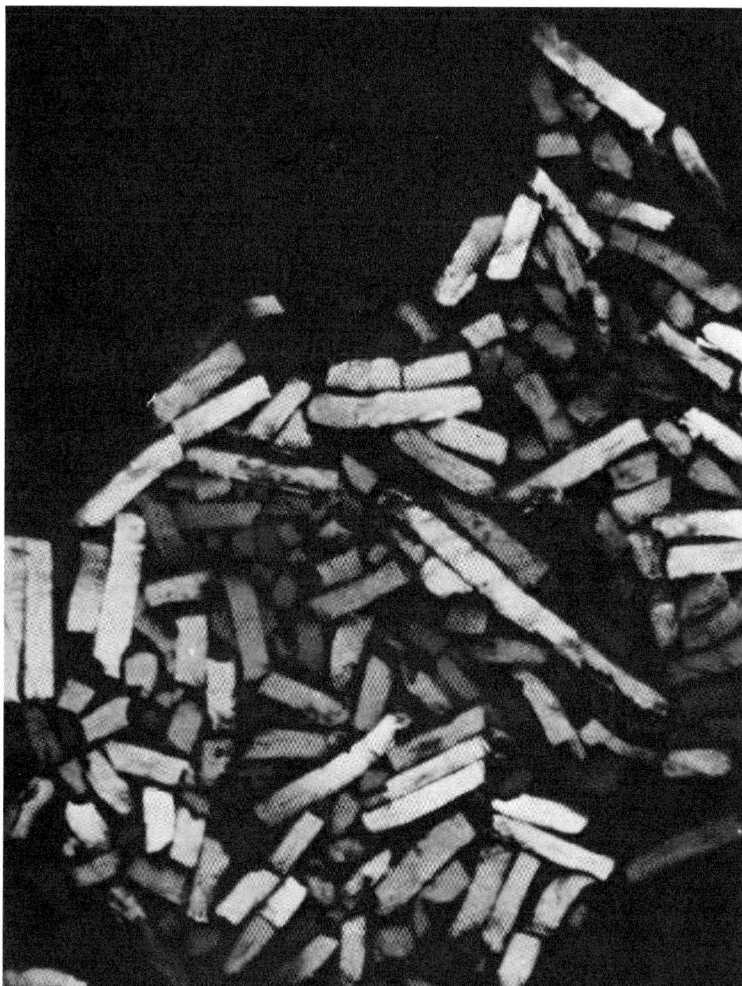

Figure 98 Microscopic view of the cross sections of a fibrillated yarn produced by film splitting with a pinned roller technique. (From Schuur and Van de Vegt, 1975.)

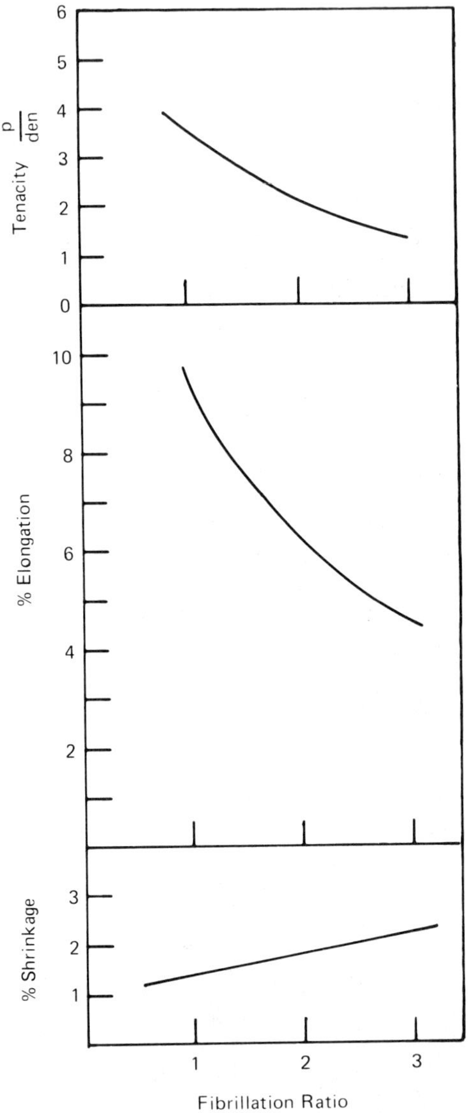


**Figure 99** Effect of the fibrillation ratio on the tenacity, elongation, and heat shrinkage (at 130°C) of split-film yarn. (Courtesy of ICI.)

fibrillated network can be controlled. Elfers and Schlegel (1974a) described a fibrillator roll with disk-shaped cutting elements embedded in rows at the bottom of grooves but sticking out from the roller surface. With this fibrillator polyamide film is also supposed to be successfully fibrillated. In another fibrillation device described by Elfers and Schlegel (1974b), the splitting tools are fine wire pieces which are one-sided fastened. The free ends are uplifted by the centrifugal force developing when rotated for fibrillation.

A patent applied for by Imperial Chemical Industries Australia, Ltd. (1978) describes a process in which heated needles are used for fibrillation of an unstretched film. The primary polypropylene or high-density polyethylene film is pressed between two heated needle plates and subsequently stretched at low temperatures in a 1:8 ratio. Fibrillated fibers with good bulkiness are said to result from this process. Droghanova and Martynova (1976) discuss in their publication the product properties and the application of fibrillated polypropylene film products. Very finely fibrillated film fibers are mainly used for such applications as twine, cord, and weaving yarns, while coarse fibers are used for cables, ropes, and thick cordage.

A patent by Shell Internationale Research Maatschappij N.V. (1972e) describes the conversion of fibrillated films into carded tows from staple fibers. The equipment used in their production consists of a needle roll fibrillator unit and an adjacent stretch-tear zone in which the fibrillated film is stretched and torn to single fibers by two pairs of rolls running with different speeds. Subsequent carding yields tow from which relatively smooth yarns can be spun. The staple length distribution can be controlled by operating conditions in the stretch-tear zone. Lambeg Industrial Research Association (1972) described a combination of fibrillation just before or after the needle-roller treatment with embossing, vacuum metal deposition, and printing.

### 3. *Methods of Uncontrolled Statistic Fibrillation of Monoaxially Drawn Films or Film Tapes*

At the beginning of the 1930s, Jacqué and his co-workers (I. G. Farbenindustrie AG, 1938) observed that monoaxially stretched films of polyvinyl chloride or of polystyrene developed an increased splitting tendency with increased stretch ratio. Such highly stretched films could be easily split into fibrous products by mechanical action, such as twisting, brushing, or rubbing.

At the beginning of the 1950s Dow Chemical Co. (1955) in the United States developed a rather similar process for the production of fibrillated film fibers from polyvinylidene chloride films. At about the same time, Rasmussen in Denmark began his pioneering work on film-to-fiber technologies. In those days the distinction was already made between the uncontrolled purely mechanical fibrillation and the uncontrolled chemomechanical fibrillation with the help of additives. In the middle of the 1960s Imperial Chemical Industries in England began with the production of polypropylene split fibers. Shell Oil Co. as well as several Japanese companies, such as Sekusui Chemical Co., Mitsui Petrochemical Industries, and Mitsubishi Rayon Co., started their intensive work in this field.

Mechanical fibrillation is a process by which a highly monoaxially stretched polymer film is split only by mechanical action into a network of still interconnected fibrillar segments having a wide statistical distribution in fiber segment thickness and in separation length. The main prerequisite for the fibrillation by purely mechanical stresses is the use of film-forming polymers with only weak intermolecular attractive forces and the application of high monoaxially applied stretch ratios.

To chemomechanical fibrillation belong all those processes in which additives are made to the film-forming polymer that introduce statistically distributed inhomogeneities into the film, contributing to higher fibrillability. These processes also result in network-like interconnected fibrous structures with varying separation length of the fibrillar segments and varying segmental thickness. In order to introduce weak spots into films by chemical means, several techniques have been suggested. The addition of an incompatible polymer to the matrix polymer is one of the methods. Very efficient also is the formation of voids in the film, such as by adding soluble salts, which are extracted after the film has been made. The same can be achieved by the addition of compounds decomposing with the development of gaseous products at the extrusion temperature. In the following section the technical possibilities for achieving simple mechanical fibrillation will be discussed.

*The Mechanical Tools for Uncontrolled Film Fibrillation.* The basic principle of mechanical fibrillation is the application of shearing forces on the highly stretched film or film tapes perpendicular or at a given angle to the stretching direction.

The simplest process to initiate fibrillation by mechanical stresses is the well known twisting technique. Twisting of highly stretched film tapes to over 1000 t/m (i.e., 25 t/in.) on a ring twister results in relative finely fibrillated yarns suitable for applications in packaging twine and rope cords. Shell Internationale Research Maatschappij N.V. (1969b) has described a special device which can be used efficiently in the fibrillation of film tapes. Figure 100 illustrates the principle of their design.

After twisting around the rotation axis of the twisting machine the twisted film tape is transported through a bended rotating tube having a length of 200 mm and an inner diameter of 12 mm. By rotation of this tube, the film tape is deflected from the axis of the tube inlet, thus introducing a false twisting action in relation to the number of revolutions of the tube. At 3000 to 10000 revolutions per minute the twisted film tape is subjected to a shear stress of approximately 10 kg/mm$^2$, resulting in a manifold splitting of the tape. The texture of the end product is characterized by Figure 4, already shown in Chapter 1. The cross sections of the fiber segments vary very much in thickness, as can be seen from the wide distribution of the titer in yarns made by twist fibrillation in comparison to products made by needle-roller fibrillation or by sawtooth-edge slitting (Chemiefaser Lenzing AG process) in Figure 101.

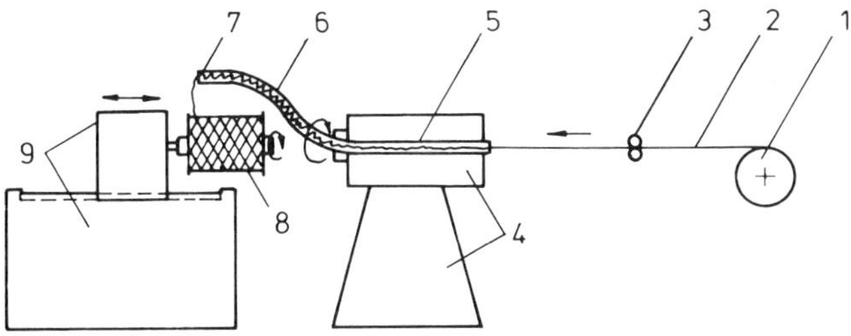


**Figure 100** Scheme of a special false twist apparatus for twist fibrillation. 1, Film tape package; 2, film tape; 3, guide and tension rolls; 4, mount, drive, and heater for false twister; 5, twist tube entrance side; 6, bended tube section; 7, twist tube exit side (parallel to 5); 8, takeup package; 9, yarn winder. (From Shell Internationale Research, Maatschapij N.V., 1969b.)

In most cases the tensile strength of fibrillated and twisted yarns for polyolefin products is almost identical with the tenacity of the original film tape. In the case of fibrillating film tapes from polymers having low splitting tendency, substantial losses in tensile strength are observed on fibrillation by twisting. The fact that there is less

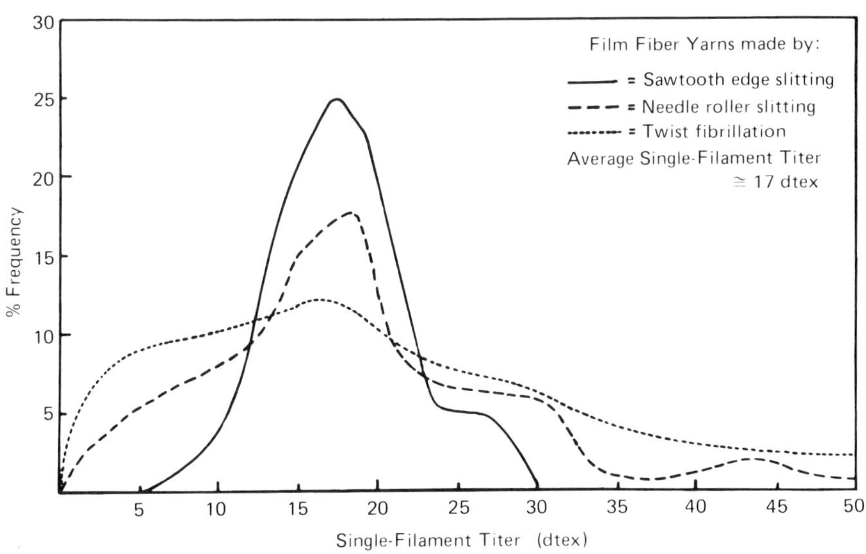


**Figure 101** Distribution of the single-filament titer for film-fiber yarns made by different film-to-fiber separation techniques. (From Krässig, 1977.)

fibrillation leads to a more uneven distribution of the stresses on fewer fiber segments. In this case, edge notches coming from defect blades in tape cutting diminish tenacity even more (see Table 18 in Sec. VI. B.1.). Other processes operating with false twisting techniques were described by Dow Chemical Co. (1965), by Société Rhodiaceta (1965, 1966), and by Plasticisers, Ltd. (1967).

Another technique for uncontrolled statistic fibrillation makes use of high-velocity air streams applied to highly stretched film tapes in "twist jets." Du Pont (1966a) described such an air jet device in which an air stream having a velocity near that of the speed of sound hits the film tape oblique. The air stream should preferably be inclined slightly forward along the line of tape movement. When the air influx takes place tangentially to the tape pass, the fibrillated yarn will be increasingly interlaced and twisted. The device suggested by Du Pont is schematically shown in Figure 102.

Another type of jet is claimed by Fiber Industries, Inc. (1972). In the process described by this company the tape is subjected to the action of two air streams, one after the other, blowing onto the tape perpendicular to its movement. The air channels are arranged in such a way that both the air streams will hit the tape tangentially with respect to its axis (i.e., first from the left side and thereafter from the right side). Thus the false twist imparted to the tape will be reversed by the second air stream. Figure 103 gives a schematic view of this pneumatic false twisting device.

To ensure good fibrillation the tape has to be kept under slight tension of approximately 0.05 to 0.20 p/tex, which can be adjusted by the feed and the takeoff rolls. The air emerges under a pressure of about 10 to 250 psi. With this process tapes of about 10 to 25 μm thickness will be split into 5 to 300 fibrils per tape. An advanced development of the same principle represents a process claimed by Imperial Chemical Industries, Ltd. (1977) in which the film tape is being heated in between and after false twisting by two air jet treatments. The first

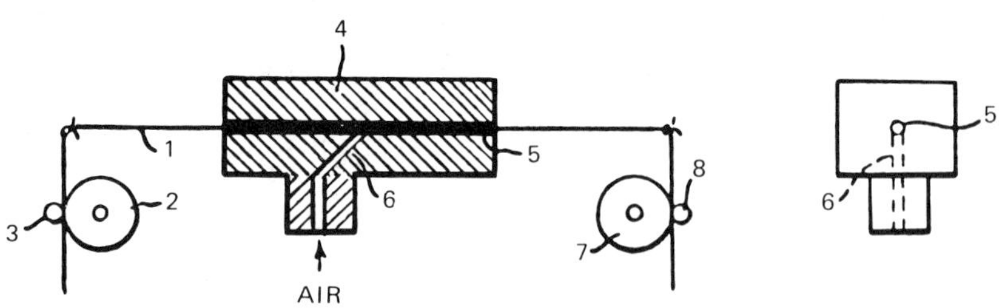


Figure 102 Scheme of a jet device for film tape fibrillation. 1, film tape; 2, 3, feed rools; 4, jet device; 5, cylindrical tape passage; 6, gas entrance way; 7, 8, takeoff rolls. (From E. I. du Pont de Nemours Co., 1966b).

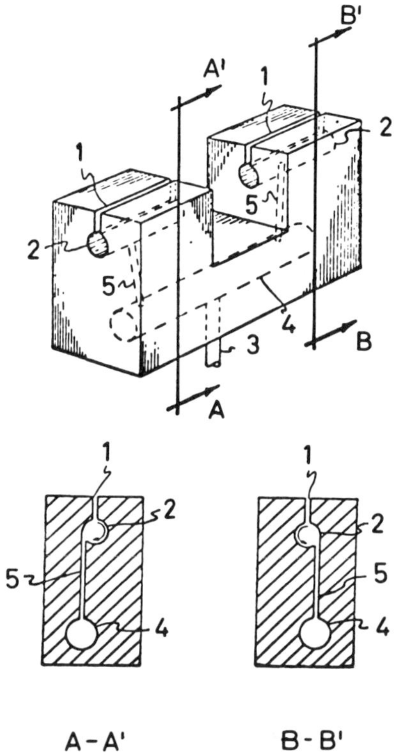


Figure 103 Scheme of a jet-twist fibrillator: 1, film tape slots; 2, film tape passageways; 3, gas entrance way; 4, gas manifold box; 5, gas passageways. (From Fiber Industries, Inc., 1972.)

false twist imparted in the first jet is set and unwound in a first heating zone. After passing the second air jet the twisted fibrillated tape is subjected to a second heat-setting treatment without tension. The result is a bulky and elastic fibrillated film yarn.

A process for splitting polypropylene tapes in a gas jet stream without twisting action was described by Chemie Linz AG (1976). Fibers of 5.5 to 11 dtex are supposed to be obtained when a highly stretched film tape is run with a speed of 400 m/min through a splitting tube in which it is subjected to the action of several high-velocity gas streams acting parallel to each other subsequently over the whole width of the tape. The bounce angle between the tape axis and gas streams should lie between 60 to 150°. Since the tapes are at least as large as the inner diameter of the splitting channel, they cannot escape and are fully subjected to the changing gas pressures in the splitting device.

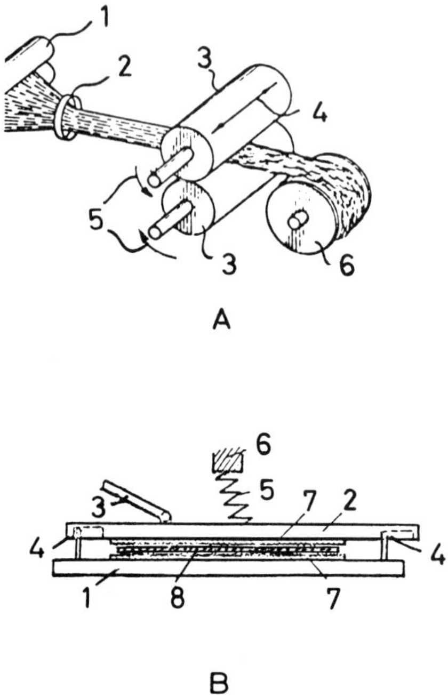


Figure 104 Schemes of film fibrillation tools: (A) with oscillatable rollers; 1, roller assembly; 2, eyelet; 3, pair of rollers; 4, oscillatable roll; 5, counter rotation; 6, takeup package. (From Phillips Petroleum Co., 1971b.) (B) with oscillating plates: 1, stationary plate; 2, oscillating plate; 3, transmission rod; 4, pins and grooves; 5, compression spring; 6, holding point; 7, rough surface portion on plates; 8, film sheet. (From Rasmussen, 1961.)

Statistic fibrillation of highly stretched films can also be achieved through exposure to ultrasonic waves.

Jacqué generated transversely acting forces by rubbing the highly stretched films or film bands between two countercurrently running rubber belts. Further splitting and parallelization can be achieved by rotating cylindrical brushes. A great many of the fibrillation methods are based on this originally suggested principle. Some of them shall be mentioned here. Phillips Petroleum Company (1971b) described a method in which the tapes are transported through a pair of rubber-coated rollers, one of them oscillating in its axial direction as shown in the upper part of part of Figure 104. The to and fro movement of the oscillating roll exerts a transverse stress to the tape, initiating the film-to-fiber fibrillation.

The same effect can be achieved with a device suggested by Rasmussen (1961) in one of his numerous patents; its principle is demonstrated in the lower part of Figure 104. The device consists of two plates having rough surfaces and pressed together by spring action. One of the plates is stationary and the other is moved to and fro perpendicular to the motion of the film tape which is run through between them.

A special form of film fibrillation using shearing forces represents the wet grinding of highly stretched films. This wet grinding process aims for the production of aqueous fiber suspensions to be applied in the manufacture of wet layed nonwovens. According to the process suggested by Leykam-Josefsthal AG (1970), the stretched polypropylene film tapes are chopped in short pieces and the shreds are mechanically fibrillated by pulping in a Hollander. A similar method has been patented by Hoechst AG (1976). The solid concentration of the film-shreds suspension in the wet beating operation ought to lie in the range of 1 to 15 g/liter, Nippon Steel Chemical Corp., Ltd. (1975a, 1975b) developed a closely related processing principle starting from tubular film made from a blend of polystyrene and polyethylene or polypropylene. In a tumble mixer, cut pieces of highly stretched film are fibrillated in the presence of water to fibers of about 30 μm in diameter and 50 to 80 mm in length. The addition of binders to the aqueous suspension improves the adhesion of the resulting split-film fibers in the subsequent manufacture of wet layed nonwovens. In this connection, reference should also be made to the publications of Berger and Schmack (1974) quoted in Sec. I.D.5, in which they tested the fibrillability of films from blends of polypropylene and polyester by means of grinding in a refiner.

As demonstrated by Rasmussen (1967a), it is possible to fibrillate a film not only by applying mechanical forces perpendicular to the stretching direction but also by exerting alternating lengthwise-acting tension on the film. He describes a fibrillation unit in which the film is stressed between two rubber belts running at different speeds, one of which may have a rough surface. The special pressure members pressing the belts together are shown on the right side of Figure 105. It is advisable to support this type of stress fibrillation by introducing incisions into the film prior to the stress treatment with the help of needle or knife blade rollers (Phillips Petroleum Co., 1971b; Polymer processing Research Institute, Ltd., 1970).

A similar technique of lengthwise stress fibrillation was patented by Soko Co., Ltd. (1971, 1973). Here the fibrillation is achieved by pressing the film between two pairs of compression rollers and by several roller assemblies against a rubber belt. Since the roller speed increases in the direction of the film movement and the pressure rolls are running faster than the rubber belt, multiple tensions are applied on the film, causing film-to-fiber separation. A scheme of the design

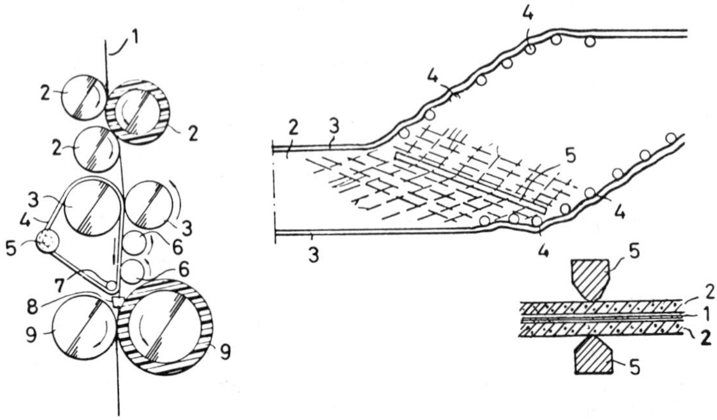

A B

Figure 105 Schemes of film fibrillation tools: (A) Rubber belt stretching devices. 1, film or film tapes; 2, holding roller assembly; 3, middle roller assembly; 4, rubber belt; 5, guide roll; 6, pressure rolls; 7, guide rod; 8, collecting funnel; 9, stretch roller assembly. (From Soko Co., Ltd., 1971.) (B) Double rubber belt fibrillator. 1, film or film tapes; 2, rubber belts; 3, belt guide; 4, pressure rolls; 5, pressure members. (From Rasmussen, 1967a.)

and the action of this fibrillation device is given on the left side of Figure 105.

Another way to initiate the fibrillation of monoaxially stretched films is to perform manifold bending actions. During such treatments the film tapes are folded lengthwise and bent at the same time so that they split along the folds. A patent taken by Allied Chemical Corporation (1969a, 1969b) shall be used to demonstrate such a process. The scheme shown on the upper part of Figure 106 illustrates the fibrillation device used. The highly stretched film tape to be split is drawn under tension through the holes of several parallel arranged plates which are alternately stationary and movable perpendicular to the processing direction. The back and forth movement of the movable plates causes the film to be bent up to 180°.

A fibrillation tool suggested by Rheinstahl Henschel AG (1971a) is shown in the lower part of Figure 106. In its use the film tape is led through a narrow funnel in which the tape is folded together. The folded tape is then wound 270° around a conically grooved roller. From this roller the tape is taken off at a sharp angle, thereby crossing the incoming folded tape and being forced to bend sharply. This device thus combines two fibrillation principles. Firstly, the tape is subjected to a transversely acting rubbing at the crossing point of the incoming and outgoing tape. Secondly, the tape is undergoing sharp bending, causing a lengthwise splitting action under high tension.

A further film-to-fiber fibrillation technique consists of compressing highly stretched tapes by treating them with expanding compressed air (Van Tilburg, 1975). The film tapes are compressed between a supply belt and a press roller. Into this pressure zone compressed air is introduced. At the end of the pressure zone the evading gas expands acting on the tapes, initiating fibrillation. In a process described by VEB Textilkombinat Cottbus (1973) the splitting of films to yarns or fibers is enforced by subjecting the film to ionizing radiation before the fibrillation treatment.

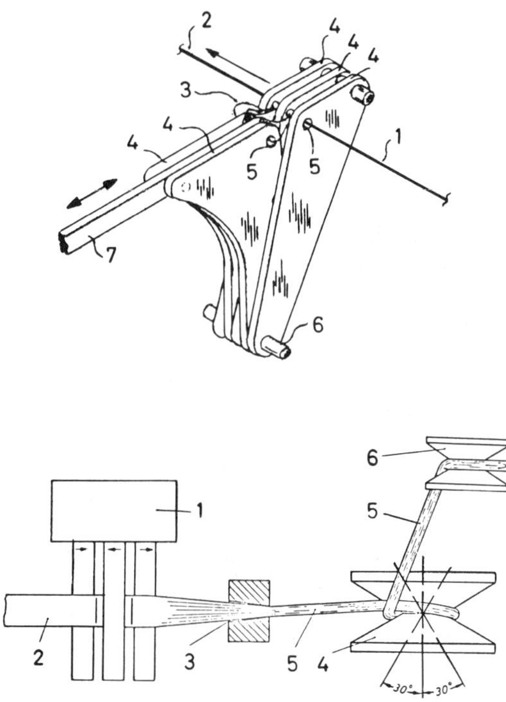


**Figure 106** Schemes of film fibrillation tools: top, bending plate fibrillator. 1, incoming film tape; 2, fibrillated tape; 3, bended film tape; 4, bending plate (two on the left-hand side are movable); 5, tape-leading holes; 6, bearing rod; 7, transmission rod. (From Allied Chemical Corp.; 1969a, 1969b.); bottom, bending roll fibrillator. 1, stretching trio; 2, film tape; 3, collecting funnel; 4, roll with conical guide notch; 5, folded tapes (being fibrillated between 4 and 6); 6, sidewise displaced 2nd roll with conical guide notch. (From Rheinstahl Henschel, 1971a.)

At extensive fibrillation treatments fine fiber dust is often formed, disturbing further processing. Such fine fiber particles stick during subsequent twisting on ring-spinning frames, especially to the thread brakes and thread guides. The same can happen during the cone winding. The deposition of broken fibers and fiber dust can partly be avoided by finishing the fibrillated yarn with an appropriate oil before twisting. A procedure for reducing the formation of fiber dust has been described by Fibron, Inc. (1977). The yarn is transported following the fibrillation through a heated zone where the temperature is high enough to melt the adherent particles, binding them solidly to the yarn. However, the temperature has to be carefully controlled in order not to damage the yarn.

Fibrillated yarns consisting of regular or irregular networks of interconnected fibrillar segments quite often show a certain filmlike appearance and luster even after twisting, when compared with staple fiber yarns. This is due to the mainly rectangular cross sections of the fiber segments. The twisted yarns often show an unsettled appearance due to the helically twisted form of the flat fiber segments. Through compression treatments in a narrow funnel, the flat fiber segments can be brought to a compact elliptical cross section, giving the fibrillated yarn more of a textile character which is of advantage in subsequent processing (Barmag-Barmer Maschinenfabrik AG, 1974).

*Chemomechanical Methods for Fibrillation.* The general characteristic for chemomechanical fibrillation methods is the introduction of weak spots into the polymer film by chemical means or the addition of chemical compounds. During stretching and subsequent mechanical action exerted by fibrillation tools the film will easily split at these weak points. Therefore it is insignificant whether the mechanical film-to-fiber separation is performed in a controlled way, such as needle-roller treatment, or uncontrolled, such as stressing, rubbing, or bending.

An important application of the chemomechanical fibrillation technique is, for example, the needle-roller fibrillation of highly stretched polymer blend films. The wet grinding of foamed polypropylene films may serve as an example for the uncontrolled fibrillation of a chemically modified film. The common feature for all these various methods is the decreased splitting energy demand owing to the chemical modification.

Rasmussen (1963c, 1965a, 1966a, 1967b, 1969c, 1970b, 1971c, 1971d) was the first to propose splitting of films from polymer blends into fibrous products. This principle has been exploited in the German Democratic Republic for a technical process in the manufacture of floor coverings. It was therefore investigated in great detail. Berger and Kammer (1978) have developed a model for understanding the fibrillation of films from polymer blends on the basis of the energy balance on a microfissure. In order to initiate a rupture on the face contact surface of two incompatible polymers, a certain amount of energy is needed. This energy consists of surface, plastic, and elastic energy. The lower the fibrillation energy of polymer blends is, the greater is the

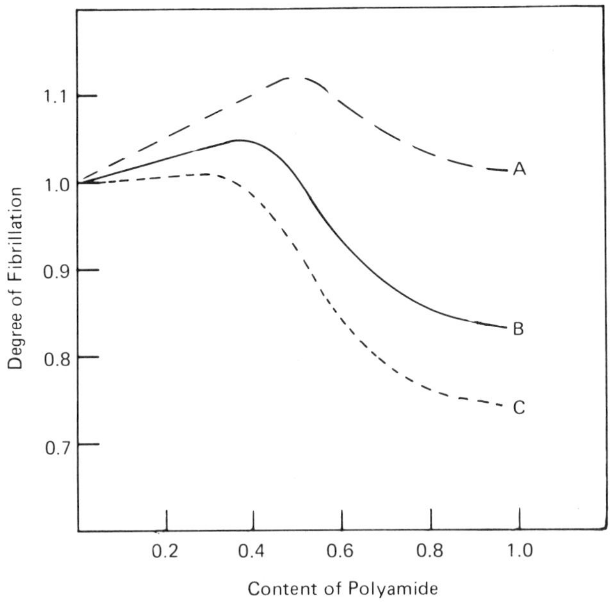


**Figure 107** Effect of the ratio of the Youngs moduli of polyamide and polyethylene on the fibrillation tendency of polyamide/polyethylene-blended films: (A), 1:1; (B), 1:0.667; (C), 1:0.5. (From Berger and Krammer, 1978.)

ratio of the Young's moduli between the disperse phase and the matrix phase. Figure 107 shows the relative fibrillation tendency of films made from blends of polyethylene and polyamide at different ratios of Young's moduli. The results of a further study on the same subject are contained in another publication by Berger and Kammer (1979). Dawczynski et al. (1978) investigated the system polyester/polyethylene in the composition of 70 vol. % polyester and 30 vol. % polyethylene. Polyethylene is embedded in the form of microfibrils in the polyester matrix. It facilitates the fibrillation by forming interfaces with low adhesion. The blend film is passed subsequently to the warp and weft system of a modified Malimo sew-knitting machine. Slitting of the warp film is achieved by slide needles, fibrillation of the weft film by a fibrillator unit, and the sewing is performed with polyester filament yarn. The textile aspects of this technique have been described by Schumann et al. (1978).

A patent of Teijin, Ltd. (1973) describes a blend of 70% polyester with 30% polyamide yielding an easily fibrillatable film after extrusion and double stretching. The splitting of the film results in fibers with titers between 8 and 19 dtex. A blend of 100 parts of polystyrene with

10 parts of polyethylene or polypropylene is recommended for chemomechanical fibrillation by Nippon Steel Chemical Corp., Ltd. (1975b). After tubular film extrusion and biaxial stretching, a film results which can be fibrillated by mechanical action in a mixer into fibers with increased oil absorption ability. Another process of this kind was described by Teijin, Ltd. (1970). From a blend of high-density polyethylene, polypropylene, and ethylene-propylene copolymer a film is made and monoaxially stretched to a high stretch ratio. The film is then heated to a temperature somewhat higher than the melting point of the component with the lowest melting point and finally fibrillated with compressed air, giving very fine fibers.

Another principle to create weak spots in a polymer film is the formation of voids or gas bubbles in the film. Rasmussen (1961) suggested the addition of soluble salts to the polymer during film extrusion which might be extracted after stretching in order to introduce voids into the film. In the same patent Rasmussen also claims the use of compounds decomposing at the extrusion temperature or during heat stretching, forming gaseous decomposition products. In this way a film with a foam texture is obtained. In the drawing treatment the voids form extended interstices acting as starting points for subsequent film splitting. Figure 108 shows a length view and the cross sections of the filaments of a split-fiber yarn made with such a technique (see also Figure 5 in Chapter 1). Figure 108 demonstrates the possibility to manufacture relatively regular fibrillated yarns.

A technically important procedure for the manufacturing of foamed films from polypropylene is the extrusion of polymer in the presence of azodicarbonamid or trihydrazino-sym.triacene at temperatures just above 190°C. Under these conditions the added compounds decompose, forming carbon dioxide and nitrogen. Under proper extrusion conditions the gas bubbles will be very finely distributed all over the melt. The foamy melt will be extruded through a flat die, water quenched, and short-gap stretched on a roller stretching unit. Figure 109 shows a flow sheet of a production line for the manufacture of foamed films. The disadvantage of the manufacturing of foamed films is the loss in tenacity. A pore volume of 30 to 40% in a monoaxially stretched foam film reduces the tensile strength by approximately 30% in stretch direction, as demonstrated in Figure 110 (Lenz, 1977).

When such a foamed film is fibrillated on a needle-roller unit the tenacity is once more reduced considerably (see Figure 99). To compensate for the reduced tenacity, fibrillated yarns made from foamed film should be given a higher twist count in subsequent twisting. Generally, foamed films can be fibrillated by all techniques of controlled and uncontrolled film-to-fiber separation. In a patent taken by Toyo Kayaku Co., Ltd. (1976) the foamed polypropylene film is fibrillated as follows. The film is first stretched at 140° by 300% in transverse direction and subsequently by 100% in machine direction, causing the

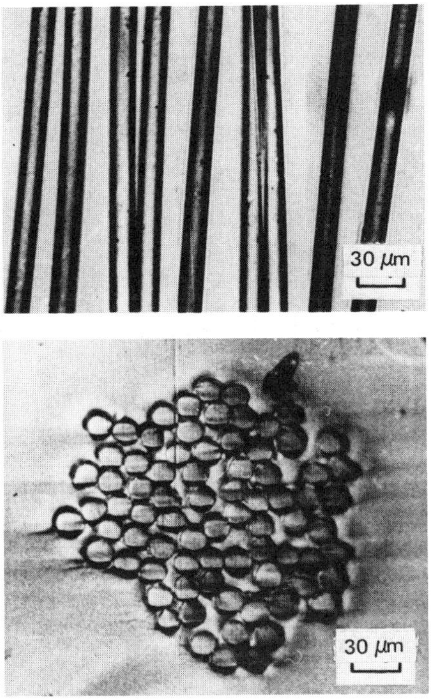


**Figure 108** Side view (top) and cross sections (bottom) of film fibers made from foamed polypropylene film tapes (uncontrolled chemomechanical fibrillation). (From Condit and Johnson, 1969.)

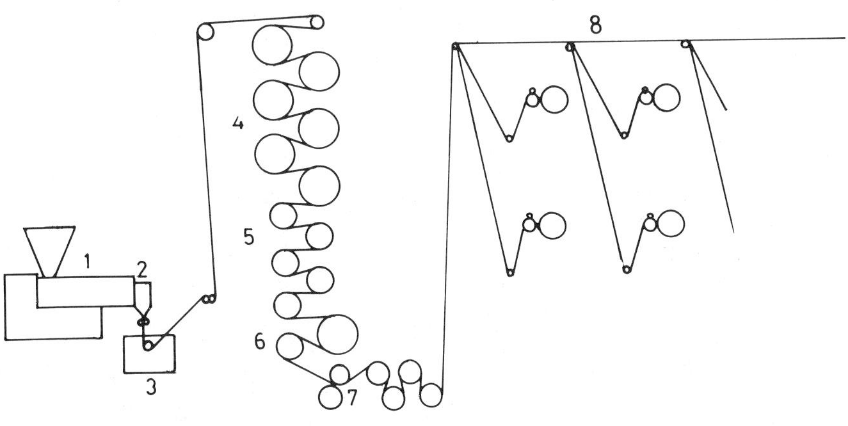


**Figure 109** Production line for foamed polypropylene film: 1, extruder; 2, extrusion die head; 3, water bath; 4, preheating rollers; 5, stretching rollers; 6, annealing rollers; 7, circular splitting knives; 8, winding unit. (From Lenz, 1977.)

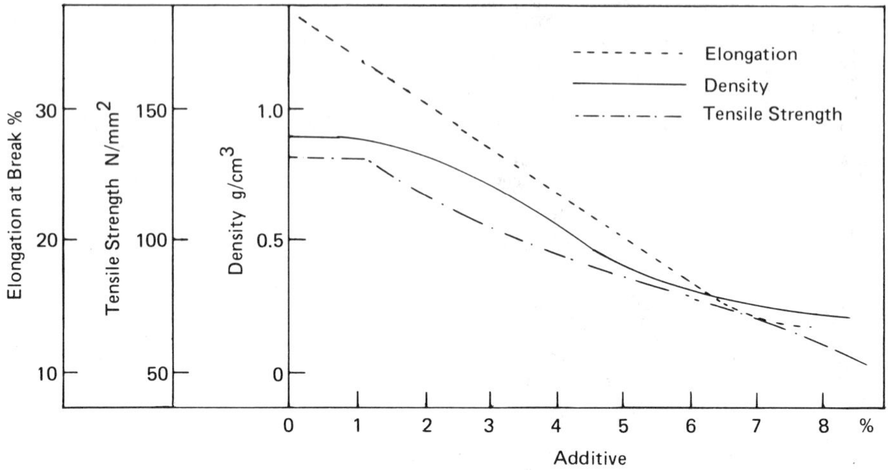


**Figure 110** Effect of the addition of foaming agent on density, tensile strength, and elongation at break of foamed polypropylene film. (From Lenz, 1977.)

splitting. In a process described by Showa Yuka K.K. (1974) the tubular film extrusion of a foamed polypropylene melt is performed at a speed ratio of 1:50 between the die exit speed and the takeoff speed. The film is said to begin to fibrillate already at a distance of approximately 3 cm from the die slot.

Idemitsu Kosan Co. (1975c, 1975d) describes a foamed polypropylene film containing 50% calcium sulfate which is extruded, stretched at a ratio of 1:5, cut to pieces of 10 cm (4 in.) length, dispersed in water, and ground in a disk refiner, giving a fiber suspension said to be applicable in paper manufacturing. Another variation, also described by Idemitsu Kosan Co., Ltd. (1975a, 1975b) is to produce film tapes with a porous layer. Two polypropylene films are coextruded with a blend of 70% polypropylene and 30% polystyrene containing 0.3% azodicarbonamide, resulting in a three-component laminate with a foamed core. The laminated film is stretched at a draw ratio of 1:10 and thereafter fibrillated to fibers consisting of a porous core and compact outer layers on both sides.

Finally, a patent taken by Asahi Dow, Ltd. (1972) shall be mentioned. In this case the foaming effect is achieved by the addition of 1% pentane to the extruded polymer. The resulting foamed film yields, on fibrillation, jutelike fibers with good tenacity in the range of 65 cN/tex. Quite a different principle for introducing weak spots into a film has been claimed by Mitsubishi Rayon Co. (1970b). A monoaxially stretched polypropylene film is first impregnated with a solution of chlorohydrocarbons in benzene and subsequently subjected to a corona discharge treatment. This treatment is said to result in an improved fibrillation tendency. Muzzy et al. (1976) reports that it is not possible to manufacture nonwovens by needle-roller fibrillation from foamed films suited for satisfactory applications in the processing of dress goods.

## VII. MANUFACTURE OF TEXTURIZED SPLIT-FILM YARNS

Fibers or multifilament yarns made by film splitting generally have a smooth and somewhat waxy feel and a filmlike luster undesirable for textile applications. In analogy to the texturizing of smooth synthetic multifilament yarns spun by conventional spinneret-spinning techniques, means were found to impart into film yarns and fibers a higher bulk, a more textile hand, and a duller appearance by introducing crimp to these products subsequent to the fibrillation.

For the texturizing of fiber or multifilament yarns from synthetic polymers the following techniques can be applied:

1. Heat relaxation or thermoshrinkage of fibers having "biocomponent" structure either in a physical or chemical respect. The prerequisite for the crimp development is different thermoshrinkage behavior of the structurally or chemically different components of the fibers (see Figure 111).

2. False twisting of multifilament yarns. A relatively high twist count introduced into the yarn in one direction is heat set at a temperature higher than the glass transition point of the fiber-forming polymer. During untwisting the material is cooled to stabilize the

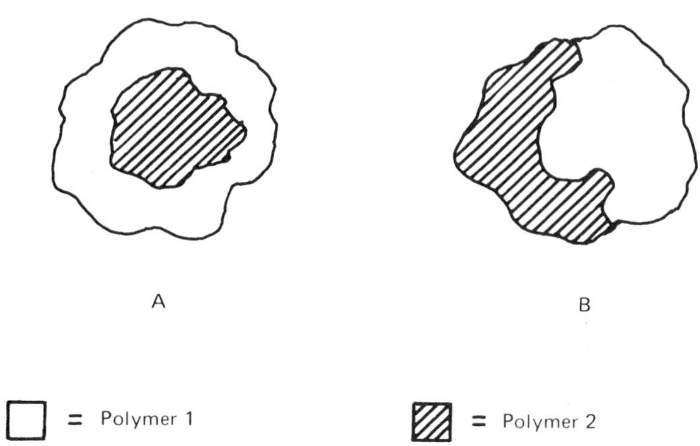


**Figure 111** "Bicomponent" fiber structures: (A) skin-core structure; (B) side-by-side structure.

introduced crimp. This technique is only applicable to fibers from thermoplastic polymers enabling the heat setting of the twist, giving the material the property of crimp recovery in relaxation of extension treatments.

3. Stuffer box treatment of tows consisting of more or less endless filaments using steam heating. This process imparts to the yarn a two-dimensional zig-zag crimp configuration. This is in contrast to the above mentioned methods which result in a helical crimp.

4. Several other texturizing methods, such as the knife-edge technique, friction texturizing, etc., will not be discussed in detail here.

The bicomponent process is the most important and most widely applied crimping technique for fibrous products made via the film. It is easier to introduce latent crimp properties into the film than to impart thermoplastic deformation to a smooth film yarn. In the case of melt-spinning techniques it is more convenient to thermoplastically deform multifilament yarns or stable fibers either by false twisting or stuffer box treatment. This is a consequence of the simpler technology of making multicomponent films than the more complex techniques of spinneret-spinning of bicomponent filaments by melt- or wet-spinning processes. Film fibers crimped using the bicomponent structure latent crimp potential have the quality advantage that their crimp has a three-dimensional configuration producing especially high bulk. Of course it is also possible to subject split- or slit-fiber yarns to stuffer box crimping treatments. However, the zig-zag crimp configuration achieved with this technique is more suitable for staple fiber products than for multifilament type yarns.

There are several ways to produce film fiber products with bicomponent, latent crimp structure:

1. By coextrusion of two polymers having different chemical structures and differing thermal shrinkage properties; a similar effect can be achieved by lamination of two films from polymers differing in structure and thermal shrinkage.
2. By coextrusion of two polymers having identical chemical structure but differing in molecular weight or molecular weight distribution.
3. By one-sided thermomechanical treatment of a homopolymer film before, during, or after the film splitting or slitting step. This technology is in some respects related to the knife-edge crimping. By the one-sided heat treatment, a fine structure difference is introduced into the film or film-fiber product, resulting in a difference in the thermal shrinkage behavior of the two sides of the film or the film fibers.

### A. Bicomponent Film Fibers from Polymers of Differing Chemical Structure

The following aspects ought to be considered in the choice of components for the production of crimpable bicomponent film fibers:

1. The two polymer layers in the bicomponent film must have good adhesion to each other.
2. Both components must be equally easily fibrillatable.
3. On heat treatment the two components ought to have different shrinkage characteristics.

In the crimped filaments the component with the higher heat shrinkage is always situated in the inner part of the crimp helix. According to Fitzgerald and Knudsen (1967) the following correlation exists between the shrinkage of the components and the crimp development:

$$C_f = kC_p(\Delta\alpha/h)$$

where:

$C_f$ = the crimp frequency, defined as the number of helical turns per length of fiber

$k$ = a constant, slightly dependent on the ratio of the moduli of the two components

$C_p$ = the crimp potential function, reflecting the relative amounts of the components and their distribution

$\Delta\alpha$ = the differential shrinkage of the two components

Of special importance in this equation is the factor K. It is essential for successful crimping that the shrinkage and modulus act in the same direction. In order to achieve this, one has to care in the choice of the two polymers, that the polymer having the higher thermal shrinkage also possesses the higher modulus in order to exert a higher stretch tension after stretching (Barbe and Sieroff, 1977). Otherwise only a weak and unstable crimp will develop in the heat relaxation treatment in spite of a substantial difference in the length change of the two polymer components. As will be shown in greater details in the next section, instantaneous crimp will develop immediately after the stretching when the two components possess a differing stretchability and relaxation behavior. Franz and Michels (1978) discuss in their publication the criteria for the crimpability of bicomponent split-film fibers from polyamide and polyester. The Figures 112 and 113 demonstrate the correlation between the degree of crimp, fiber fineness, the ratio of mass distribution of the components, and shrinkage difference. Increasing the proportion of polyamide having a greater relaxation ability will increase the development and degree of crimp.

It is advantageous to split or slit the bicomponent film into very fine single filaments in order to achieve fine crimp curvatures with small radius of curvature. Furthermore, it is important to perform

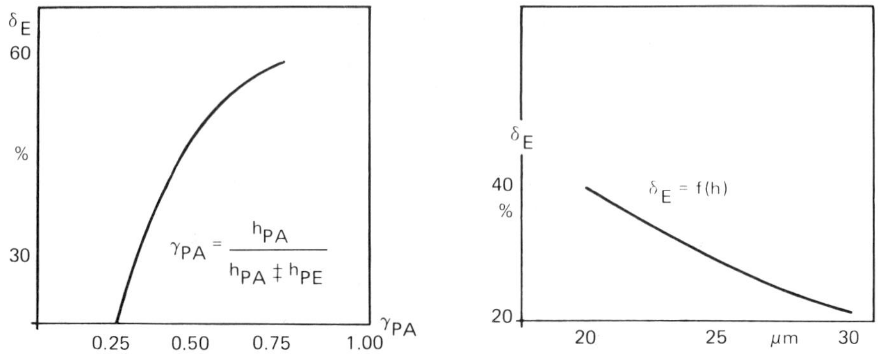


**Figure 112** Effect of bicomponent polymer composition and fineness on the ability of crimp development. Left side: correlation between the content of the more shrinking component (PA) and the total crimp ($\delta_E$). Right side: effect of film or fiber thickness (h) on the total crimp ($\delta_E$). (From Franz and Michels, 1978.)

the heat relaxation treatment, developing the crimp under weak tension. Rasmussen (1966a, 1971b, 1971d) was the first to recognize the possibility for manufacturing texturized film fibers from bicomponent films. Besides this, he demonstrated other possibilities for the modification of split-film fibers using multicomponent films with laminary layer structure. Using a combination of polyamide and polyester allows one to achieve not only a high degree of crimp, but due to the presence of the polyester component also a very good crease resistance, and due to the presence of the polyamide component excellent deep dyeing with acid dyestuffs. A combination of hydrophilic and hydrophobic polymer components gives fabrics made from bicomponent film fibers agreeable wearing comfort and easy drying property. The textile hand of film-to-fiber textiles is markedly improved by partial lamella separation of the bicomponent layers through frictional forces in use.

The laminary coextrusion technique in the manufacture of texturized film fibers can also be combined with the false twisting of highly stretched bicomponent film tapes from polypropylene and polyamide. In a friction texturizing process of the tapes, splitting and partial lamella separation is achieved. In the course of subsequent heat treatment intense crimping is obtained (Rasmussen, 1972b). In another patent Rasmussen (1979) describes a special effect yarn. The false twisted bicomponent split-fiber yarn is bound together in distances of 2 cm by the partial molten polypropylene component. In this way a yarn with intermittently interrupted bulk is obtained.

There are a great many processes for the manufacture of texturized film fiber yarns based on the bicomponent concept described in the literature or in patents. Here some of them may be listed: Rasmussen (1971b, 1971d), Allied Chemical Corp. (1969b, 1972, 1973), Eastman Kodak Co. (1969), Freedman (1970), Smith & Nephew Polyfabrik, Ltd. (1970), Smith & Nephew Research, Ltd. (1971), Mehta (1971), Mitsubishi Rayon Co. (1971b), Shell Oil Co. (1971), Briston (1972), Shell Internationale Research Maatschappij N.V. (1972c), Berger and Schmack (1974).

The prerequisite for success in these processes is, as mentioned before, reasonably good adhesion between the components in the bicomponent film structure. However, unfortunately, this does not often occur. An appropriate fibrillation tendency of both film components is also quite often difficult to achieve. Franz and Michels (1975) suggest for the case of polyamide/polyester bicomponent film

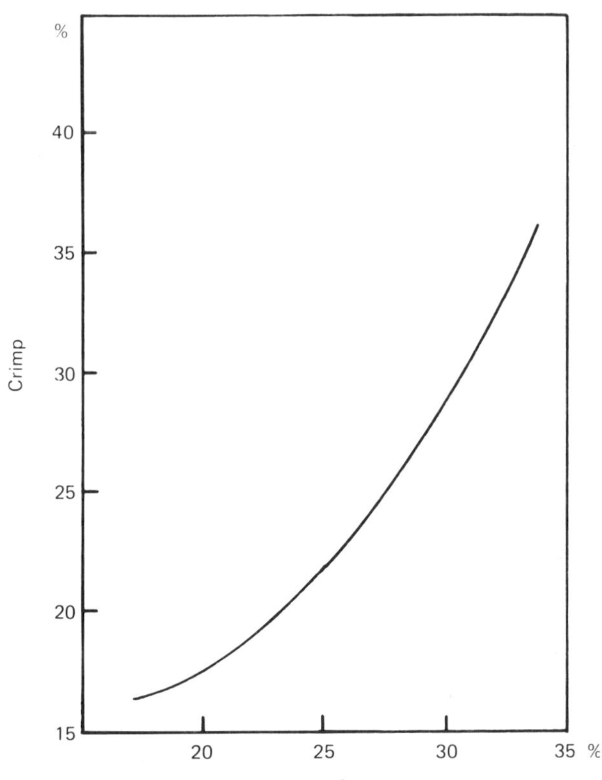


**Figure 113** Effect of the shrinkage difference of the two polymer components on the crimp of a bicomponent fibrillated film tape yarn. (From Russek and Glowacka, 1978.)

to use in the coextrusion not the pure polymers but blends of them to minimize the disadvantages of differing split energies and of low adhesion. The process they describe uses two polymer blends in the coextrusion, each consisting of one of the two main polymers blended with one or more polymer added. Whenever the main blend component in both layers of the coextruded film is identical, the adhesion as well as the fibrillation tendency by the island/matrix structure is markedly improved. The two components might, for instance, consist of polyester and high-density polyethylene in one layer and of polyester and polyamide in the second layer. The blend ratio in both components is 70 parts of polyester to 30 parts of high-density polyethylene or polyamide. The coextrusion of these two polymer blends and the monoaxial stretching results in a bicomponent film which can be easily fibrillated with needle rolls. The crimp will be introduced under high-temperature conditions during dyeing with dispersion dyestuffs. For the manufacture of crimpable split-film fibers Russek and Glowacka (1978) suggest using a bicomponent film made of two polymer blends: one consisting of 20% polypropylene and 80% polyester, the other of 80% polypropylene and 20% polyester. This bicomponent film shows good adhesion of the two layers, and at 140°C an optimal crimp will be achieved for the split-film yarn made therefrom.

Another method of producing latent crimp potential based on the bicomponent concept is the use of copolymers, such as those consisting of 95% polypropylene with 5% polyvinyl acetate for one film layer and of 97% polypropylene with 3% polyethylene for the second layer. Fibrillation and heat setting of the bicomponent film results in split-film fibers of good tenacity and elongation properties combined with a bulky crimp (Institute Textile de France, 1976).

Mitsubishi Rayon Co. (1975) suggests the lamination of stretched and unstretched films which are subsequently split and crimped in a heat treatment. For this purpose an unstretched film made from homopolypropylene is laminated together under pressure with a stretched film made from a copolymer consisting of propylene and ethylene at 135°C. Splitting of the resulting bicomponent film and heat setting leads to highly crimped fibers. This combination was also investigated by Neveu (1975). In his work he especially examined the effects of the drawing ratio, the thermal treatment, and the relaxation ratio in heat setting. Mehta (1971) suggests that split-film fibers consisting of polypropylene and high-density polyethylene combined in a bicomponent structure can be used in the manufacture of needle-fortified nonwovens. The resulting nonwovens can be heat set at 120°C, whereby the high-density polyethylene component melts and further fortifies the nonwoven structure.

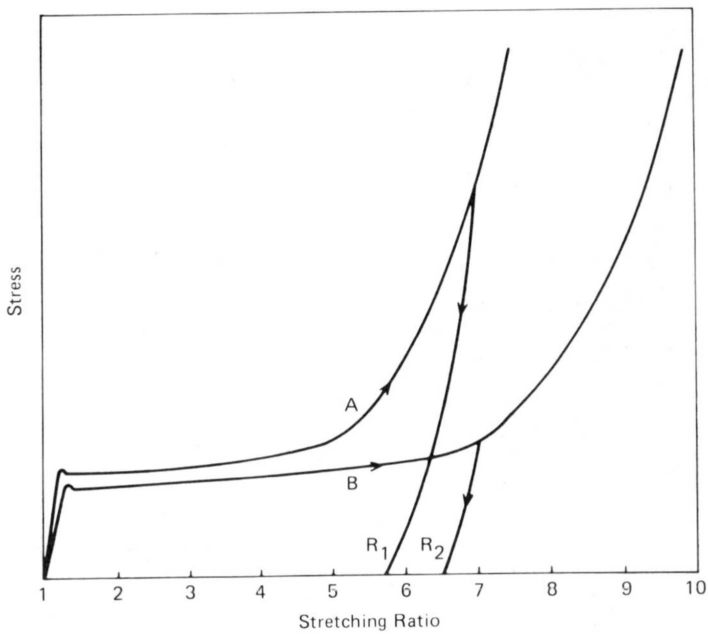


Figure 114 Schematic representation of the stress-strain curves of a preoriented high molecular weight layer (A) and a less oriented low molecular weight layer (B) in a laminated film during stretch and relaxation. (From Schuur and Van de Vegt, 1975.)

## B. Bicomponent Film from Identical Polymers Differing in Molecular Characteristics

In bicomponent films from polymers with different chemical structure the adhesion between the two film layers is quite often problematic. Therefore it is advantageous to combine bicomponent layers from identical polymers differing in their molecular characteristics. The coextrusion of polymers with identical chemical composition but widely differing molecular weights solves the problem of good adhesion of the layers in the bicomponent film and gives excellent latent crimp potential since the two layers demand different stretch tensions at the same stretch ratio, thus resulting in differing internal stresses. As shown in Figure 114 the orientation in the higher molecular weight layer has already achieved a high level at stretch ratio of approximately 1:6, while the lower molecular weight layer is still in its flow and nonelastic extension

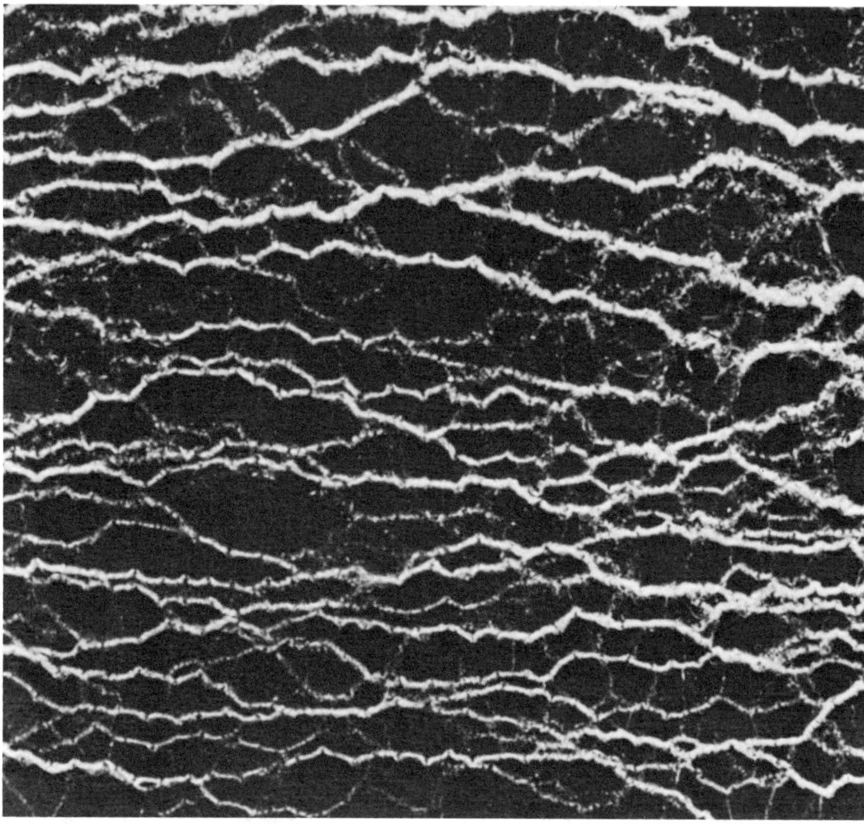

**Figure 115** Microscopic length view of a fibrillated bicomponent film yarn crimped by stress relaxation. (From Schuur and Van de Vegt, 1975.)

region. This results in higher internal stresses in the higher molecular weight layer, giving a higher relaxation tendency. The difference in relaxation ability between the two layers of the bicomponent freshly stretched film leads to crimp initiation in the heat relaxation treatment after fibrillation. A helical crimp is spontaneously formed in which the higher molecular weight layer lies on the inner side of the helix structure. In this case the crimp is initiated mainly because of the difference in elastic tension. Mitsubishi Rayon Co. (1973) exploits this principle in the following way: A laminated film from two polypropylene components with different melt indices is cut into tapes, stretched to a high stretch ratio, fibrillated, and subsequently heated in the relaxed state. The results are highly crimped film fiber yarns. Figure 115 demonstrates the crimp achieved by the application of this method.

Not only do differences in average molecular weight cause latent crimp potential in bicomponent film products, but also differences in molecular weight distribution. The layer with the broader molecular weight distribution will be oriented to a higher extent at a given stretch ratio, probably due to the plasticizing effect of the lower molecular weight portions. This results in a higher elastic recovery with the release of internal stresses.

In order to achieve optimum crimp development the processing parameters must be properly balanced throughout the whole process in order to avoid counteracting tendencies. For example, in tubular film extrusion the high molecular weight compound has to be introduced on the outer part of the tube forming in coextrusion, since this side cools faster and has less time to relax extrusion orientation. Friedman and Vlasov (1974) present in their publication a theoretical survey dealing with the extrusion of two polymer melts with different melt viscosities to a bicomponent tubular film. Mitsubishi Rayon Co. (1973) describes the manufacture of multifilament film fiber yarns from polypropylene bicomponent film using two polymer grades with intrinsic viscosities of 1.5 and 2.5, respectively. Air jet interlacing of this yarn with a monocomponent split-film yarn from polypropylene having an intrinsic viscosity of 2.5 results, upon heat relaxation, in an especially bulky yarn product.

### C. Crimped Bicomponent Film Fiber Products Using One-Sided Thermomechanical or Chemical Treatments

The texturizing of synthetic mono- or multifilament yarns is also possible by a one-sided thermomechanical treatment known as the "knife-edge" treatment (Taslan process). The thermoplastic yarns are hereby drawn under high tension at a certain angle over the edge of a heated blade arranged perpendicular to the processing direction. The yarn side which is in contact with the heated bar is substantially altered in its fine structure and thermal shrinkage behavior so that upon heating the difference in thermal response toward the other yarn side is sufficient to initiate crimping. The combination of this effect with the film splitting operation leads to film fibers with strong latent crimp potential, as shown by processes developed by Hercules Inc. (1970) and by Chemiefaser Lenzing AG (1970; see Harms et al., 1971a, 1971b, 1973a, 1973b; Krässig, 1972, 1975a, 1976).

The latent crimp potential is introduced into slit fibers through simultaneous edge bending ov a heated bar, the slitting operation, and stretching as described in detail in Sec. VI.A.4. The formation of a quasi "bicomponent" structure in this multifunctional one-step treatment is the cause of the latent crimp potential which can be triggered in a hot air relaxation step. Figure 116 shows in polarized light the cross section of single capillaries of a slit-film filament yarn produced

Figure 116 Illustration of the "bicomponent" structure of film fibers made by the film splicing process of Chemiefaser Lenzing AG. (From Krässig, 1977.)

by the Chemiefaser Lenzing AG's film yarn process. The bicomponent structure caused by the one-sided heating on the cutting tool is clearly visible.

Further proof of the existence of a bicomponent structure in the one-sided heat treated film is given by the results of model experiments on double layered film drawn under identical conditions over a heated flat bar. Light scattering studies, x-ray diffraction, and infrared dichroism using frustrated multiple internal reflection (FMIR) technique show that the film layer exposed to the heated bar has a better crystallinity and lower degree of orientation than the layer not in contact with the heated bar. Table 20 records the results of these studies (Harms et al., 1973b; Krässig, 1976).

This bicomponent structure resulting from fine structure differences leads, in the course of a hot-air treatment in an oven or in a jet-tube channel, to a helical crimp of the polypropylene slit-film fibers, as demonstrated in Figure 117. Another process applying a one-sided modification treatment of film in split-film fiber production which leads to highly crimped fiber products by thermal initiation was developed

**Table 20** IR Absorption of the Two Layers of Polypropylene Film Which Underwent a Heat Treatment on a Heated Bar Simulating the Conditions of the Lenzing Slit-Fiber Process (FMTR-Technique using Polarized and Nonpolarized IR Radiation)

| | Tacticity % | Order index $J_{1455}/J_{840}$ | Orientation by dichroitic ratio | | |
|---|---|---|---|---|---|
| | | | 1168 $cm^{-1}$ | 995 $cm^{-1}$ | 840 $cm^{-1}$ |
| Polypropylene film | | | | | |
| side touching the heated bar | 70 | 1.21 | 1.40 | 2.07 | 2.48 |
| side away from the heated bar | 75 | 1.08 | 1.55 | 2.27 | 3.02 |
| Ratio heated to nonheated side | 0.93 | 1.12 | 0.90 | 0.91 | 0.82 |

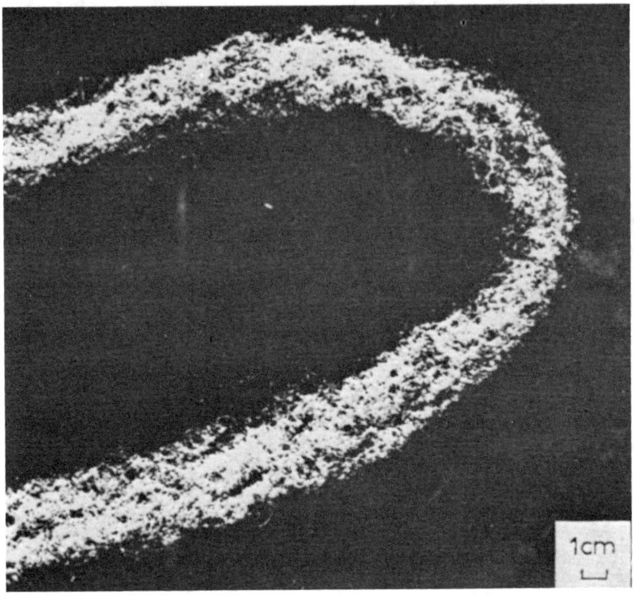


Figure 117 Demonstration of the crimpability of a polypropylene split-fiber yarn produced according to the film splicing process of Chemiefaser Lenzing AG. (From Harms, Krässig, and Sasshofer, 1971b.)

by VEB Textilkombinat Cottbus (1973). According to this process the film, prior to film splitting or slitting, is exposed to powerful high-energy electron irradiation. A web of split fibers made from film modified in such a way is texturized when subjected to treatment at elevated temperature. At the same time the surface of the fibers is molten and the web is solidified by fibers sticking together at their crossing points.

### D. Thermomechanical Procedures for Production of Crimped Film Fiber Yarns

The majority of all procedures for the manufacture of crimped film fibers makes use of the bicomponent principle. In the literature no reference can be found of the application of conventional methods for the crimping of spinneret-spun multifilament yarns, such as false twisting, friction texturizing, and stuffer box crimping. Only relatively few patents deal with other possibilities of thermomechanical crimping. Sekusui Chemical Company (1971) claims that tension-inducing contact of a highly stretched polypropylene film tape with a fast rotating heated roll having a coarse surface produces crimped split-film fibers. A process described by Mitsubishi Rayon Company (1971c) works with a pair of rolls, one having a metal surface and the other being rubber coated, rotating with different circumferential speeds. During the passage through this set of rolls a highly stretched polypropylene film is split into fibers of approximately 20 dtex having about 22 crimps per cm (i.e., 9 crimps/in.).

For the sake of completeness, one other principle for the production of crimped elastic split-film fiber yarns should be mentioned. When a fibrillated polypropylene film tape is drawn through a heated aperture in a metal plate having a temperature above the melting point of polypropylene, adhesion centers are formed between single filaments or filament segments, in the accumulation zone at the entrance of the aperture, resulting in a bulky yarn with good elastic properties (Mitsubishi Rayon Co., 1972a). However, it should be stated that yarns made from bicomponent films whose components possess good adhesion and sufficiently different thermal shrinkage properties have, in most cases, a markedly superior crimp stability than the crimp in film fiber yarns produced only by thermomechanical methods.

## VIII. THE SIGNIFICANCE OF POLYMER ADDITIVES IN FILM FIBER PRODUCTION

In the preceding sections it was quite often mentioned to what extent the processing and properties of film tapes and of film fibers could be altered by blending different polymers. In this section the significance and potential of low molecular weight additives for altering certain characteristics of polymer film products shall be discussed. Certain inorganic and organic additives can change the response of the film forming polymer towards chemical and physical influences, such as oxidation, heat deterioration, or light degradation. Generally, the modification of polymer characteristics with the help of such additives always means an upgrading and improvement of the use potential of the products made from them.

The numerous polymer additives offered on the market shall be examined in terms of their significance in the improvement of film tape and film fiber products with respect to the following characteristics: color, luster, UV-light stability, heat and flame resistance, stability against oxidation, affinity towards dyestuffs in printing and bath dyeing, prevention of electrostatic charging, reduction of friction and blocking effects, changing splitting tendency, or adhesion properties to other substances. Regarding the effect of additives, not only is their chemical nature important but so also is their concentration and evenness of distribution in the polymer. For this reason the techniques of incorporation will be discussed first.

## A. Techniques for the Incorporation of Additives into Polymer Melts

Additives are usually dry and finely pulverized powders which have to be mixed with the polymer which is mostly in granulate form. The aim of the mixing technique is, in the first instance, to achieve in the extrusion a constant concentration per time unit of the additive throughout the polymer during the continuous production process. The second goal is the achievement of a constant concentration of the additive in the polymer melt over the whole die width or die circumference. Thirdly, the mixing technique has to assure a very fine and even distribution of the additive throughout the polymer melt, especially in cases where it is not soluble in the polymer.

The latter requirement poses special technical problems to be solved. Due to the high viscosity of polymer melts, high shearing forces are necessary for a proper dispersion of additive agglomerates. A wide particle size distribution leads to disturbances in the process, such as clogging of the filters, formation of die encrustations, and ruptures of the film in stretching due to inhomogeneities. Moreover, the tensile properties of the resulting film fiber yarns are adversely effected by a wide particle distribution in the additive or by uneven distribution of the additive in the film product. If the additive is soluble in the polymer, incorporation is less problematic.

The simplest incorporation technique is "dry-blending" which consists of mixing the polymer granulate with the additive powder in a rotating drum mixer. As a result of the different electrostatic charges of the two components the additive powder is evenly distributed over the whole granulate surface. Under such conditions every granulate particle is covered with a layer of additive powder. This method is successful, especially at low additive concentrations, and when the additive is soluble in the polymer. For the incorporation of color pigments this simple method is quite often not suitable since pigment agglomerates are in most cases not sufficiently dispersed under the shearing forces effective in the extruder.

In these cases, a two-step mixing technique of the polymer with the pigment additive is to be recommended. In the first step the pigment powder is dry-blended with the polymer granulate and extruded to a granulated pigment-polymer compound with a high pigment concentration, the "master batch." In the second step the pigment master-batch granulate is blended down to the desired pigment concentration with the polymer granulate. The double extrusion leads to a more even distribution of the pigment particles in the film. Nevertheless, this double extrusion method demands added manpower requirements and additional energy. Furthermore, double extrusion of the master-batch polymer includes the danger of heat degradation.

For the incorporation of polymer-insoluble pigment, therefore, a third processing technique has been developed. This method comprises the blending of polymer granulate with a powdery pigment preparation in which every pigment particle is encapsulated in a thin layer of polymer. These encapsulated pigments can, for example, be produced by precipitation of the polymer from a solution containing dispersed pigments under high turbulance. Thus the pigment particles are embedded in a thin layer of a polymer which is compatible with the polymer to be pigmented. The polymer used for the pigment encapsulation acts in the extrusion as a kind of carrier.

The concentrated pigment master batches can also be produced by intensive mixing aggregates, such as Banbury mixers, pinmills, or double-screw extruders. The pigment concentration in the masterbatch formulations is generally in the range of 25 to 60%. The pigment concentrates are offered in the form of granulates, pellets, chips, or powders. The mixing of pigment powders and polymer granulate may be accomplished continuously or discontinuously. In the discontinuous procedure the components are weighed in the desired proportion, placed in the drum of a drum-mixing machine, and mixed by rotation. Subsequently the well dispersed mixture is fed into the extruder through the inlet funnel. In the continuous procedure the inlet funnel of the extruder is equipped with a gravimetric or volumetric metering device which feeds the single components in any desired proportion into the extruder feeding zone where a mixing tool provides an even distribution. With both procedures care has to be taken that vibration of the inlet funnel does not cause demixing of the components.

A normal one-screw extruder, by the nature of its function, is no ideal mixing device but more a conveying machine. Its mixing effect, however, can be markedly improved by a special screw design. Extruder screws with a particularly good mixing effect have a specially designed shearing section after the conveying zone which cares for intensive mixing of the melt components. This shearing or mixing section of the extruder screw carries, in most cases, naps adjusted to the screw core instead of the conventional screw threads. These naps divide the polymer melt stream many times over. The use of such shearing or mixing sections, however, involves the danger of thermal damage to the polymer due to possible local overheating. Figure 118 shows such a shearing or mixing section of an extruder screw.

Another possibility for the homogenization of polymer melts with additives is the use of static mixing devices. Static mixers have no moving parts; they work on the principle of alternately dividing and rejoining the polymer streams with specially designed chicanes built into a tube section. In relation to the desired mixing effect, an optimal number of stream-dividing chicanes have to be used. The number of mixing elements is also dependent on the pressure decline occurring with increased flow resistance. The more the pressure declines the better is the mixing effect. For static mixers several special designs were developed, such as by the firms Kennex, Sulzer, or the multiflux

**Figure 118** Shearing and mixing section of an extruder screw. (Courtesy of Chemiefaser Lenzing AG.)

system of Barmag-Barmer Maschinenfabrik AG. Figure 119 shows a length cut through such a static mixer.

Filtration of the pigmented polymer melts deserves special attention. Every filter screen clogs after a certain time and causes an interruption of the extrusion process. To save work cost and waste connected with such interruptions it is advisable to use filtration systems whose filter material renews itself automoatically through pressure sensing devices recording the pressure increase resulting from the mounting clogging. There also exist double-screen systems whose filters can be interchanged (Balk, 1978). Figure 120 presents a photograph of a quick-change double-screen filter unit. For the control of the pigment particle distribution in the film or in the film fiber product, microscopic observation can be used. If possible, no particles or agglomerates of particle should appear greater than 5–10 μm.

## B. The Pruduction of Dull or Colored Film Fiber Products

The film-to-fiber technologies are applied mainly to polyolefins. A significant drawback of polyolefins is that in the unmodified state they can hardly be dyed in conventional bath dyeing with any known dyestuff because they lack functional groups with dye affinity. The production of colored split- or slit-film fibers for polyolefins, therefore, was only possible up to now through melt extrusion incorporation of colored pigments. Through the suitable choice of pigments or of blends of pigments almost any color shade or depth can be achieved. The melt extrusion pigmented film fiber products stand out particularly for their very good stability against light, laundering, and dry-cleaning treatments. High color fastness values can be achieved in any respect. By the use of white pigments in the melt extrusion, such as titanium dioxide, film fibers can be dulled. Film fiber products show quite often a glossy appearance not common to textile products produced from spinneret-spun fibers. The dulling of the split-film fibers suppresses this shiny appearance most efficiently.

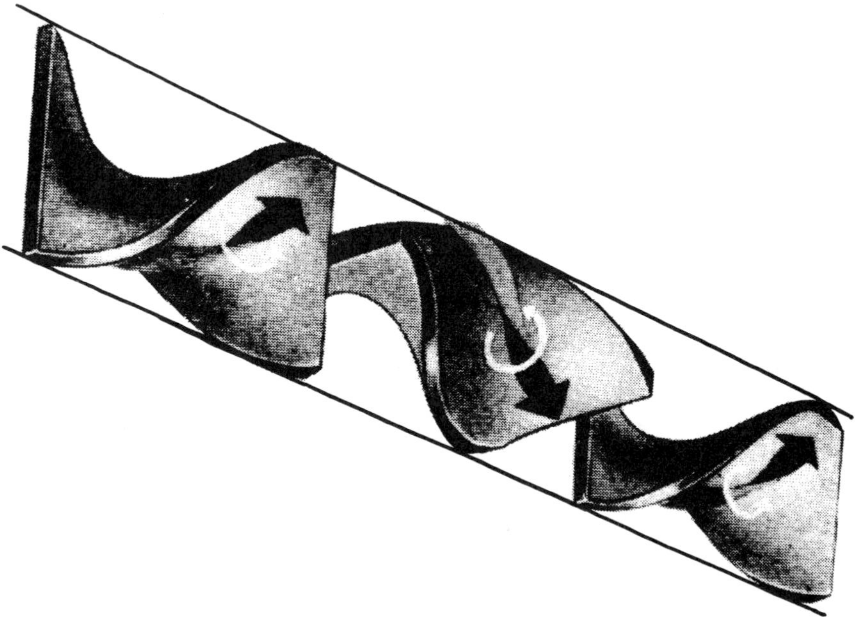

Figure 119 Demonstration of the mixing action of a static mixer. (From Schaab and Stoeckhert, 1979.)

Figure 120 Quick-change double-screen filter unit. (Courtesy of Chemiefaser Lenzing AG.)

For the sake of completeness it should be mentioned that film tapes can be given a less glossy appearance through mechanical surface treatments. The choice of pigments is subject to the following criteria:

1. Sufficient temperature resistance to withstand exposure to the extrusion temperature
2. Good light fastness and weathering resistance
3. Sufficient resistance against chemical influences
4. Desired shade and color depth
5. Film transparency and opacity
6. Costs

The following pigment types are available for the pigmentation of polyolefins:

*Inorganic pigments*

Yellow shades: lead chromate, cadmium sulfide
Orange shades: lead molybdate, cadmium selenide
Red shades: mercury selenide, cadmium selenide, iron oxide
White shades: titanium dioxide
Black shades: carbon black (lamp black)

*Organic pigments*

Yellow shades: pigments of the azo- and diazo type
Orange shades: diazo pigments
Red shades: perylene pigments
Blue shades: pigments of the phthalocyanine type
Green shades: pigments of the substituted phthalocyanine type.

As already mentioned, pigments can cause many processing difficulties during the production of pigmented films and the melt extrusion of dyed film fibers. This point has to receive special attention mainly in the choice of inorganic pigments. In the manufacture of tapes or film fibers the fact that these pigments are generally markedly harder than organic pigments has to be considered. They often cause a quicker blunting of cutting tools. Titanium dioxide of the rutile modification in particular shows serious abrasive effects. Besides quick wear of the cutting blades, pigments sometimes form deposits at tape and thread guides which interfere with smooth processing.

In the past, extensive efforts were made by polyolefine producers to make bath dyeing with conventional dyestuffs possible. The patents dealing with this problem are quite numerous. For the solution of this problem the following routes were taken:

1. Copolymerization of the olefin monomer with other monomers carrying dye-receptive functional groups. Suitable monomers are: acrylics, acrylonitrile, vinylpyridine, etc. Furthermore, grafting of polyolefins with such monomers can serve the same purpose.
2. Incorporation of small amounts of polymeric polar substances carrying functional groups with high dyestuff affinity. Suitable are additions of polyvinyl pyridine, polyvinyl pyrrolidone, or copolymerizates of propylene with aminoalkyl acrylate.
3. Incorporation of organic metal compounds, primarly of nickel, aluminum, or zinc.
4. Chemical modification of the polyolefin or the polyolefin product prior to the extrusion dyeing of the product made from it, such as halogenation, sulfonation, or phosphorylation.

Since the copolymerization of monomers of different kinds with stereospecific catalyst systems is quite often difficult to achieve, graft copolymerization modification has been considered for some time as a realizable method to solve the dyeability problem for polyolefins. High-energy radiation-induced graft copolymerization was studied quite extensively. However, so far no practically applied process or marketable products have been derived from such efforts. More successful have been the efforts based on adding dyestuff affine substances to the polymer melt in the course of extrusion processing. The best system actually practiced is based on the incorporation of organic metal compounds, especially those containing nickel. For this the "nickel quenchers" which have been used for a long time as UV stabilizers of polyolefin, mainly of polypropylene, can be used. In dyeing with chelating mordant disperse dyes sufficient migration of the dyestuff into the polypropylene fiber product and fixation by complex formation is obtained. This method yields good fastness properties combined with high ultraviolet stability (Botros, 1979).

The incorporation of polymers containing basic nitrogen groups allows the use of a broader range of water-soluble acid dyestuffs (Mitsubishi Rayon Co., 1970). However, in most cases the fastness properties achievable in this way are so far unsatisfactory. In this context, the blending of polypropylene with polymers, such as polyester, polyamide, polycarboxylic acids, or polyvinyl chloride was also recommended. The insufficient compatibility of these polymers with polyolefins, however, together with the mostly unsatisfactory light fastness of colorings of such blends causes great problems (Ivett, 1968; Roberts, 1971; Lee, 1975). A slight increase in the dye affinity of polypropylene split fibers can be obtained by grooving or profiling the film prior to the fibrillation step (Shell Internationale Research Maatschapij N.V., 1971).

## C. Additives Influencing the Splitting Tendency of Films

The splitting tendency of films or film tapes, respectively, can be a desired or an unwanted property. Film tapes for weaving or rascheling applications should show low splitting tendency since splitting interferes with fabric processing. Also, the split weaving of stretched polypropylene films on looms is adversely affected by high splitting tendency.

For the production of split- or slit-film fibers or yarns, however, a high splitting tendency is understandably advantageous and important. In Sec. VI.B.3. different ways for the introduction of weak spots enhancing and controlling the splitting of film have been discussed. Here it should be remembered that one way to introduce weak spots is the introduction of voids into the film (e.g., by the addition of azodicarbonamide during extrusion). This compound is blended in the dry state either as powder or as a concentrated master-batch formulation with the polymer chips and extruded under conditions of elevated temperature under which the azo compound decomposes, forming gaseous nitrogen and carbon dioxide. To achieve a perfect distribution of the gas-forming additive in the melt, resulting in evenly distributed micro-size gas bubbles, an exact coordination between screw revolution, melt temperature in the metering zone, and melt pressure in front of the extruder filter screen is essential. By water-immersion cooling of the broad die extruded flat primary film and stretching in the small gap of a roller stretching unit, an excellent foamed film with a density of

0.5 to 0.6 g/cm$^3$ can be made. This foamed film splits easily into a fine fibrous network when passed through a needle-roller fibrillation unit. Patents of Mitsubishi Rayon Co. (1971d, 1972b) describe such processes using different ways for forming polypropylene films with azodicarbonamide. As discussed in Sec. VI.B.3, the splitting tendency can be enhanced with other additives too. Several independent inventors have recommended the incorporation of high percentages of calcium carbonate as a means for improved fibrillation tendency (Shell Internationale Research Maatschapij N.V., 1971a; Nippon Ekika Seikei K.K., 1973).

It is surprising that there also exist additives which diminish the splitting tendency. Thus the Standard Oil Company (1978) recommends the addition of 1% of a mixture of 30% finely distributed silicium dioxide and 70% low-density polyethylene to polypropylene in order to minimize solitting tendency and enhance efficiency in the weaving of flat film tapes. The use of this additive is claimed to lead to smoother and softer tapes which split less and at the same time the number of lost and ruptured wefts in catcher loom weaving is decreased enormously. For nonsplitting packaging tapes from polypropylene the addition of 5–20% talcum is recommended (Veba Chemie AG, 1978).

## D. Additives for the Improvement of Heat and Light Stability

The extrusion of thermoplastic polymers to films is done at temperatures which lie considerably above the melting temperature of the polymer. As a result of this heat exposure a certain thermo-oxidative decomposition of polymer molecules is mostly unavoidable. In Sec. II.A, the correlation between extrusion temperature and the melt-flow index of differently stabilized grades of polypropylene was discussed. Results of such studies given in Figure 33 show that the melt-flow index change is largely dependent on the degree of stabilization of the polymer with antioxidative acting additives.

One distinguishes two types of antioxidants: primarily acting stabilizers and secondarily effective stabilizers. Sterically hindered phenols belong to the primary antioxidants. Typical secondary antioxidants are, for example, the thioesters of long-chain fatty acids. Between these two types a synergistic effect is possible. The chemical mechanism of the stabilization effect is supposedly based on an extinction of radicals formed in the homolytic thermal dissociation of the polymer. The decomposition radicals normally react with oxygen, forming peroxides which, in turn, cause the disruption of the polymer chain molecule. This degradation reaction is stopped by the added stabilizer.

The stabilizer addition may be done during manufacture of the polymer or by the converter before extrusion. In case of polypropylene the necessary stabilizer concentrations lie in the range of 0.05 to 0.1%.

**Table 21** Heat Stability of 1000-Denier Weaving Tape Made from Different Grades of Polypropylene

| Polymer type | Time for tenacity to fall to 50% at 120°C (hours) |
|---|---|
| Packaging-twine grade | 120 |
| General-purpose grade | 320 |
| Extra-heat-endurance grade | 900 |
| Very-highly-stabilized grade | 1500 |

The commercially available polymer grades normally contain sufficient amounts of heat stabilizers. The stabilization is not only a protection against thermal oxidation during processing of the polymer into films and film fibers, but also serves the prolongation of the lifetime of end products made from them. In Table 21 the thermal stability of differently stabilized polypropylene film tapes at a testing temperature of 120°C is listed. According to degree of stabilization, the half-life of the tensile strength of film tapes made from polypropylene lies between 120 and 1,500 hr.

Figure 121 shows the correlation between the half-life of the tensile strength and the exposure temperature for a well stabilized all-round polypropylene film tape. For the assessment of heat resistance of film tapes the presence of pigments, the tape thickness, and environmental factors such as the content of film tape sacks or the kind of latex binder for carpet backings, have to be taken into account. Humid heat, especially treatments at the boil, are more damaging than dry heat.

Like other polymers, polyolefins are also decomposed under the influence of light in the presence of oxygen. The photolytic degradation of polymer molecules is caused by ultraviolet light in the wavelength range of 200–400 nm (1 nm = $10^{-6}$mm). This near-ultraviolet radiation range equals approximately 5% of the complete sunlight radiation. Compared with the bond energy of carbon-carbon bonds or of carbon-hydrogen bonds, the energy of this UV radiation is very high. Especially sensitive to the exposure of UV radiation is the bond between hydrogen and a tertiary carbon atom. Under the effect of UV radiation the hydrogen atom splits off, and under the influence of oxygen a hydrogen peroxide group is formed, initiating the degradation. For this reason the heat and light stability of polypropylene is considerably lower than that of polyethylene. Since fabrics made from film tape or film fibers have a relatively large surface through which UV radiation can penetrate the polymer, a very good light stabilization of these products is of utmost importance (Broatch, 1968; Brittain, 1968; Fitton and Gray, 1971; Campbell and Skoroszewski, 1972; Weber, 1977).

Stabilizers, particularly for polypropylene, come from the following groups of chemical compounds: benzophenone derivatives, benzotriazole derivatives, substituted benzoic acid esters, organic nickel complexes, and sterically hindered amines. Furthermore, the utilization of o-hydroxybenzophenone and of nickel chelates, as well as mixtures of both, is common practice (Leu et al., 1975; Laus, 1975). Benzophenones and benzotriazoles act as UV filters by absorbing UV radiation and transforming it into heat. The maximum of absorption of most commercial UV-filter stabilizers lies in the wavelength range of 300–400 nm. In this way UV radiation, which is dangerous for polyolefins is prevented from penetrating into the product and thus from setting off the radical decomposition mechanism, degrading the polymer molecules. The nickel complexes, however, owe their stabilizing action to the

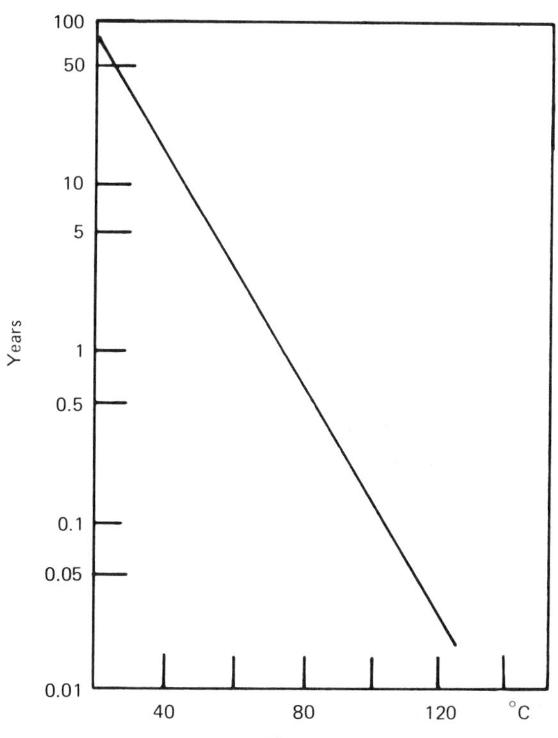


**Figure 121** The heat resistance of polypropylene weaving tape. (Courtesy of ICI.)

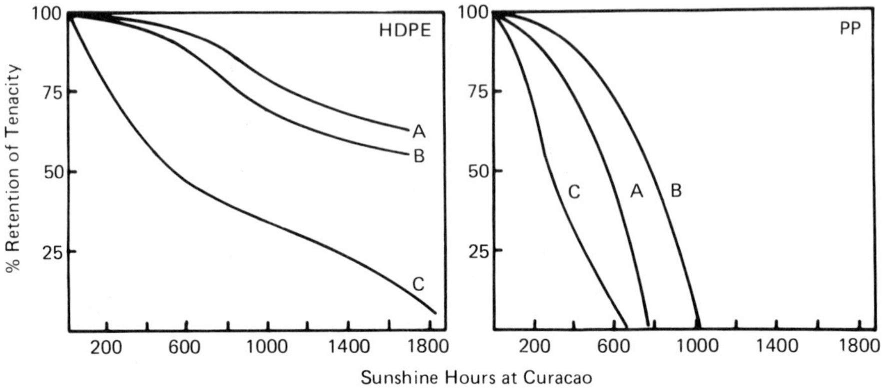


**Figure 122** Comparative effect of light stabilization systems in high-density polyethylene and polypropylene weaving tapes: (A), 0.3% benzophenone derivative/0.3% trisnonyl phenol phosphite; (B), 0.3% nickel complex stabilizer; (C), pure polyethylene or polypropylene. (From Fitton and Gray, 1971.)

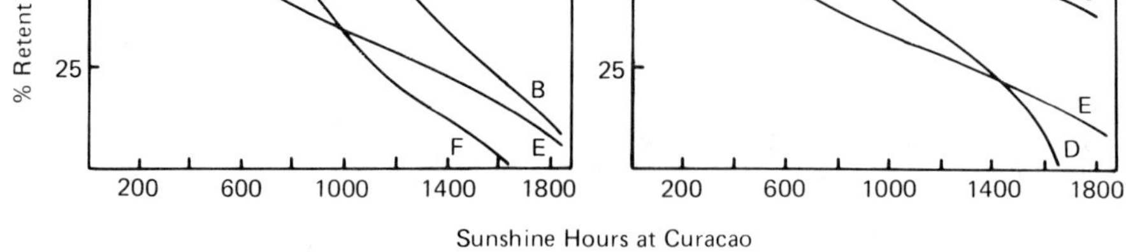


**Figure 123** Antagonistic effect of rutile titanium dioxide on the light stability of stabilized polyethylene weaving tape: (A), 0.3% nickel complex stabilizer; (B), 0.3% nickel complex stabilizer/1% rutile titanium dioxide; (C), 0.3% benzotriazole derivative; (D), 0.3% benzotriazole derivative/1% rutile titanium dioxide; (E), pure high-density polyethylene; (F), pure high-density polyethylene/1% rutile titanium dioxide.

ability to extinguish excited states in the polymer molecules. Through a complicated energy transfer mechanism the activated energy state initiated in the polymer molecule by the UV light is dissipated, preventing the splitting of chemical bonds leading to molecular degradation.

Figure 122 gives a comparison of the efficiencies of benzophenone and of nickel complex quenchers in stabilizing polypropylene and high-density polyethylene, respectively. The efficiency of nickel complex quenchers and of benzotriazole stabilizers in high-density polyethylene is weakened considerably, especially at longer exposure times by the presence of 1% titanium dioxide of the rutile type, as demonstrated by the experimental results presented in Figure 123. Such an antagonistic effect is also ascribed to phthalocyanine blue and cadmium sulfide pigments (Ivett, 1968, Fitton and Gray, 1971). Many other pigments, however, have the supporting effect of UV filters.

Carbon, in the form of carbon black is the best known UV stabilizer. Carbon black incorporated into high-density polyethylene surpasses in its stabilizing action even the common light stabilizers, as shown by Figure 124. This superiority of carbon black becomes particularly apparent at light exposure times exceeding half a year. Figure 125 demonstrates that most other pigments have a smaller stabilizing effect than common UV stabilizers. However, most have the ability to prolong the lifetime of high-density polyethylene products considerably. Stabilization of polyolefin products against different environmental influences is relatively complicated. In order to protect polypropylene properly a combination of various stabilizers has to be applied, such as (1) an antioxidant of the primary or secondary type, (2) UV stabilizer of the filter or nickel complex type, and (3) an organic phosphite improving the light fastness of pigments.

## E. Flame Retardant Additives

In the last two decades the demand for reduced flammability has found growing attention. In the United States and in Great Britain laws and regulations are setting stringent standards for certain textiles with respect to flammability and burning behavior. The same applies to materials used in the construction industry and in public utilities, such as trains, buses, airplanes, theaters, schools, old-age homes, etc. Fabrics made from film tape or film fibers used in those applications have to conform with existing security regulations.

Polyolefins are generally very easily flammable and combustible and also melt under the influence of excessive heat. They therefore have to be stabilized and protected by blending with additives retarding flammability and burning. The test specifications applied in determining flammability and burning behavior are quite different for various application areas and from country to country. They are normally

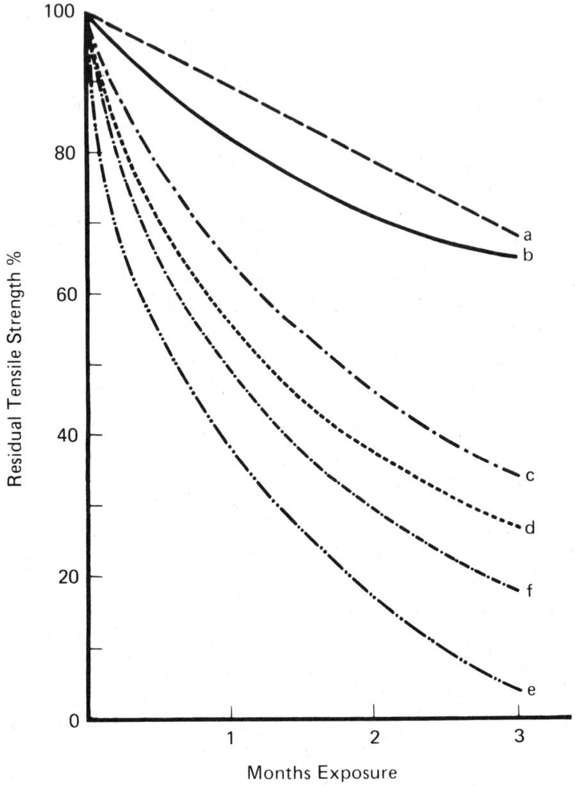


Figure 124 Ultraviolet stability of stabilized and unstabilized polyethylene and polypropylene film tapes: a, polyethylene black; b, polyethylene UV-stabilized natural; c, polyethylene unstabilized natural; d, polypropylene stabilized natural; e, polypropylene unstabilized natural; f, jute. (From Hossack, 1971.)

contained in national material standards, such as ASTM for the United States or DIN-norms for Germany, or in manuals of supraregional industrial or standards institutions. It would go far beyond the scope of this book to list even part of the existing regulations and test methods.

The standard procedure for reducing the flammability and burning of polyolefin products consists of the incorporation of halogen-containing organic compounds into the polymer in the course of extrusion. The action of the organic halogen compounds is based on the interaction of halogen atoms split off at elevated temperature with radical sites created in the polymer by the pyrolysis processes. This interaction prevents or reduces chain-splitting reactions, leading to combustible gaseous decomposition products which initiate inflammation and maintain burning. Halogen atoms furthermore react under recombination with radical decomposition products, forming noncombustible halogenated compounds.

The following halogen compounds find application as flame retardant additives for polyolefins: chlorinated or brominated paraffins, tetrabrom-phthalic anhydride, tetrabrom-diphenyl oxide, etc. Chlorinated compounds have the disadvantage that at elevated temperature they split from hydrogen chloride which can cause corrosion of processing equipment, especially of the extruder. Generally, brominated organic

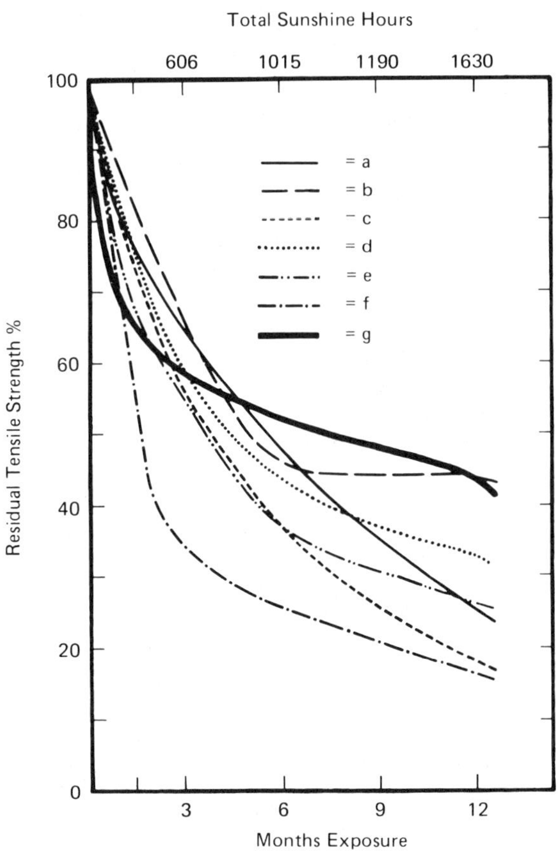


Figure 125 Ultraviolet stability of colored polyethylene tape yarns (700 denier) in a fabric: a, natural stabilized; b, black (101); c, red (401); d, yellow (501); e, green (701); f, natural unstabilized; g, blue (801). (From Hossack, 1971.)

compounds are more efficient in reducing flammability than corresponding halogenated compounds.

A synergistic intensification of the effectiveness of halogen compounds can be achieved when they are being used in combination with antimony trioxide in the lower percentage range and with proper adjustment of the halogen to antimony ratio. At elevated temperature antimony trioxide readily forms antimony trichloride or tribromide which have lower decomposition temperatures than the halogenated organic flame retardant additives. They therefore give off free halogen atoms earlier than the organic additives. Unfortunately, the use of antimony trioxide can cause some problems in the production of films, film tapes, and film fibers, such as filter clogging, blade blunting, stretch ruptures, etc. The application of properly micronized antimony trioxide grades is advisable.

A class of very effective flame retardants are phosphorous-containing halogenated organic compounds, such as brominated organic phosphoric acid esters. These compounds can be dispersed easier and more finely throughout the polymer melt due to their solubility in the polymer. Thus they do impair the film processing much less. Mitsubishi Rayon Co. (1971e) claims, in a patent, the use of an addition of 5% tribromo-trispropyl-phosphate in the presence of 0.3% of a synergist to polypropylene. By double stretching of the extruded primary film at 180°C, easily fibrillatable flame-retardant film tapes are obtained. With halogenated phosphorous organic compounds, however, the danger of decomposition during extrusion cannot be excluded. Furthermore, these organic phosphorous-containing compounds are prohibitive for many applications due to their high price. For film fibers, formulations of chlorinated paraffins with antimony trioxide find wide application (Bash, 1978). More recently, aluminum oxide hydrate has found growing attention as a flame retardant. During heat exposure these compounds set water free. This consumes heat, thus retarding polymer decomposition. The problem of flame retardation will become more and more important in the future. For this reason great efforts are being undertaken everywhere in order to find and introduce new and better flame retardant compounds (Mitsubishi Rayon Co., 1971f, 1972c; Moorwessel and Pfirrmann, 1975).

## F. Additives for the Reduction of Electrostatic Charging

The tendency for electrostatic charging of synthetic polymers through friction causes great problems during the production process as well as during the use of final polymer products. During the various production steps electrostatic charging can be avoided by use of radioactive decharging devices or by electrical air ionizers, enhancing the leakage of the electrical charge from the film or film fiber surface. Nevertheless, it is necessary to apply finishes which are able to diminish the tendency for electrostatic charging in further processing and in later use of film fiber products. Very effective is the application of cation-active antistatics added to the spinning finish. This method, which is used widely in textile processing, is also applied with good exerpience in film fiber production and processing. It is advantageous to add to the antistatic also an emulsified oil for the improvement of slipping ability. However, one has to take into account that paraffinic oil penetrates into the polyolefin and makes it swell. Therefore only such oils (e.g., organic phosphoric acid esters) may be used which are insoluble in the polymer. The finishing of film fiber products with such emulsions does not only reduce the tendency for electrostatic charging but diminishes also the abrasion of fiber fragments and the appearance of polymer dust, disturbing the further processing of split-film yarns in twisting and weaving. Moore (1979) gives some information on the selection of appropriate finishes. Furthermore, it is possible to impart to film fiber products antistatic properties by incorporation of additives (McMeekin, 1968; Evans, 1971), such as fatty acid amides. These incorporated additives migrate steadily to the film or film fiber surface and there prevent the development of electrical charges in frictional actions. Their drawback, however, is the reduction of the transparency of the film.

Antistatically finished polymer grades are available on the market. Their processing into films, film tapes, or film fibers, however, is to some extent problematic. Films from these special polymer grades can be produced only with the circular die blowing technique or with the flat die chill-roll method. Flat die extrusion with water cooling cannot be applied due to water solubility of antistatic agents incorporated into these special polymer grades. Even very low concentrations of the antistatic additives in the range of 0.1% increase water adhesion to the film so much that it can not be dried sufficiently before stretching by the usually applied air blowing technique. At higher concentrations of the antistatic additives, even deposits on the chill roll may occur, disturbing further processing. Also, the formation of blisters on the film was observed in such cases, leading finally to unstable yarn packages. Film tape layers tend to glide away from one another since the antistatic additives do not often mix homogeneously with the film-forming polymer but form separate phases which tend to migrate to the surface, interfering with the necessary adhesion of the tape layers in the package.

## IX. SURFACE TREATMENTS OF FABRICS MADE FROM POLYOLEFIN TAPE YARNS

For some applications it is necessary to print, coat, or laminate the polyethylene or polypropylene film tape fabrics. When such fabrics are being used for packaging purposes the fabric surface often has to serve the designation of the content or as advertizing space. For this, it has to be printable. Through extrusion coating the fabrics have to be made dense and watertight when being used for sacks to package bulk goods. Through laminating the tape fabrics with aluminum foil, paper, or plastic foams, composite materials with completely novel characteristics and for a wide variety of applications can be obtained. In all these cases special surface treatments are essential to overcome the low surface tension and poor adhesion properties of the polyolefin tape fabrics in order to achieve good printability or good adhesion in caoting or in lamination processing. This goal is generally not attainable with additives. A certain exception are nickel complexes, as claimed by Lee (1975).

Good compatibility of the tape fabric with print pastes, coating melts, or laminating adhesives can normally be achieved by high-frequency corona treatment of the film or fabric. This treatment applies the discharging of a sinus-shaped alternating current of 12–18 kW at frequencies of 20–40 kH over a system of electrodes onto the film or fabric surface. The generator producing the alternating current, together with the electrode system forms the corona unit. Through automatic adjustment of the generator to the operation state of the electrode system a maximum energy transmission at lowest loss level is guaranteed. When the corona is ignited the emitted electrons are being accelerated in the electrical field, delivering their energy to the surface of the polymer film or fabric, creating chain ruptures and roughening of the surface on a microscale. Through subsequent reactions of the radical sites created in chain rupture with other products formed in the corona treatment, such as ozone, nitrous oxides, and hydroxyl radicals, functional groups are formed (i.e., peroxide groups, ozonide groups, carboxyl groups, ketone groups, aldehyde groups, etc). These functional groups act through formation of electrostatic double layers as adhesion centers for printing pastes, hot melts, glues and other substances.

Naturally, good adhesion is also dependent on other parameters enhancing contact, such as squeezing pressure, temperature, and duration. They influence the adhesion process through the response of the rheological behavior of materials involved in coating or laminating treatments. A very important application of corona treatments of film tape fabrics is the pretreatment prior to coating with low-density polyethylene. The coating is done in such a way that the preheated and corona treated fabric runs into a gap formed by a water cooled roller and a pressure roller covered with silicon rubber. From a flat die of the width of the fabric to be coated, low-density polyethylene with a melt index of more than 20 g/10 min is extruded into this gap. The fusible film is squeezed under a pressure of several bars onto the fabric where it solidifies immediately, forming a good adhering bond between the coat and the fabric. With a 10–20 μm-thick low-density polyethylene layer the usability of tape fabrics is considerably increased.

In order to increase adhesion between film product and the polymer coat, in extrusion coating ozone electrodes also in use. They transform the oxygen of the air into ozone which is blown against the molten film of the coating polymer. By strong oxidation action functional groups are formed, similar to those created in corona treatments. These functional groups act as adhesion centers between the film tape fabric and coating layer. However, to date it is not yet completely clarified whether surface oxidation is the only reason for the improvement of the adhesive strength between the two polymeric substrates. In ozone and corona treatments a roughening of the surface may also contribute to improved adhesion. The fact that the effect of corona treatment prior to coating or lamination fades slowly away when the corona treated material is stored lets us realize the importance of electrical surface charges improving adhesion as well (Kim et al., 1971). In contrast to this, Owens (1975) has brought forward some evidence for the improvement of adhesion by hydrogen bonds between enolized ketone groups on the corona or ozone-treated film or film tape surface with carboxyl groups on the coating substrate.

Another possibility to pretreat tape fabrics prior to extrusion coating for better coat adhesion is the application of "primers." However, the application of primers is complicated and expensive. The finishing of film tapes or film fiber yarns with oil emulsions in order to improve their slipping behavior in twisting and weaving operations and to decrease their tendency for developing electrostatic charging has been discussed before. It should be mentioned, however, that the utilization of such auxiliary finishing agents interferes with coating or lamination because their presence dramatically reduces adhesion. They have to be either avoided, when subsequent coating or lamination is intended, or they have to be removed prior to further processing.

# 3

# Machine Systems for the Manufacture of Film Tapes, Film Yarns, and Fibers

In this chapter a survey of the machine systems will be given which are used in the production of films, film tapes, film yarns, and film fibers. In addition, the machinery necessary for the manufacture of film tape or film fiber products will be shortly reviewed. Due to the limited size of this book, however, the discussion of these subjects will have to be restricted to the basic principles to be applied for proper equipment selection. Construction details of individual machines offered on the market cannot be described thoroughly. However, it is intended to give the less experienced reader practical recommendations which may help him when confronted with tasks connected with the manufacture of film tapes, film fibers, or products derived therefrom.

## I. FILM EXTRUSION EQUIPMENT

As discussed in Sec. II of the preceding chapter, the most important operation in the manufacture of film tapes, film yarns, and film fibers is forming a primary unstretched film by melt extrusion. As outlined in the introduction of Sec. II, three basic methods are available:

1. The extrusion of the polymer melt through a flat die and the solidification of the extruded film by water bath immersion
2. The extrusion of the polymer melt through a flat die and the solidification of the extruded film on chill rolls
3. The extrusion of the polymer melt through an annular die and the solidification of the extruded film by air cooling.

Schematic illustrations of the principles of these film forming methods were given in Figures 31 and 32.

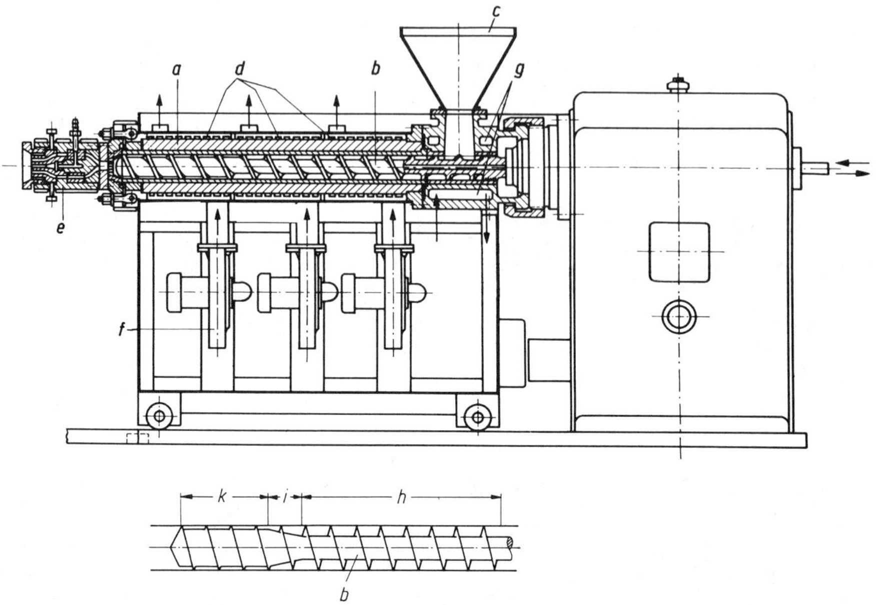


Figure 126 Sectional view of a single screw extruder: a, cylinder; b, screw; c, feed hopper; d, heating elements; e, die; f, cooling fan; g, cooling channels for the feed throat; h, feed section; i, compression zone; k, metering zone. (From Schenkel, 1963; Knappe, 1975.)

The machine type generally used for the plasticizing and melting of the thermoplastic polymer and for conveying the melt through the die is the screw extruder. It can be either a single or a double screw extruder in horizontal or vertical arrangement. The principle of the single screw extruder is illustrated in Figure 126. In the cylinder (a), which can be heated by heating elements (d) grouped in several zones, a specially designed screw (b) is single-side mounted. The polymer granulate or powder is introduced through a feeding hopper (c). The rotating and deeply grooved feeding section of the screw, which is in most cases close to the hopper cooled with water to prevent sticking of the polymer, conveys the latter along the cylinder toward its heated sections (d). There the polymer is plasticized and melted by the heat coming from the heating elements and by frictional heat. In order to control the latter, most extruders are equipped with additional outside cooling facilities for the cylinder. In some extruders the screw can also be cooled with a liquid flowing through a central duct. The rotational speed of the screw can be varied either in progressive steps by transmission or continously variable by a suitable electric drive.

The basic structure and design of a double screw extruder is in principle very similar to that of a single screw extruder.

The advantage of a double screw extruder lies in its better mixing characteristic, which makes it superior for the manufacture of pigment master-batch granulate, and in its better performance in the processing of polymer powders. On the other hand, single screw extruders are quite satisfactory for the processing of polymer granulates. Furthermore, they are lower in cost and cheapter in maintenance. Extruder screws are commonly characterized by the ratio of length to diameter (L/D). In the processing of films from high-density polyethylene, somewhat higher L/D ratios can be used than in the processing of the less heat-stable polypropylene. Higher L/D ratio means a longer residence time and more heat exposure of the polymer in the extruder. On the other hand, screws with higher L/D ratios give improved heat transfer conditions and more uniform distribution of coloring or stabilizing additives.

From the mathematical treatment of the extrusion, for which good summary reviews were given by Meskat (1955), Paton et al. (1959), McKelvey (1962), Schenkel (1963), and Knappe (1975), one can derive

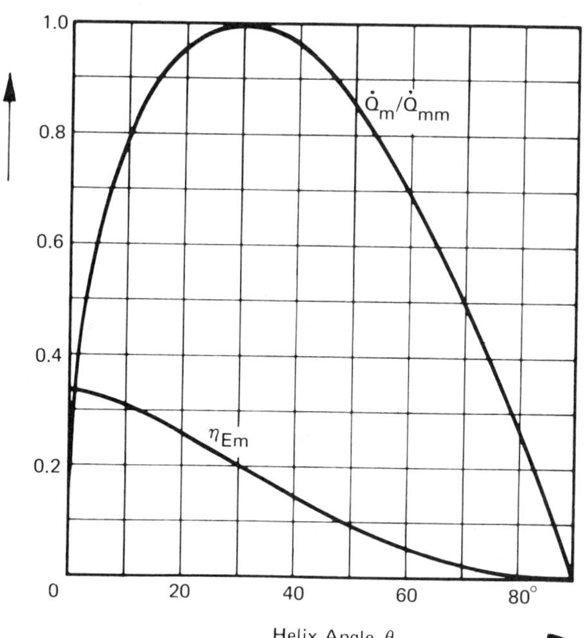


Figure 127 Pumping efficiency ($\eta_{Em}$) and relative throughput ($\dot{Q}_m$/$\dot{Q}_{mn}$) in their relation to the helix angle of the screw. (From Knappe, 1975.)

that, in the case of compounds with structural viscosity behavior, the highest pressure increase is achieved with screws having constant helix angle. With decreasing helix angle the pressure increase can be enhanced. The relations between helix angle ($\theta$) on the one hand, and the pumping efficiency ($\eta_{Em}$) or relative throughput ($\dot{Q}_m/\dot{Q}_{mm}$) on the other hand, are illustrated in Figure 127. The data in Figure 127 indicate that a helix angle for screw threads of 10-20° seems to be the best compromise to give a reasonable pumping efficiency at a high relative throughput.

In the construction of screws the design of the discharge zone is relatively easy on the basis of a desired maximum throughput. However, it is somewhat more difficult to properly design the remaining screw zones, namely the feeding zone, compression zone, and possibly necessary shear and mixing zones. Another important screw characteristic is the "compression ratio," namely the ratio of the thread volume in the feeding zone and the thread volume in the discharge zone of the extruder screw. Since in most cases screws are used for which the thread widths are constant along the whole screw, the ratio of the thread depths ($x = h_f/h_d$) can be used to characterize the compression ratio. Generally, compression ratios of 2 to 5 are in use. A good guide for the proper selection of compression ratio is the ratio of the bulk weight to the melt density. The compression ratio should not be much higher, since under such conditions overfeeding of the discharge zone occurs and leads to unstable pressure conditions.

The following Figure 128 schematically presents the most commonly used extruder screw designs. The screw design designated as (a) is the "One-step screw," characterized by a short compression zone. The flow conditions in the discharge zone are in this case only slightly influenced by feeding and melting. One-step screws with short compression zones are favorably applied in the processing of polymers with a narrow melting range and low melt viscosity, such as polyesters or polyamides. Screws designed according to scheme (b) are one-step screws with extended compression zones. This screw design has shown some advantages in the processing of polyolefins (Dowd, 1963). Maddock (1960) suggested a one-step screw design (c) characterized by an extended discharge zone. This screw design gives excellent performance with respect to constancy of throughput and high uniformity of the extruded products (Badonsky, 1960). Screw design (d) with progressively decreasing thread depth along the screw is used especially in the processing of polyvinyl chloride. Compared with one-step screws this design is relatively sensitive in operation towards fluctuations in rotational speed and temperature. There are also extruder screws in use which carry at the discharge end "mixing torpedoes" (e). These shear and mixing sections can either be smooth, forming a narrow high shear gap to the cylinder wall, or are grooved for a combination of shearing and mixing actions. Such mixing or shearing sections can

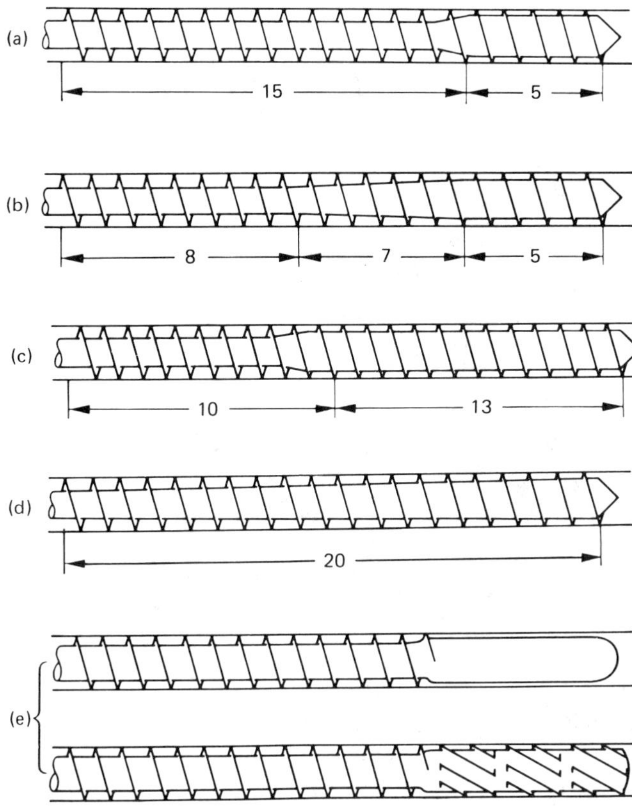


Figure 128 Commonly used screw designs (Knappe, 1975): (a), single-stage screw with short compression zone; (b), single-stage screw with extended compression zone; (c), single-stage screw with extended metering zone; (d), screw with constantly increasing root diameter; (e), screws with mixing torpedo (homogenizer).

also be incorporated somewhere in the extruder screw. The photograph shown in Figure 129 gives examples for shearing and mixing sections.

The variations in screw design applied by various machine manufacturers, especially with respect to achieving optimum kneader action, are innumerable. It would go far beyond the scope and size of this book to refer to more than the fundamental principles. However, the screw design suggested by Maillefer (1963), shown in Figure 130, is worth mentioning. The Maillefer screw is characterized by adjacent threads in the feeding zone having different cross sections. The

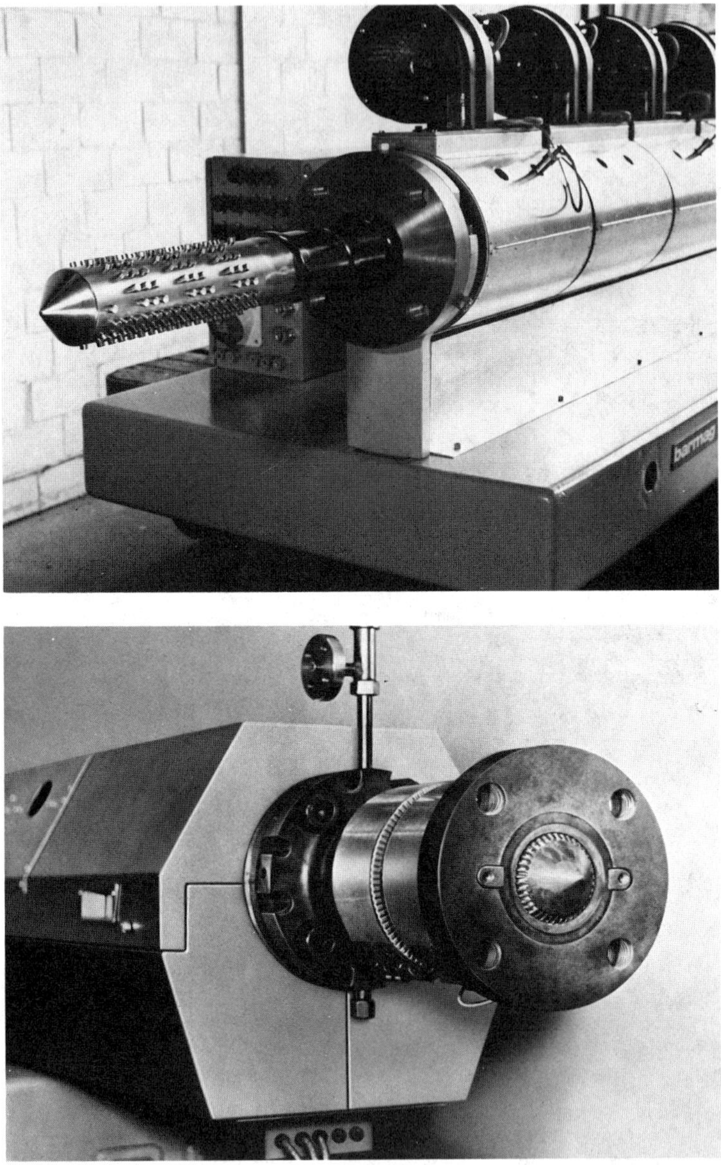


A

Figure 129 (A), extruder screw with LTM-mixing section of BARMAG design (top); BARMAG extruder with superposed dynamic mixer (bottom). (Courtesy of Barmag-Barmer Maschinenfabrik AG.)

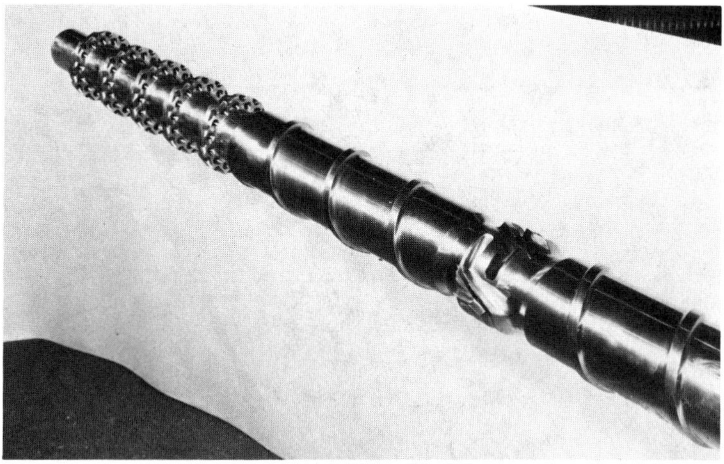

B

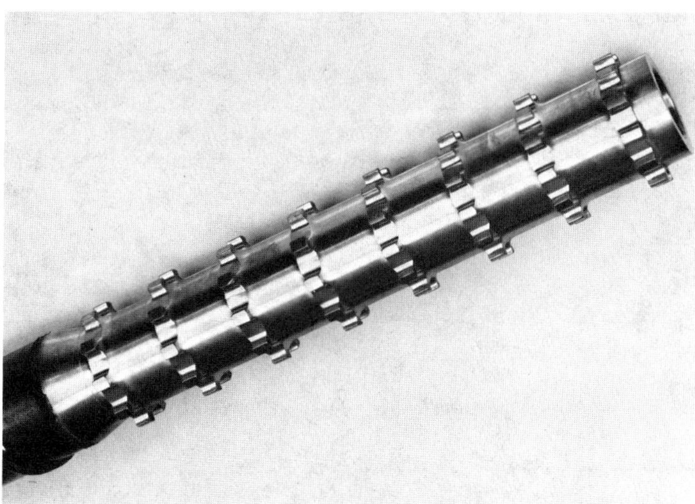

C

Figure 129 (continued) (B), Screw end with shearing and mixing section (design Windmöller & Hölscher). (Courtesy of Windmöller & Höllscher.) (C), Screw mixing section (design Reifenhäuser GmbH & Co.). (Courtesy of Reifenhäuser GmbH & Co.)

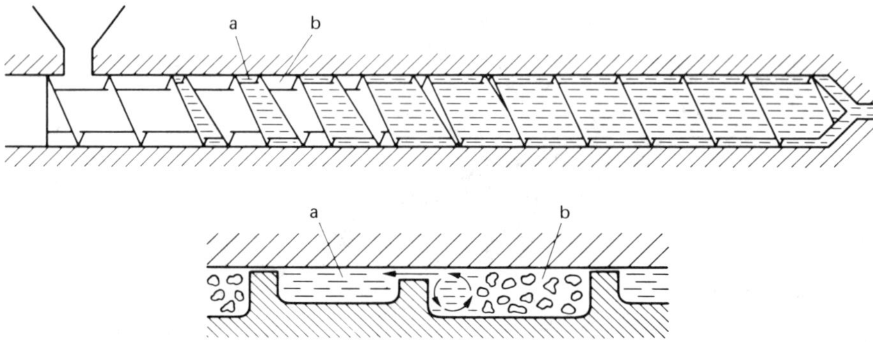


Figure 130 Double flighted Maillefer screw. (From Maileffer, 1963.)

thread channels are separated by somewhat lower crosspieces which allow only already plasticized material to enter from the deep thread channel (a) into the shallow thread channel (b).

Quite often, especially in processing hydrophilic polymers, polymers with air enclosures, or those containing low boiling plasticizers, it is necessary to remove gaseous components before the melt enters the die. For this purpose, "decompression screws" are applied. Figure 131 shows the design principle of such a screw. The screw design determines a very important extruder characteristic, namely pressure

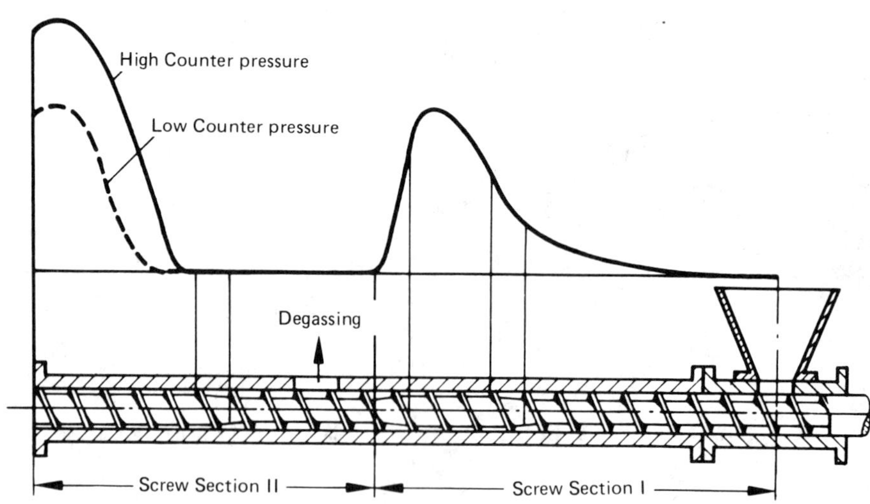


Figure 131 Scheme and pressure profile of a vented extruder. (From Krämer, 1979.)

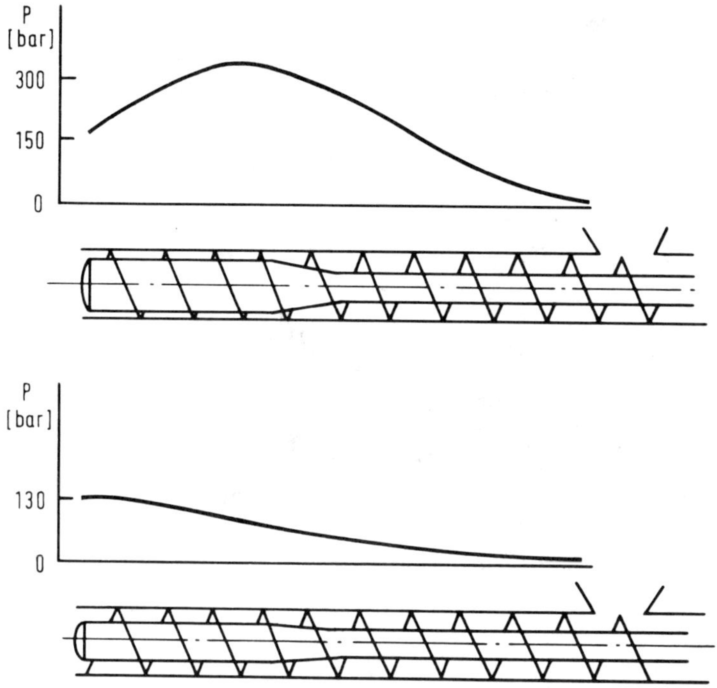


Figure 132 Pressure buildup in conventional extruders with screws having metering zones with low (top) or large (bottom) flight depth. (From Krämer, 1979.)

buildup in the cylinder. The regularity of extrusion and the uniformity of the extruded product are largely dependent on pressure conditions. The pressure built up in extruders with one-step screws depends mainly on the depth of the screw threads in the compression and metering zones, as shown in Figure 132.

In Table 22, small, medium, and large size extruders are listed with their main characteristics, namely the screw diameter, effective screw length (L/D ratio), the range of rotational screw speeds, their output, and also the commonly supplied input power. The weight of extruders ranges from approximately 1 metric ton for small machines to approximately 10 metric tons for very large ones. The floor space required is not only determined by the extruder and its special design features, but to a much greater extent by the whole system: the extruder, the die, and the film cooling and solidification unit.

A

B

**Table 22** Characteristic Data of Small, Medium, and Large Size Extruders

| Extruder size | Screw diameter D in mm | Effective screw length L/D | Revolution range n/min | Maximum output kg/h | Driving power kW |
|---|---|---|---|---|---|
| Small | 20-45 | 15-30 | 0-500 | 2-15 | 60 |
| Medium | 50-90 | 15-30 | 0-400 | 10-50 | 300 |
| Large | 100-400 | 10-25 | 10-100 | 20-300 | 5000 |

## II. FILTER SCREEN EQUIPMENT

In the production of uniaxially stretched film, film tapes, or film fiber yarns the absence of impurities is of utmost importance. Therefore the polymer melt coming from the extruder screw head has to be properly filtered before entering the die.

For this purpose, generally metal screen filters with up to 10,000 meshes per $cm^2$ (64,500 meshes per $in.^2$) are used. There are numerous systems offered on the market by a large number of filter manufacturers. The principle systems can be classified as follows:

1. Filter candle units, mostly used in pairs to allow rapid change-over
2. Revolving filter screen cartridges for easy filter changes
3. Continuous metal filter belt units, also allowing longer operation without interruption of the film extrusion
4. Long-life screen filters with very large filter areas up to several $m^2$, allowing continuous film extrusion over time periods of up to one week.

Figure 133 gives photographs of a continuous filter belt unit used in a film tape line and of a long-life filter unit.

**Figure 133** Photograph of a continuously operating wire fabric screen filter: (A), "auto-screen" system (patented by Mr. Kalman, Process Development, Ltd.) (Courtesy of Chemiefaser Lenzing AG); and of a "non-stop filter" of BARMAG with dismantled filter elements (B). (Courtesy of Barmag-Barmer Maschinenfabrik AG.)

## III. FILM FORMING DIES

Two basically different die concepts are in use, namely the slit die applied in flat film extrusion, and the annular die used in film tube blowing.

### A. The Slit Die

Wide slit dies must be constructed in such a way that the polymer melt coming from the discharge zone of the extruder is distributed evenly over the whole width of the die and leaves the slot in uniform amount, thickness, and average speed. In order to achieve such conditions the die head entrance has to be specially designed. In most cases wide slit dies have a central feed line through which the melt coming from the extruder discharge and from interposed filters enters the die head. Very important is the achievement of an even distribution of the melt over the whole width of the die slot. This is done with the help of a

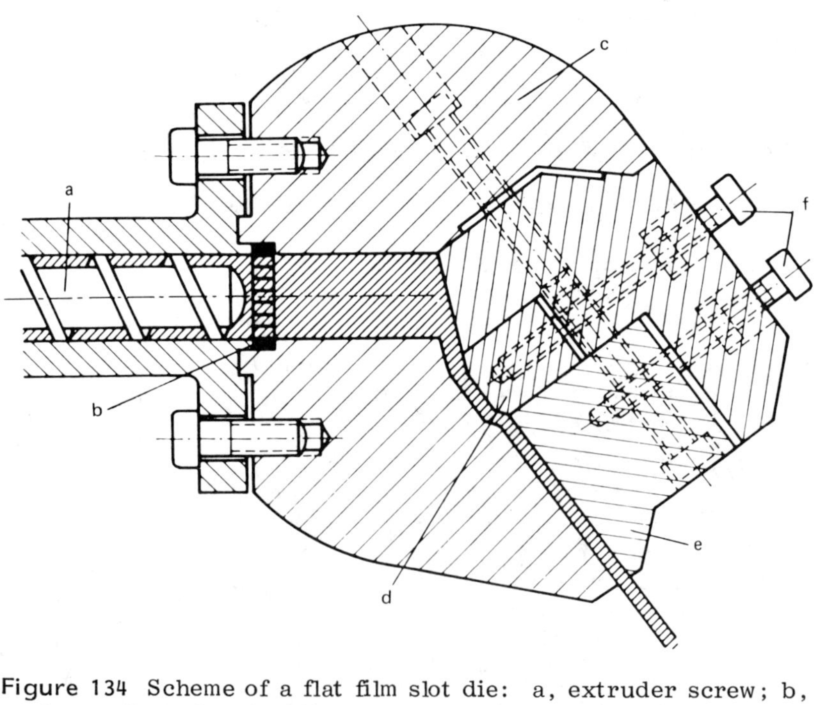


Figure 134 Scheme of a flat film slot die: a, extruder screw; b, screen pack; c, die body; d, adjustable restrictor bar; e, adjustable lip; f, adjustment screws. (From Domininghaus, 1960.)

A

B

C

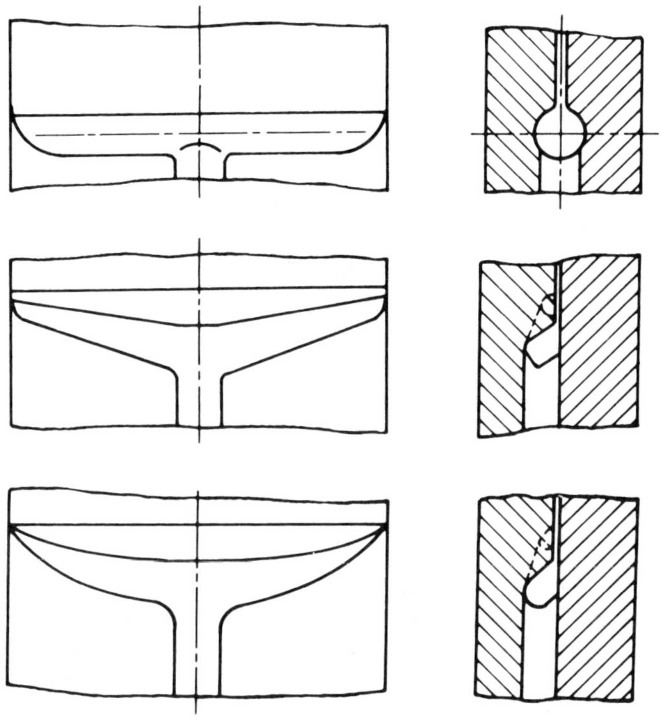

Figure 135 Manifold designs: (A), T-shaped manifold with constant diameter; (B), coat-hanger manifold with decreasingly rectangular diameter; (C), coat-hanger manifold with rheologically proportioned decreasing diameter.

manifold supply of the melt and of flow restriction zones and restrictor bars.

Figure 134 gives a sectional drawing of a slit die for flat film extrusion (Domininghaus, 1960). The design of the manifold is of utmost importance. A large number of suggestions have been made. Typical ones are shown in Figure 135. Common for all these manifold designs is the T- or coat-hanger-shaped spreading channel. Since the pressure decreases with increasing distance from the central die inlet, the flow velocity of the melt decreases correspondingly. Decreasing flow velocity of polymer melts with non-Newtonian flow properties passing through narrow channels means higher flow viscosity. All this would logically result in reduced melt discharge towards both outer sides of the die slot. In order to correct for this, the manifold opens in a "flow restriction zone" in which the melt passes a narrow slot path which is shorter the further the distance is from the central die inlet. Figure 136 illustrates this in a schematic manner.

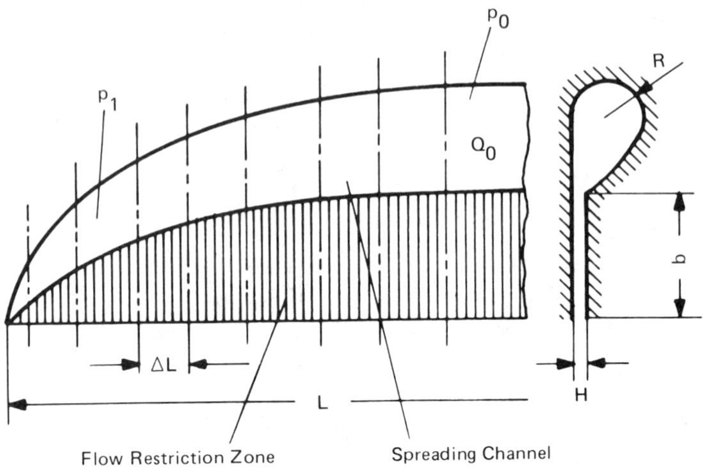


Figure 136 Die manifold geometry. (From Krämer, 1979.)

In the narrow slot of the flow restriction zone, pressure fluctuations can cause material deformations which lead to alterations in melt output, especially apparent in the die slot center. To compensate for this, most slit dies have, in addition, "flow restriction bars" adjustable with setting screws and which allow further narrowing in a very controlled manner of the center exit of the flow restriction zone. Similar fine adjustments to achieve even flow over the whole width of the die slot can be obtained with "flex-lips." Figure 137 shows schematically some die designs with flow restriction zones, restrictor bars, and adjustable flex lips.

Very important for the proper performance of slit dies is the temperature constancy. It is advisable to equip the die body with several independently regulated electric heating zones and to provide both outer edges of the die with well proportioned higher heating units. Slit dies are available in extreme cases up to operating widths of 2500 mm (100 in.). However, for film tape production lines one should not use overly large operating widths; the operators need safe manual access to all tape positions. The die lip opening can vary for slit dies in the range of 0.5 up to 20 mm. At slit openings of 0.5 mm, operating pressures up to approximatley 300 bar are common.

## B. Annular Dies

Besides slit die extrusion, film formation with annular dies has gained increasing importance during the last two decades.

The design of annular dies for blown film extrusion is decisively determined by the selected processing direction. Blown film extrusion can be performed either horizontally, vertically downwards, or vertically upwards. Horizontal blown film extrusion is seldom applied in practice though the production line concept is substantailly simpler. However, the disadvantages (i.e., gravity deflection of the film tube and the temperature difference between the bottom and upper side of the tube caused by the updraft of hot air) have hindered a more frequent application.

Vertical downward operation has so far only been applied in the production of relatively small tube diameters. The internal cohesion of the polymer melt has to carry the weight of the downwards hanging film tube, a condition which can only be fulfilled in the manufacture of thin and not-too-wide tubes. On the other hand, the downward operation has certain advantages, such as the easier application of additional water cooling during the film solidification and less interference by updraft of air heated by the extruder operation. The most commonly used operating method for the processing of blown film is the vertical

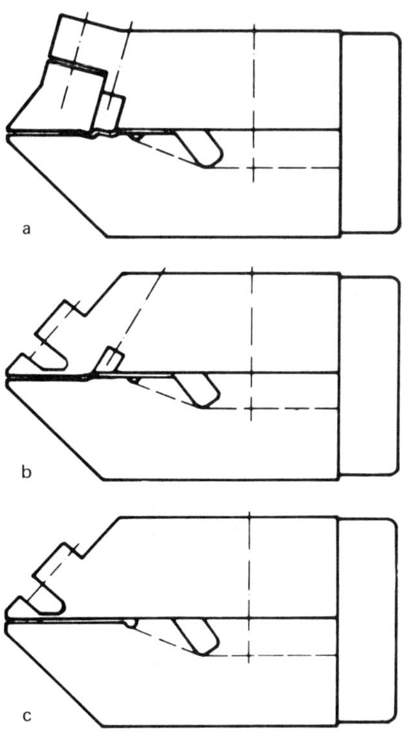


**Figure 137** Various types of flat film slot dies: a, die with restrictor bar and adjustable lip; b, die with restrictor bar and "flex-lip"; c, die without restrictor bar and "flex-lip." (From VDI-Gesellschaft für Kunststofftechnik, 1978.)

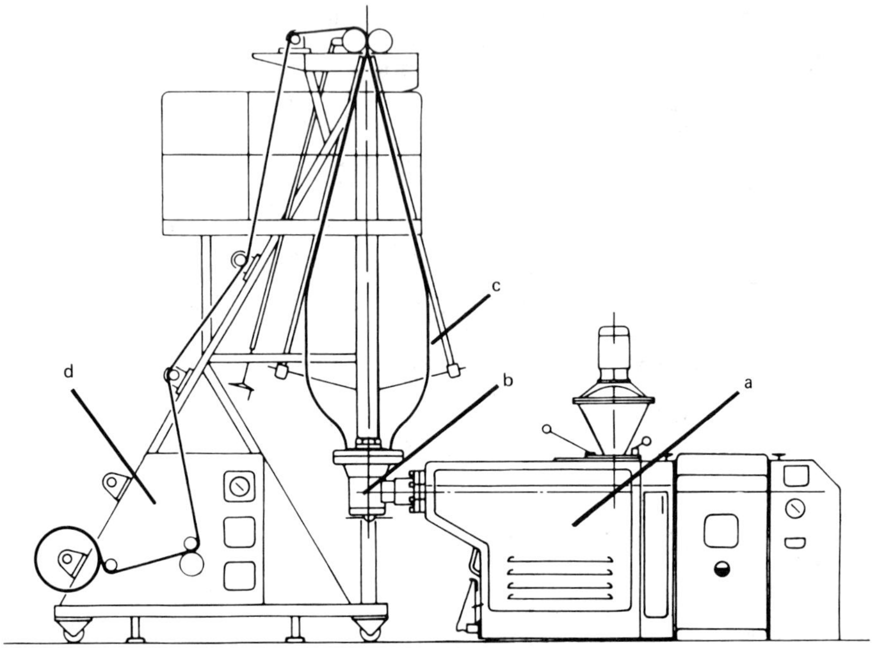


Figure 138 Vertically upwards blown film extrusion: a, extruder; b, blown film die; c, film bubble; d, takeoff station. (Schenkel, 1963; Knappe, 1975.)

upwards film blowing method which is illustrated in Figure 138. In this technique the film tube is carried by the take-off rolls; it allows the use of annular dies with 1500 mm slot diameter and more.

The annular dies used in vertical upwards operation can either be "straight-through" or "crosshead" dies. The first ones have center feed, the second ones side feed. The straight-through die head can be connected to the vertical extruder or in combination with an elbow adapter also to horizontal screw extruders. The crosshead dies, however, is only used in combination with horizontal screw extruders. Basically there are three design types for annular die heads: (1) the side-feed die design, (2) the center-feed die design, incorporating a spider or a breaker plate to support the mandrel, and (3) the "spiral mandrel" die design.

The side-feed die heads belong to the oldest blown film die constructions. The melt flows around a mandrel which is mounted in the die body. To prevent the melt from flowing directly up to the die lip after entering the die head, channels are formed in such a way as to increase flow resistance on the inlet side. With increasing distance from the melt feed the channels are widened to reduce the flow resistance. In this way a uniform axial flow pattern is achieved. Spider lines in the film tube on the inlet side are unavoidable. Figure 139 illustrates the construction features of this die head design principle, which is mostly used in film tube extrusion with not too great diameters. In center-feed die heads the mandrel is fastened to the die body with spider legs or perforated disks. The latter also act as breaker plates to give even melt distribution. Their use, often in combination with spider leg mandrel attachments, largely improves film thickness uniformity over the tube circumference. The spider lines, which are otherwise a problem, can thus nearly be eliminated, as indicated in Figure 140, by

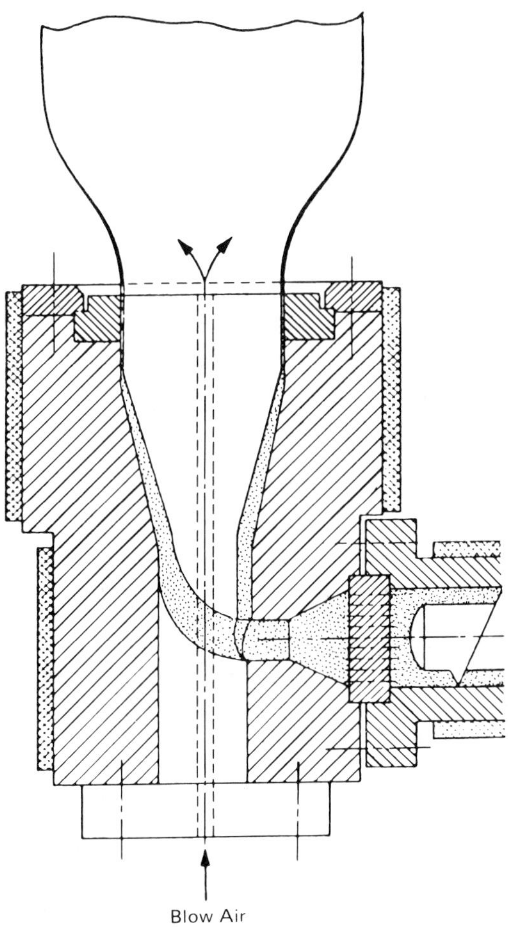


Figure 139 Scheme of a side-fed film blowing die. (From Krämer, 1979.)

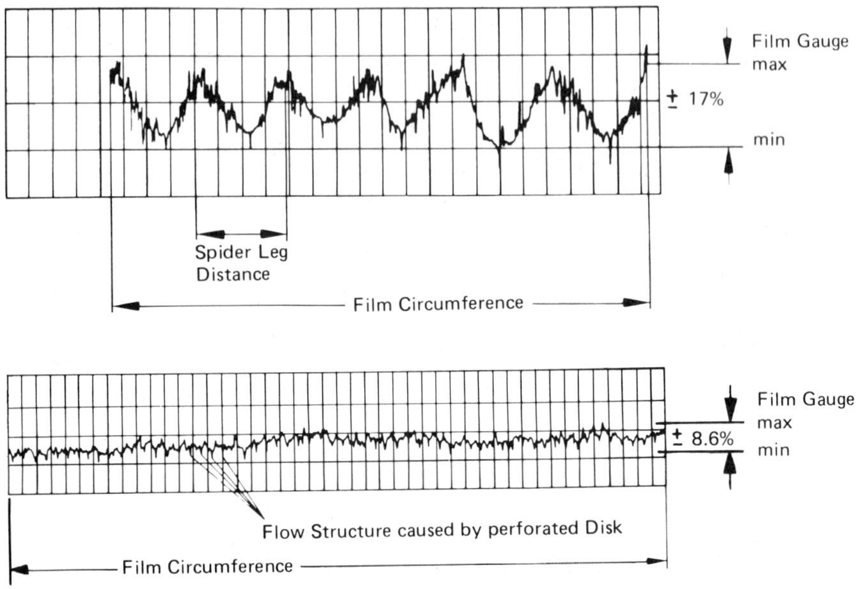


**Figure 140** Film gauge profile of film blown with a spider-type die without (top) and with (bottom) breaker plate mandrel support. (From Krämer, 1979.)

corresponding results of profile measurements on film blown without and with perforated breaker plates.

Another possibility to eliminate spider marks is given by the use of "smear devices" (i.e., threadlike grooves cut into the mandrel or the die body, directly positioned behind the spider leg or disk attachment). While thickness variations with simple spider-type center-feed dies is normally in the range of ±12-18%, thickness uniformity is improved to ±7-10% by the application of breaker plates and/or smear devices. The construction principle of center-feed dies with spider leg mandrel mount followed by a smear device section is given in Figure 141.

In spiral mandrel die heads the melt is introduced into the die through a center feed. The mandrel has on its inlet side a number of entrance holes through which the melt enters and is directed radially to the outer circumference of the mandrel into spiral channels. Through the special design of the latter the melt is evenly distributed over the whole circumference of the mandrel, which in this die construction is held in place by attachments to the die head body on the feed side, not interfering with the melt flow. Proper dimensioning and positioning of the melt overflow nips and of the flow channels is of utmost importance. The thickness distribution along the circumference of the film tube can be directly related to the perfection of the melt distribution. Examples of spiral channel design and of the improvements in film thickness uniformity possible are illustrated by Figure 142.

With spiral mandrel die heads, profile uniformities in the range of 4-6% can be obtained. This die head design is especially suited in the processing of polyolefins. Figure 143 gives an example of the spiral mandrel die head design.

Multilayer films are gaining increasing importance also for specialty film fiber products. The construction design of die heads suited for the extrusion of two or three film components is naturally more complicated, since for each polymer separate feed and distribution systems are necessary. Such multilayer film extrusion die heads are in use for bicomponent film processing from high- and low-density polyethylene and polyamide. In the latter case the presence of an adhesive interlayer calls for a die head which allows three components to extrude. Each of the components requires a separate melt-flow system in the die

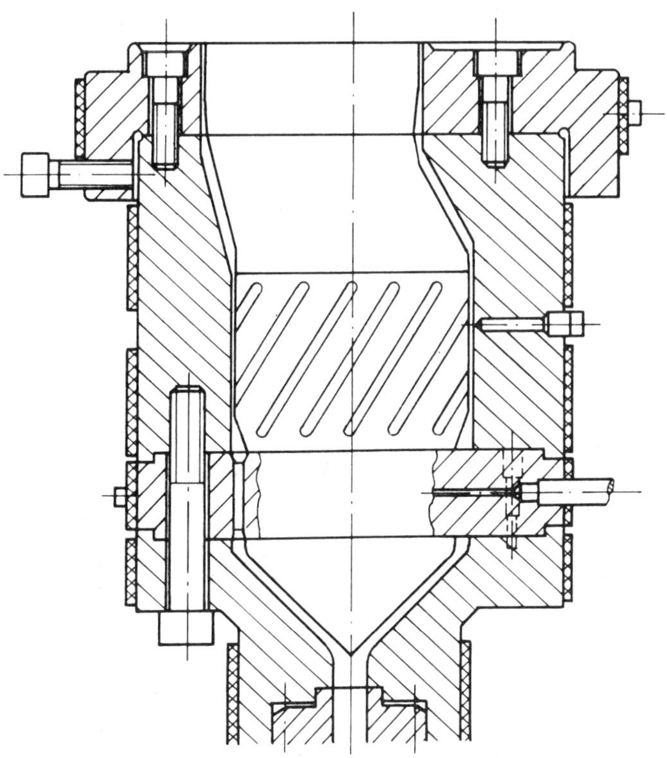

**Figure 141** Spider-type blown film die with smear device to obliterate spider marks in the melt flow. (From Krämer, 1979.)

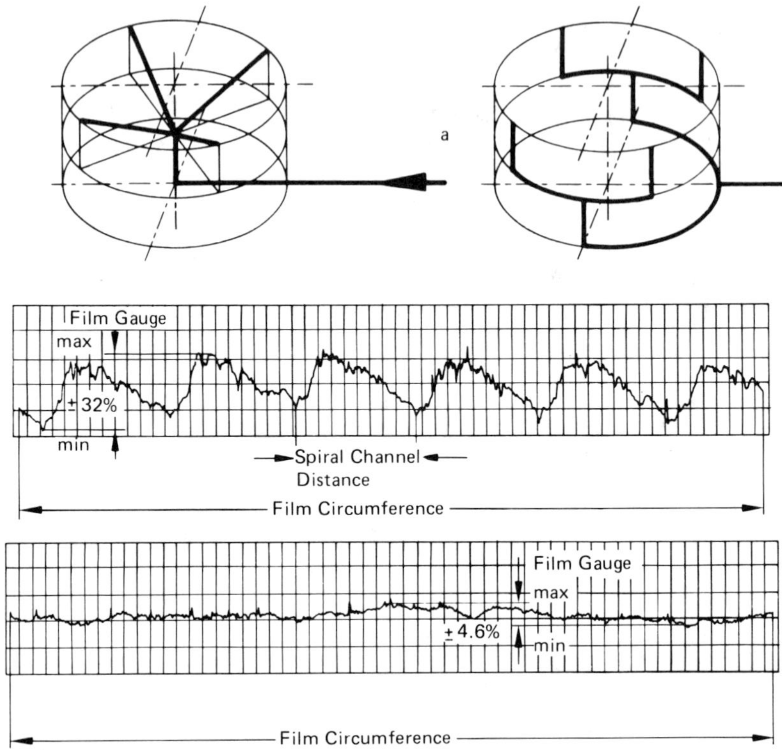


Figure 142 Scheme of the feed systems in spiral mandrel blown film dies (top) and the film gauge distribution obtained with a poorly designed (middle) and a properly designed (bottom) spiral mandrel die. (From Krämer, 1979.)

head. In case of very large differences in the melting point of the components, spearate heating systems are desirable. However, this can only be realized in large-size die heads. Figure 144 gives a longitudinal section through a two- and three-layer coextrusion die.

Multilayer blown film extrusion requires separate extruders for each of the components.

## C. Rotational Systems Eliminating Thickness Variations

One of the major problems in blown film extrusion with side-feed or center-feed annular die heads in which the mandrel is mounted in the die head body with spider legs is the occurrence of spider lines (i.e., longitudinal thin lines placed along the length of the film tube. Similar

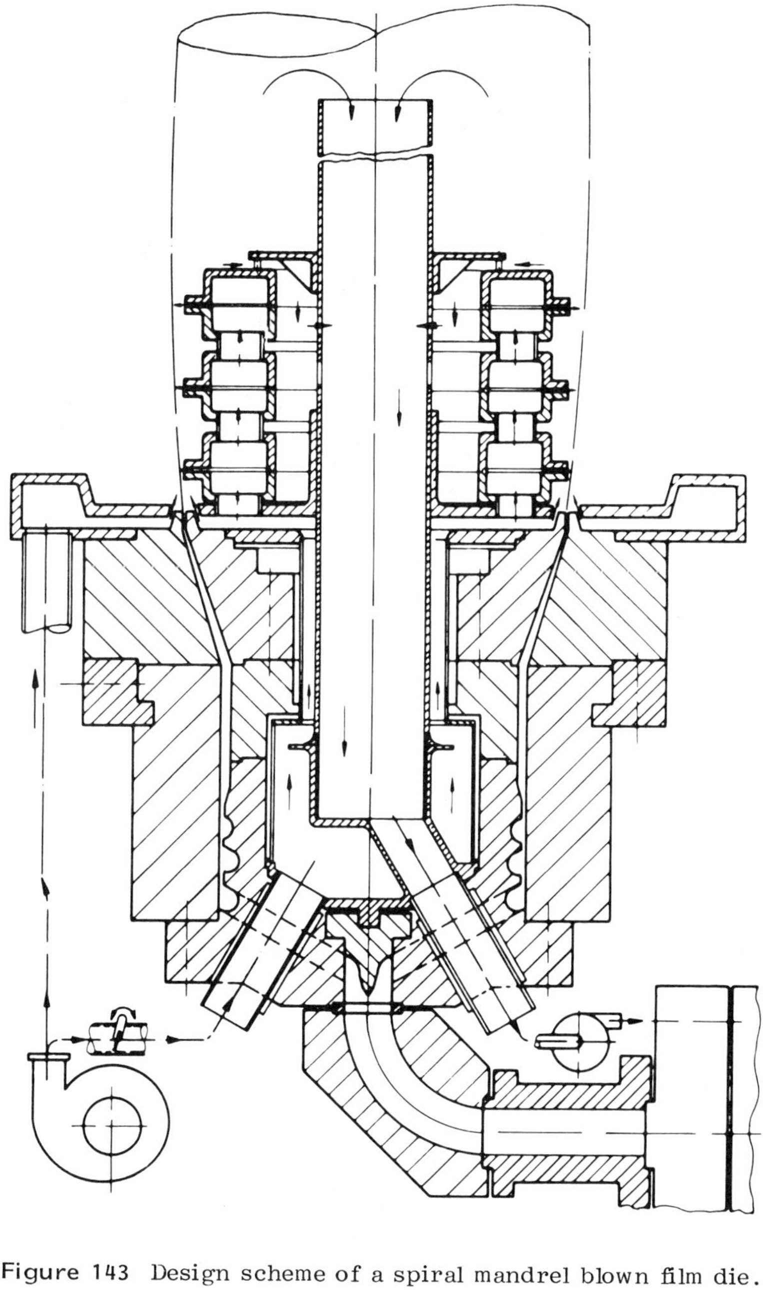

Figure 143 Design scheme of a spiral mandrel blown film die. (From Krämer, 1979.)

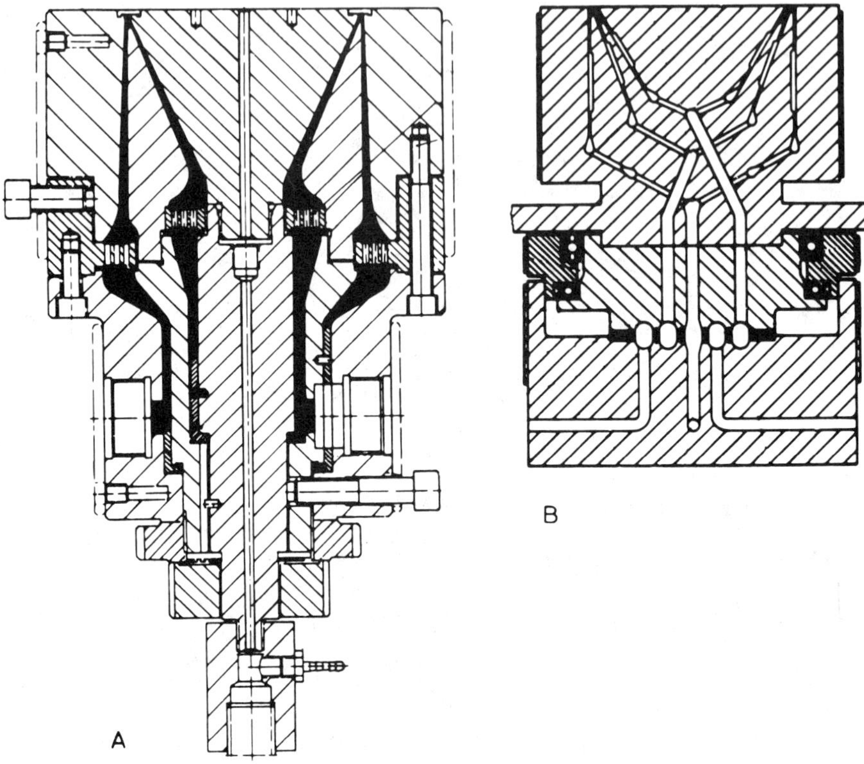

A

B

Figure 144 Two-layer (A) and three-layer (B) coextrusion dies. (From Krämer, 1979.)

irregularities can also be caused by flow disturbances, by uneven temperature gradients in the melt, by failures in the die heating system, by uneven film cooling, or by drafts in the operating area. With stationary blown film extrusion equipment these gauge variations appear always in the same position along the film tube and lead to "hoops" in the reeled film or to occasional weak tapes in tape production. With the use of perforated breaker disks, smear devices, or spiral mandrels, spider lines and some of the other irregularities can be partly eliminated.

Other possibilities to eliminate such thickness variations in blown films are the use of rotating extruders, rotating die heads or cooling rings, and reversion take-off units combined with turnover lever systems. For extruder rotation only vertical extruders with direct central-feed dies are well suited. The extruder is mounted on a turntable on which extruder, die head, and cooling ring oscillates 360° in alternately changing direction of rotation. Figure 145 shows a photograph

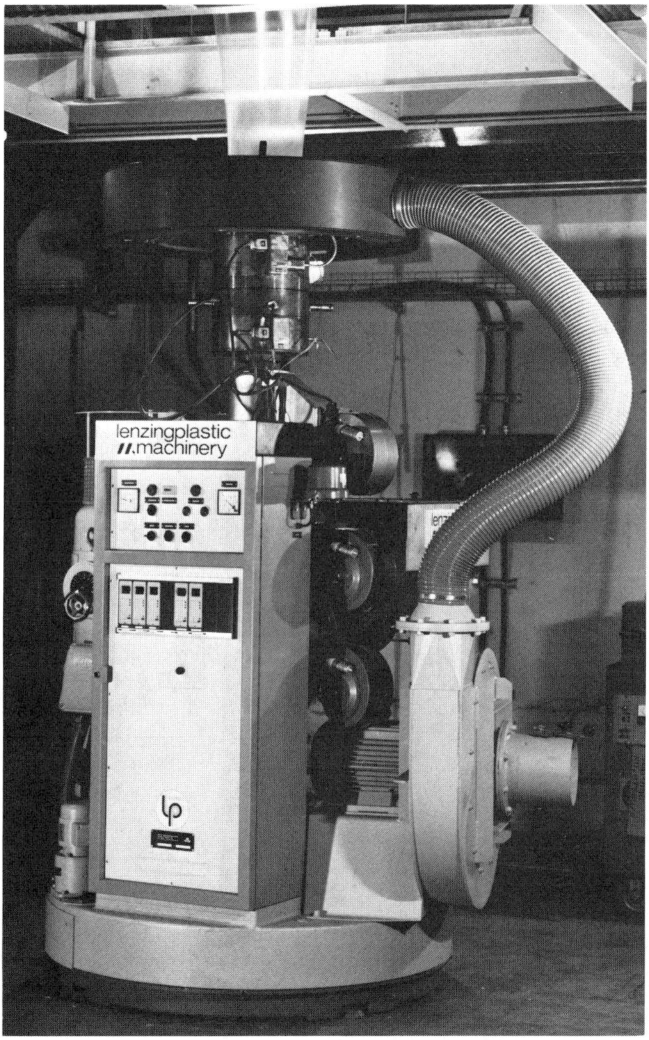


Figure 145 Photograph of a rotational film blowing extruder "Rotatruder" Oelikon-Lenzing System). (Courtesy of Chemiefaser Lenzing AG.)

of such an extruder unit. This system is limited to extruder sizes of 90 mm diameter and not very high screw lengths up to approximately 30 D.

Rotating die heads have the advantage of being technologically more simple than rotational extruder units. However, they are not so effective in the elimination of gauge variations. Relatively good results can

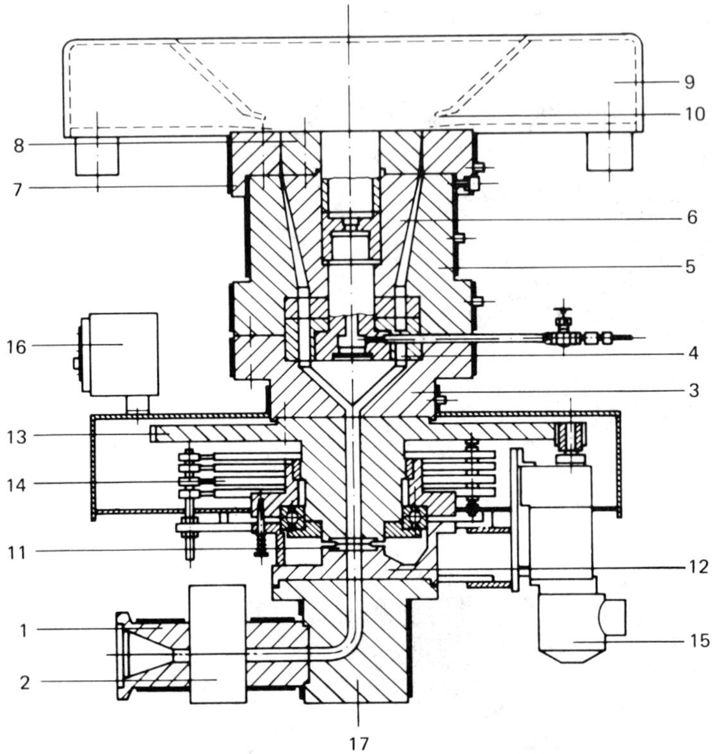


Figure 146 Scheme of a rotating blown film die with rotating cooling ring: 1, adapter; 2, screen pack changer; 3, adapter ring; 4, mandrel support; 5, housing; 6, mandrel; 7, die; 8, mandrel crown; 9, cooling ring housing; 10, lip elements; 11, raceway; 12, lower part of turning mechanism; 13, upper part of turning mechanism; 14, ring wheel body; 15, drive unit; 16, control units; 17, elbow adapter. (From Krämer, 1979.)

be achieved when die head and cooling ring are rotating simultaneously. Figure 146 shows the design of such a rotating die head assembly. Reversion take-off units connected with turnover lever systems, such as offered, for example, by the German machine manufacturer Windmöller & Hölscher, allow elimination of most of the shortcomings of gauge variations and can also be applied on large extruders and very large film widths.

## IV. FILM SOLIDIFICATION EQUIPMENT

In flat film extrusion two methods of film solidification are in use, namely (1) water immersion cooling, and (2) the chill-roll cooling technique.

In blown film extrusion the film cooling and solidification is generally achieved by contact with the outside air and by the inflation air. Only small losses occur through the takeup of "air rings" or "cooling rings." In downward extrusion of blown film, water cooling can also be applied in exceptional cases. In Chapter 2, Secs. II.B and II.C the various effects of different film cooling and solidification techniques on the fine structure and properties of films of film-derived products were discussed in detail. Therefore, the discussion here will be restricted to construction and design features of film solidification devices.

### A. Water Bath Design

Water bath cooling and film solidification is of interest mainly in the manufacture of thicker films and heavy film tapes. The water bath design has to pay special attention to appropriate layout and construction of the film guide rolls, guide bars, or guide plates. The proper placing design of the water inlet and outlets is very important in order to assure an undisturbed water surface, an equal water and temperature level, and to avoid the formation of turbulence.

Essential for smooth operation is the proper distance between the die orifice and the water surface. With some thermoplastic polymers the distance has to be very narrow; for polypropylene, preferably in the range of only 5-10 mm (0.2-0.4 in). A wider air gap between the die outlet and the cooling water surface can cause "surging pulsation," whose origin is physically not yet completely understood. On the other hand, too narrow a distance between the die lip and the cooling medium surface leads often to difficulties due to turbulence in the water and to water level differences between the film inlet and outlet side which, at high production speeds, are hard to control even with appropriate flow guides or breaker plates. The level differences between film inlet and outlet side can lead to wetting of the die lips when, for any reason, the production speed has to be suddenly reduced or when film rupture occurs.

Earlier, the favorable effects of the rapid and equal water bath cooling and solidification of freshly extruded film, such as the simultaneous cooling of both film sides, formation of a "smectic," paracrystalline structure accommodating subsequent stretching, and fast development of good mechanical properties, were emphasized. However, in spite of these advantages, water bath cooling is only applied for heavier film gauges and for special cases. This is due to the following disadvantages: (1) the unavoidable large width recess, (2) the often observed rippled film surface caused by water turbulence, and (3) the dragging along of water which limits production speed and has to be carefully removed by stripper rolls. In using water bath cooling, attention should also be given to the absence or presence of additives in the polymer. Lubricants and antistatics often present in polyolefin granulates can cause increased water uptake even at moderate line speeds. Furthermore, water used for water bath cooling should not contain too large an amount of calcium salts, since these can cause speckled film surfaces or film opacity.

## B. Chill-Roll Design

The application of chill-roll units in flat film cooling and solidification has found wide application. Chill rolls act not only as a means for film cooling but also as a takeoff unit. In current use are units with one, and also two rolls.

In the design of chill rolls the intended line speed is of importance in selecting the proper diameter of the rolls. At high line speeds, rolls with diameters up to 1200 mm (48 in.) are in use. In applying high flow rates of the cooling medium and by proper design of the cooling circuit inside the rolls, temperature uniformities of ±0.5-1.0°C over the width of the rolls can easily be achieved. An example of the cooling system of a chill-roll is schematically shown in Figure 147. In

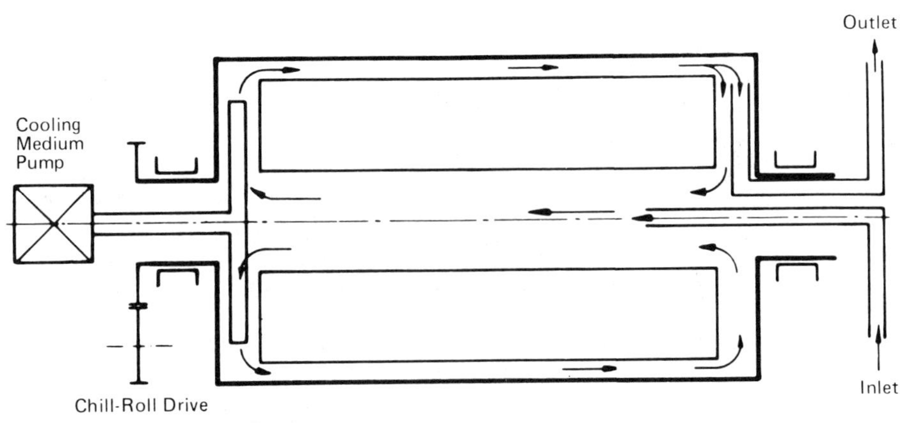


**Figure 147** Scheme of the cooling system of a chill roll with additional water circulation developed by Reifenhäuser KG. (From VDI-Gesellschaft Kunststofftechnik, 1978.)

polyolefin flat film extrusion the distance between the die slot opening and the surface of the chill roll should be relatively small (e.g., in the range 20-50 mm (0.8-2.0 in.). The chill-roll temperature for polyethylene can be in the order of 50-70°C, while the polypropylene lower chill-roll temperatures of about 20°C are advantageous. However, the surface temperature should in any case be well above the dew point, depending on humidity and temperature conditions in the operating area.

Very important for smooth operation and for good film quality is the surface finish of the chill rolls. Best performance is achieved with highly polished chromium plating. At higher line speeds mat chromium finished rolls show certain advantages with respect to reduced film sticking. In chill-roll operation, perfect contact of the still-molten film with the chill-roll surface is essential. Due to the thicker film edges—coming from a certain unavoidable width reduction—air cushions can be entrapped between the film and the chill-roll surface. This happens especially at high line speeds. The entrapped air hinders proper heat transfer in film cooling and thus causes structural nonuniformities in the end product. In order to avoid this effect, chill-roll units are generally equipped with "air-knife" systems. Several designs were developed, for example, those blowing compressed air against the still-molten film and pressing it over its whole width firmly onto the chill-roll surface (air knives or air noggles), or suction devices mounted behind the die orifice sucking the film onto the chill-roll surface, simultaneously removing any air entrapment. In Figure 41 (Chapter 2, Sec. II.B) we have already shown the various systems schematically.

## C. Film Cooling in Blown Film Extrusion

Blown olefin films can be cooled and solidified either with air or, in exceptional cases, also with water. The latter method, however, can only be applied in downward extrusion.

The still-thermoplastic hot film tube must be cooled (normally with air) in the bubble expansion zone. In order to stabilize the film tube, one has to impinge it with cooling air equally from all sides around its circumference. The cooling air is produced with the help of turboblowers and coolers fed with water and applied by cooling air rings (see Figure 44, Chapter 2, Sec. II.C.). Older designs of these cooling rings achieved the desired airflow uniformity by leading air through a labyrinth system of built-in flow channels. More modern air ring designs use flow channel constructions in which the cooling air is constantly accelerated in a controlled way up to the nozzle opening, carefully avoiding any turbulent flow. Some air ring designs have adjustable nozzle lips allowing proper control of airflow and speed. With a nozzle lip opening of 2 mm, air speeds up to 20 m/sec are commonly used. Wider slot openings allow a more moderate film tube cooling at

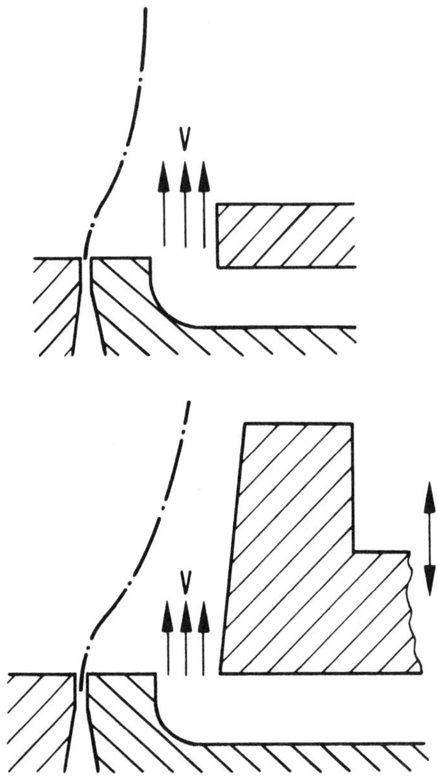


Figure 148 Demonstration of the effect of the position of the upper lip of the cooling ring (Venturi effect). (From Krämer, 1979.)

lower air speeds. A proper adjustment of the air speed to the distance between the air ring slot opening and the film is essential. At excessive air velocities and at minimum distance, fluttering of the film tube occurs, leading to undesired gauge nonuniformities.

Air velocity should be selected so that the air hitting the film has a certain turbulence in order to intensify the heat transfer and to enhance the cooling effect. The air ring lip design determines the angle of impact of the air with the film tube. Normally, the radially incoming air stream is deflected by 90°, hitting the film tube at a relatively small angle. In the processing of polymer melts of low viscosities (polymers with high melt-flow index) it is sometimes necessary to support the still plastic tube by applying the air stream at a 45° impact angle through the use of corresponding cooling ring nozzle designs.

The form of the flexible upper lip of the air ring nozzle influences—due to the enforced air velocity—the efficiency of the cooling. When the upper lip is distinctly superelevated with respect to the lower lip, the plastic film tube is sucked against the upper lip by the "Venturi effect" until pressure equalization is reached. This effect is schematically illustrated in Figure 148. A similar effect can be achieved with an "iris diaphragma" ring which is adjusted a distance of approximately 200 mm above the air ring nozzle, leaving only a narrow gap to the film tube. The possibility of a slight decentralization of the air ring further allows equalization of smaller gauge differences as experienced sometimes with side-feed annular die heads or caused by spider leg attachments of the mandrel. Thin sections of the film tube are cooled somewhat better so that they are solidified faster and can be less extended by the inflating pressure.

The effect and regularity of cooling by the air ring can be observed on the position and contour of the "frost line." The cooling air

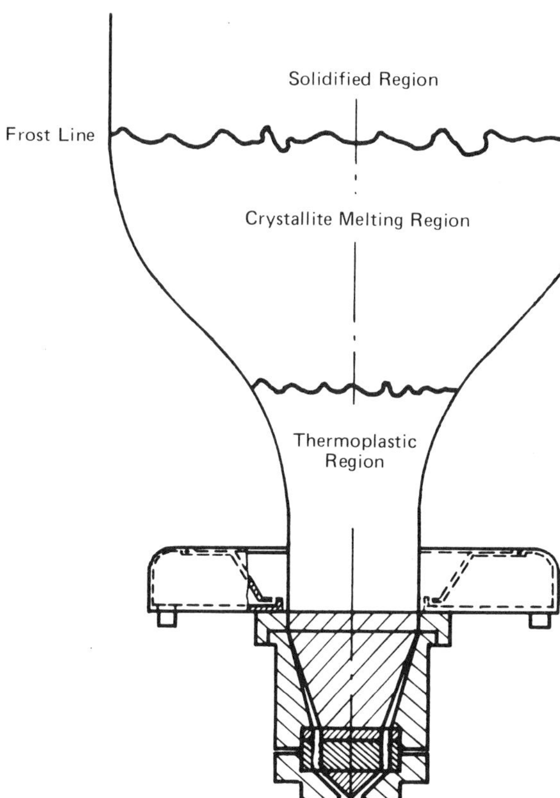


**Figure 149** Schematic demonstration of the film solidification in film blowing. (From Krämer, 1979.)

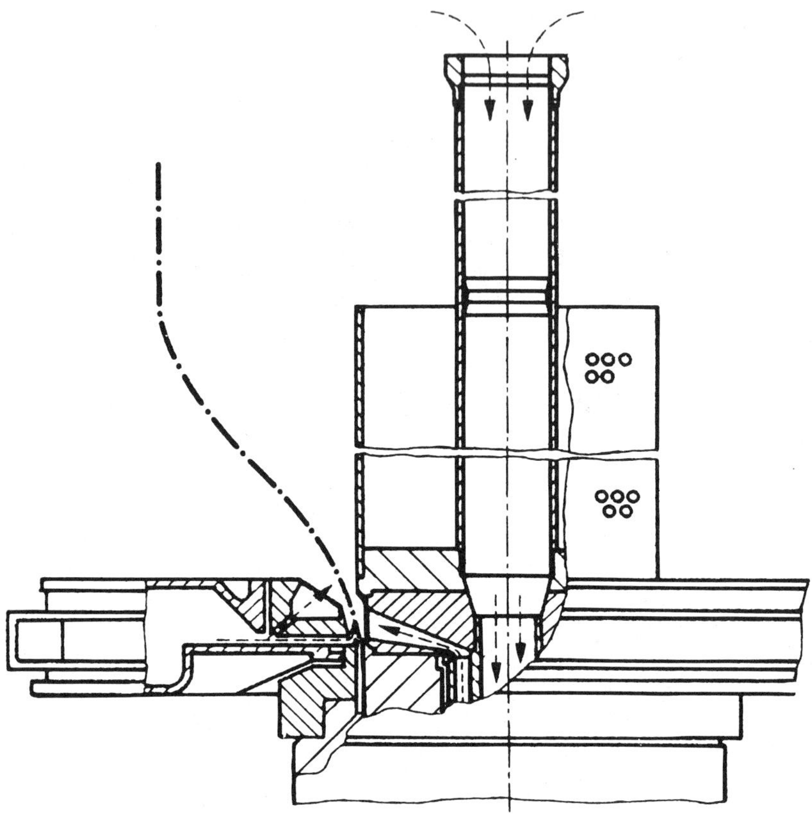

Figure 150 Scheme of a film blowing die head with combined inner and outer cooling of the film bubble. (From Krämer, 1979.)

influences gauge uniformity, the formation of the crystalline structure, optical and mechanical properties, and the further processing behavior of the primary film on its whole course, from leaving the die slot up to the final solidification at the frost line. This is shown in Figure 149. By application of combined inside and outside cooling air, it is possible to enhance the cooling capacity considerably. Supplementary inside cooling allows a substantial increase in the throughput of a given film extrusion line. Figure 150 gives an example of a blown film extrusion die using combined inside and outside film tube cooling.

## V. FILM STORAGE DEVICES

In film production, and even more important in direct film tape beaming, film storage units are an essential processing addition to any film or film tape line. In film production the use of film storages is recommended when in-line conversion, such as printing or bag forming, is performed in direct succession to film extrusion. The same is the case in direct tape beaming lines. Whenever operational troubles occur or necessary changes have to be performed, such as the occurrence of tape ruptures in direct beaming operations or changes of printing rolls, film stacking in storage units of more or less sophisticated designs are used.

The most simple design of a film storage unit is illustrated schematically in Figure 151. In such storage containers the film is laid down by tilting devices in such a way that one film layer is laid down from one side of the film storage to the other side on top of the preceding layer. By proper control, shorter tilting action may be applied from time to time in order to enhance storage capacity. More sophisticated designs use multiple roller systems in which an upper row of rollers can be altered in its distance against a lower row of rollers, thus creating place for additional lengths of film. (See Chap. 2, Fig. 47.)

## VI. STRETCHING AND HEAT-SETTING EQUIPMENT

Polymer films, film tapes, and film fibers obtain their final mechanical properties, such as desired tensile strength and elongation at break, only after drawing to a certain stretch ratio at elevated temperature.

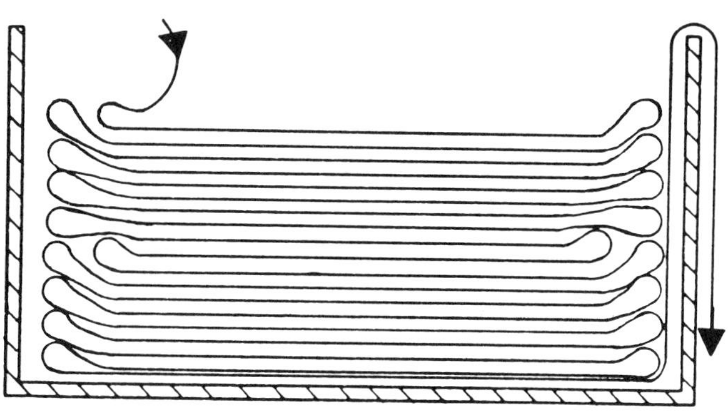

Figure 151 Scheme of a film storage container. (From Krämer, 1979.)

Figure 152 Photograph of a roller stretching unit for uniaxial film stretching. (Courtesy of Chemiefaser Lenzing AG.)

In the case of polyolefin products, stretch ratios of 1:3 up to 1:12 are commonly applied.

To perform the stretching operation on films, film tapes, and film fibers three principle methods are in use:

1. Drawing in infrared, hot air, or steam heated ovens (see Figure 72)
2. Drawing on hot plates (see Figure 73)
3. Drawing on short-gap roller stretching units (see Figure 74).

For the uniaxial stretching of films one currently prefers the short-gap roller stretching technique. Figure 152 shows a photograph of such a roller stretching unit.

Although heat is evolved in the drawing by crystallization enthalpy and by internal molecular friction, it is always necessary to preheat the film or film tapes at least above the glass transition temperature of the film-forming polymer. Therefore, the film first passes on roller stretching units a number of preheating drums. Constant temperature distribution over the full width of these preheating drums is essential. The heating is in most cases done with liquid heating mediums, allowing good temperature uniformity across the width of the rolls with variations less than ±1 to 2°C. Important also is the proper positioning of the preheating rollers and proper selection of their diameter in order to achieve the necessary cohesion, preventing film slippage.

After passing the preheating rolls the film enters the actual stretching device consisting of two pairs of pressure rolls, forming a short gap of a few, up to 20 mm. The two roller pairs run at different speeds, the second one at a faster speed corresponding to the desired stretch ratio. On the outlet side of the short-gap stretch roll systems follow heat-setting and cooling rolls. They also act as additional traction units, supplying the drawing force for film or tape stretching. For this reason the positioning and roll diameter has to be as well adjusted as that of the preheating rolls. Speed differences between the heat-setting roll directly following short-gap stretching and the subsequently mounted cooling roll allowing a certain degree of heat relaxation result in better dimensional stability of the films or film tapes.

Electrically coupled direct current (dc) motor drives, whose speed can be controlled and synchronized against each other, have to be selected in agreement with the high stretching forces and peripheral roller speeds. The drive control system should be designed in such a way as to allow change in the line speed in a continuous manner and to maintain any selected stretch factor. This is of special importance for facilitating startup procedures. Principally, such short-gap roller stretching units can also be used for film tape stretching. However, since the refeeding of broken tape ends is relatively difficult with two-sided supported rolls, one would prefer in film tape or slit-film yarn stretching, the use of drawing units consisting of one-sided supported roller systems, or "godets." These roller units are constructed with three, five, or seven rolls ("trios," "quintets," or "septets") which are placed in a staggered manner in order to achieve high angles of encirclement around individual rolls. Figure 153 shows a photograph of such a roller unit with seven rolls illustrating the general design features and the placement of rolls with respect to each other. Often these roller units are equipped with pressure rolls—sometimes rubber-coated—to counteract tape or film yarn slippage.

Generally, two such multiple roller godets form a stretching unit. The first one acting as a feeding and holding device (and in some cases to preheat the tapes or the slit-film yarn facilitating the stretching) and the second one running at a higher speed in relation to the desired stretch ratio, producing the stretch force. Usually, the stretching operation is facilitated by the use of hot-air-heated ovens. The air heating is either performed with direct infrared or indirect steam heating. For the same purpose electrically or liquid heated hot plates, "stretching irons," are in use. These heating facilities are placed between the feeding and drawing roller godets.

In film tape and slit-film yarn production, hot plate stretching is currently given distinct preference, since direct contact of the film with the hot plate surface guarantees better heat transfer. Furthermore,

Figure 153 Photograph of a septet. (Courtesy of Chemiefaser Lenzing AG.)

hot plates allow easier refeeding of broken tape ends. In the design and construction of hot plates, however, attention has to be given to the proper layout of the heating system in order to assure an even temperature distribution over the whole width. The maximum tolerable temperature variation should not be more than ±3–5°C. Figure 154 shows a photograph of a stretching unit using hot plate equipment. Hot air ovens are mostly constructed as twin-circuit, circulating hot air units. They are generally built for use in temperatures up to 300°C. To obtain an equal temperature distribution throughout the whole oven, the air speed should be between 10 and 20 m/sec. To facilitate startup operations or the refeeding of broken tape or yarn ends, the hood can normally be opened partly or completely. In the use of hot air ovens the fact has to be taken into consideration that in stretching, tape thickness and width are reduced in relation to the stretch ratio, while in hot plate stretching the tape width is somewhat less affected.

Equipment similar to that used for film tape or slit-film yarn stretching, namely two heated multiple roller godets or those combined with an interposed hot air oven, are also used for heat relaxation and heat-setting treatments. Since such treatments are usually performed under conditions which allow a certain degree of longitudinal shrinkage, the godet on the inlet side runs at a somewhat higher speed than the second godet on the outlet side of the annealing zone. Here it is also desirable to have electronically controlled and synchronized motor drives enabling the operator to change line speed continuously without altering the degree of relaxation in the annealing treatment, facilitating startup procedures or intervention in case of operating troubles.

## VII. SPLITTING AND SLICING MEANS

The most important processing step in the manufacture of film tapes, split-film yarns, and film fibers is the splitting or slicing operation. Depending on the type of technique and end product, this operation is performed either before or after stretching. In the case of film tape manufacture, tape cutting is normally performed before the stretching operation, except with the application of split weaving or knitting techniques where tape cutting from uniaxially stretched film

Figure 154 Photograph of a hot plate stretching unit. (Courtesy of Chemiefaser Lenzing AG.)

is partly carried out on the fabric processing equipment. In the case of uncontrolled mechanical or chemomechanical fibrillation of film in its entire width or on tapes, the splitting operation is generally performed after the uniaxial stretch, since the splitting is strongly facilitated by high anisotropy of strength produced in the stretch. In the case of controlled splitting or slicing techniques, the separation of film into individual, more or less endless, separated single filaments can either be done before, simultaneously, or after the film or film tape stretching, depending on the type of process applied. Since the application techniques of such splitting or slicing tools has already been described earlier in (Chap. 2, Sec. IV), reference will frequently be made to avoid duplication.

### A. Means for Fibrillation

Purely mechanical fibrillation can be achieved by applying shearing forces which act on highly stretched film or film tape perpendicular, or at a given angle, to the processing direction. Such shearing forces can be produced by a *brushing* action using stationary or rotating brushes of various designs, mainly carrying steel needles or bristles.

Similar forces can be exerted on uniaxially stretched film or film tape by *rubbing* devices. Rubbing action on uniaxial film can be exercised, for example, by a pair of countercurrently oscillating rubber-coated rollers (Phillips Petroleum Co., 1971b), or by pairs of plates oscillating transversely and having rough surfaces (Rasmussen, 1961). The application of such rubbing devices were described in Chapter 2 and illustrated in two examples given in Figure 104. Film-to-fiber fibrillation can also be caused by *stress* action, such as that produced by double-rubber-belt arrangements combined with pressure members, creating compression zones in the film moving between the belts (Rasmussen, 1967a). Also, passage of the uniaxially stretched film or tapes through pairs of rollers or belts with rough surfaces, rotating at different speeds (Soko Co., Ltd., 1971, 1973) initiates fibrillation through stress action. In this connection reference is made to Figure 105, which was presented earlier.

Stress action initiating uncontrolled mechanical fibrillation can also be produced by *bending* operations. Bending plate or bending roll fibrillator units, as shown earlier in Figure 106, were developed by several companies (Allied Chemical Corp., 1969a, 1969b; Rheinstahl Henschel AG, 1971a). The most widely used fibrillation techniques is based on *twisting* action and is performed on normal ring-spinning frames or on false-twisting machines of special design (Dow Chemical Co., 1965; Société Rhodiaceta, 1965; Plasticisers, Ltd., 1967; Shell Internationale Research Maatschapij N.V., 1969b). The design of the special false-twisting device by Shell Internationale and its operational features were described in Chapter 2 and illustrated schematically in

Figure 100. *Gas-jet* treatments are another technique for the fibrillation of uniaxially stretched film tapes (du Pont, 1966a; Fiber Industries, Inc., 1972). Examples of fibrillator tools operating with gas jets were illustrated earlier in Figures 102 and 103.

## B. Means for Controlled Film-to-Fiber Separation

Controlled mechanical film-to-fiber separation techniques based on profiling film or film tapes, either in extrusion or thereafter by using specially *profiled dies* or *profiled embossing rollers*, call for initiation of fibrillation along longitudinal grooves, through the same method described in the preceding section; namely brushing, rubbing, bending, twisting, gas-jetting, and other types of stress application. This film-to-fiber processing technique has already been described in great detail in Chapter 2 (see pp. 148-154).

Additionally, in the film-to-fiber processes based on the use of *needles* or *edge rollers* to introduce holes, cuts, or incisions into the film or film tape, final splitting is done by stress action through similar mechanical treatments as described in the preceding section. The manufacturers of needle- or edge-roller fibrillators have developed a wide variety of such tools. Some photographs showing various fibrillator roller and complete fibrillator units were already given in Figures 92, 94, 95, 96 and 97.

In general, the basic rule gained from processing experience indicates that the combination of small needle-roll diameter with short contact length between film and pinroll gives strong irregular networks with random small meshes consisting of rather coarse interlinking fiber segments. The application of needle rollers with large diameters at long contact lengths with the film results in weak fibrillates having an irregular long mesh network. The mesh length, the degree of film slitting, and the average titer of the resulting fiber segments are controlled by the contact length between film and fibrillating roller, the film or tape speed, the surface speed of the roller, and the pin density per unit area of the roller surface. The mathematical treatment of the interrelation between these process variables and the product characteristics has already been described in Chapter 2 (see pp. 164-166). The use of this mathematical treatment allows estimates of the expected fiber titer, as presented in Table 19.

For performing controlled film-to-fiber slicing by *defined cutting* methods, a number of special tools were developed by different inventors. The first process for film conversion into more or less continuously separated filaments using controlled slicing was described as early as 1956 by Dow Chemical Co. They had developed a cutting device consisting of a pair of rolls, one of which carries cutting edges parallel to its periphery, the other one a resilient backing roll. The grooves of the profiled roller are separated from one another by sharp ridges capable of scoring or slicing the stretched film. The distance of the sharp ridges are in the range of 1 mm. The scheme of the grooved roll, as given in the corresponding patent of Dow Chemical, was shown in Chapter 2 in the upper part of Figure 88. In a slicing process described by Mirsky (1967), stationary, rotary, or oscillatory blade assemblies having lengths of their sharp edges of 2 mm or less are used for slicing the film into continuously separated single filaments. Figure 88 shows, in its lower part, the design of this film-to-fiber separation tool.

Hercules, Inc. (1969) developed a film slicing tool for the production of filament type film yarns from film longitudinally striated either in extrusion or by embossing. It consists of a stationary block, carrying a series of serrations along its entire width. The distance from the sharp tip of one serration to the tip of an adjacent serration is such as to conform to the distances of longitudinal grooves in the striated film. The process application of this slicing tool has been described in Chapter 2 and its design illustrated in Figure 86. A somewhat similar process, unique, however, in other features of process performance, was developed independently by Chemiefaser Lenzing AG (1970) and makes use of a sawtooth-like cutting tool over which an unstretched or only partly stretched film is drawn under tension. The sharp edges of the cutting tool, which is heated to facilitate the film slicing, are spaced between 0.08 and 0.500 mm from each other.

The heated cutting tool serves several purposes in this process: (1) heating of the film, (2) film-to-fiber separation, (3) creating the holding effect for generating the stretching force, (4) defining the position of the neck-down point, and (5) edge bending at elevated temperature introducing latent crimp properties. Continuously separated filaments drawn from the sawtooth-like cutting tool with line speeds exceeding the feed speed by the desired stretch ratio can be produced, with titers down to approximately 7 to 10 dtex. A scheme of the slicing tool used in this process is shown in Figure 89 of Chapter 2.

## VIII. WINDING EQUIPMENT

Films, film tapes, or film fiber yarns have to be wound up in the final step of their production to facilitate their shipment and/or their further processing.

### A. Film Winding Units

The films, independent of whether they are produced by flat film or annular film extrusion, are obtained as continuous strips which have to be would for storage, shipment, or further processing. For this purpose film winding systems have been developed. Since further processing, such as printing, bag forming, welding, or splicing makes great demands on windup uniformity, the winding equipment has to be properly designed and constructed in order to allow perfect control of the rotational speed and of the winding tension independent of the steady increase of the reel diameter during the winding operation.

The principle prerequisites for a good processing film reel are: cylindrical shape, absence of hoops and creases, perfect edge alignment (deviations less than ±2 mm), and even reel tension throughout the package. Sidewise dislocation or blocking tendency are intolerable conditions. Basically, there are two types of film winder designs: the "pressure roll contact winders" and the "winders with central drive units".

For pressure roll contact winders the drive power acts not directly on the shaft of the takeup spool, but on a pressure roll. The film coming from the stretch or annealing unit is taken over by the pressure roll. The film reel rests in close contact on the pressure roll and the incoming film web. At the contact point the film reel is driven through friction by the movement of the pressure roll. For smooth operation, an exact parallel adjustment of the winding mandrel with the pressure roll shaft and perfect constancy of the contact pressure at all diameters of the film reel is essential. In order to meet these conditions the winding mandrel has to be mounted flexibly, allowing a circular arc movement with increasing film package diameter controlled by a rack and pinion drive. Through the controlled inclination of the film inlet the steadily increasing film reel weight can be compensated. The principle of a pressure roll contact winder and of a central drive winder is shown schematically in Figure 155.

These relatively simple winding units are suited mainly for relatively soft films. Up to winding speeds of about 50 m/min, transverse cutting and reel changes can be performed by hand. At higher winding speeds semi-automatic or fully automatic cutting and film reel changing devices are required. For winders with central drive units the winding mandrel is driven directly. Technical requirements for the drive system and automatic controls are much more complicated than for pressure roll contact winders. The automatic control system has to adjust the drive of the winding mandrel to the steadily changing film reel diameter in order to properly control reeling tension. This is achieved by starting the reeling with high rotational winder mandrel speed and constantly slowing down with each additional film layer.

Winders with central drive system are controlled for reel-diameter-dependent torque in most cases with the help of "Alquist drives," with hyperbolic or tensor roll control systems. However, with none of these systems can the ideal linear relation between torque and rotating speed be fully achieved. A realistic approximation is, however, possible by proper selection of the operating conditions. In the case of Alquist drive systems adjustment can be achieved with the use of adjustable transformers influencing magnetization. Control of drive systems with

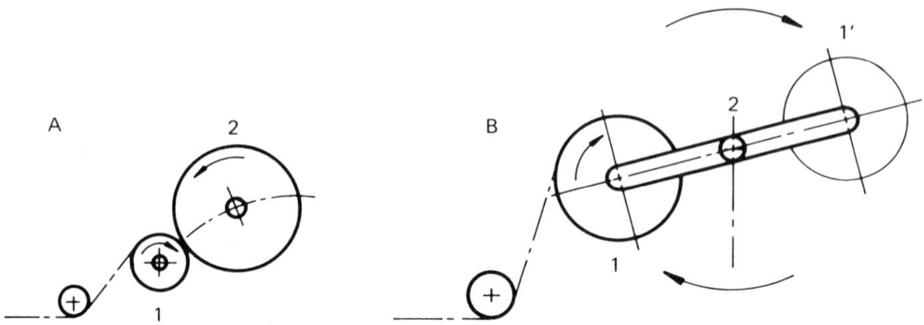


Figure 155 Functional scheme of a contact winder (left) and a central drive changeover winder (right).

(A) Contact winder (pressure roll winder): 1, pressure roll directly driven by a torque-controlled direct-current motor; 2, reeled film driven indirectly by contact with the pressure roll.

(B) Central drive changeover winder: 1, film reel directly driven by a torque-controlled direct-current motor acting through reel-diameter-dependent torque-control system; 1', film winding changeover position with independent direct-current motor drive; 2, axle of changeover bar carrying the two directly driven film winding positions.

hyperbolic characteristics is achieved with transistor or tyristor controls. Tensor roll control system operates through the tensor position which influences the controls of the winder shaft drive. In addition, a tensor control system also fulfills a storage function, allowing more sensitive film reeling.

In order to equalize film gauge variations, which even in flat film extrusion are not completely avoidable, and to prevent the formation of hoops in the film reel, extruders are sometimes placed on irridescent foundation platforms. The edge trimmings are enlarged by such measures, but the commonly used edge trimming recycling systems avoid polymer losses, and additional technical installation pays off quickly with improved quality of output.

## B. Film Tape Winding Equipment

For film tape conversion into woven or knitted fabrics of good quality, perfect tape reels are important. In order to achieve high quality tape spools, film tapes are wound up on tension-controlled cross-winding heads with continuously variable speed adjustment. Special attention has to be given to constant tape tension during winding, proper tape guides, and perfectly operating iridescent devices. There are tape winding machines on the market which allow windup of two film tapes per winding head simultaneously on one reel or side-by-side on two reel cores. In most cases modern tape winding units have two or three rows of winding heads—one on top of the other—arranged at a suitable distance (approx. 450-500 mm apart). In most cases two winding frames are placed together in such a way that the winding heads are face to face and allow for servicing by the same operator. Figure 156 shows a photograph of a modern film tape winding station. Winding tension ranges between 30 and 300 N, the tape count from approximately 500 to 5000 dtex, and the reel package weight from 5 to 10 kg.

## C. Film Tape Beaming Equipment

In weaving or knitting operations one can principally work with tape warp beams, with tape spools from a creel system, or with film reels and tape cutting systems attached to the weaving or knitting looms (split weaving or split knitting).

The use of tape beams requires the time consuming beaming operation of warp tapes from tape spools on a creel unit. The beaming of

Figure 156 Photograph of a tape winding unit. (Courtesy of Chemiefaser Lenzing AG.)

flat tape is much more delicate and problematic than the beaming of round yarns. Furthermore, the application of tape creel delivering systems for the warp inlet to beaming unit calls for additional room and investment for installations. On the other hand, direct weaving or knitting from uniaxially stretched film in the split weaving or knitting technique has certain shortcomings with respect to strength reduction by cutting defects coming from damaged or blunted cutting blades. For this reason direct tape beaming on the tape line is interesting in all cases where high quality of mechanical properties is essential for woven or knitted tape fabrics. Direct tape beaming units consist of flat or tubular film extruder combined with a film storage device, allowing continued operation in case of tape ruptures in stretching that requires tying together broken tape ends, and subsequent film stretching, annealing, and beam winding units.

Such a direct beaming line for film tapes allows direct winding of low titer high-density polyethylene or isotactic polypropylene film tapes, avoiding the tedious and time-consuming (with tapes often problematic) beaming operation from tape creels. Since in these direct beaming lines the tapes have normally undergone stretching and annealing treatments, the occurrence of warp tape ruptures in subsequent weaving or knitting operations is reduced to a relatively low level, resulting in a high unit capacity factor. There are also tape beaming lines which operate from film reels of partially or unstretched films combining the tape cutting, stretching, annealing, and beaming operations. These lines avoid some of the possible operating difficulties connected with the necessity of film storage in direct beaming in line with film extrusion (see Figure 157).

## D. Split- or Sliced-Film Yarn Winding Equipment

As in the production and processing of continuous filament, flat or texturized yarns made by conventional fiber spinning techniques from polymers such as polyethylene-terephthalate or polyamides-6 or -6,6, the furnishing of film fiber yarns on precision wound packages (cylindrical cross-wound tube, standard low angle or pineapple cone) is commonly used. The same advantages of good quality control and high economic efficiency in further application apply here the same as with conventionally spun continuous filament yarns. Here also the desirable characteristics asked for from a proper wound package are (1) a uniform package density, leaving the yarn in a relaxed condition but wound firm enough to assure safe shipping without sloughing or crumpling, (2) uniform shape without ridges or bulges and other deformations at the ends of the package, (3) absence of broken filament ends, (4) equal distribution of oil or emulsion sizing throughout the package, (5) knots, whenever used, placed recognizably on the nose of the package, (6) the

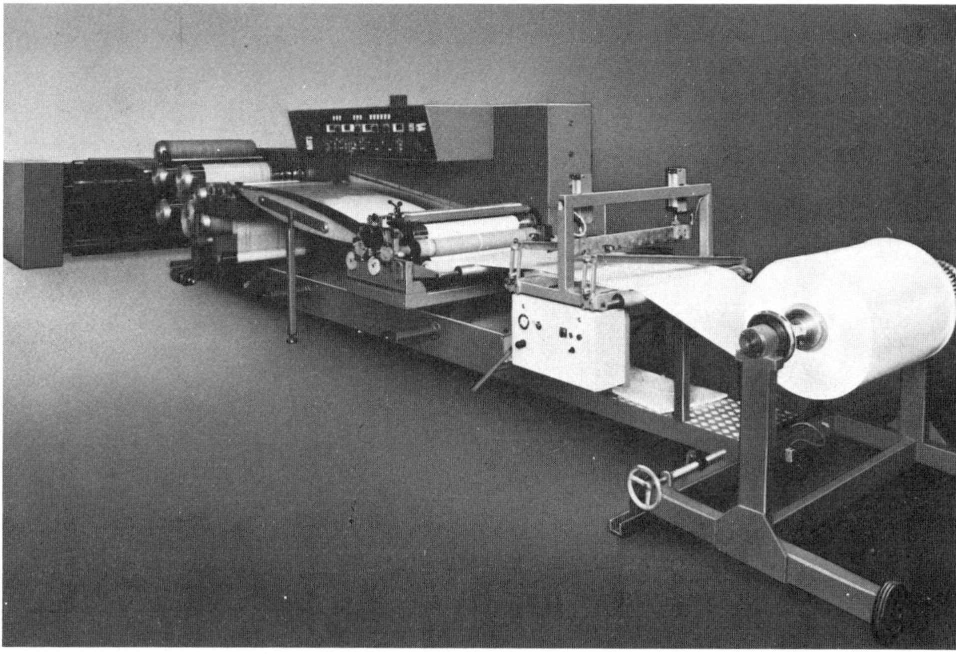

Figure 157 Photograph of a film-tape beaming line. (Courtesy of Chemiefaser Lenzing AG.)

tie or transfer end conveniently located at the cone base, and (7) perfect winding quality to allow free delivery in processing without plucks and jerks.

In order to achieve such standards, machine manufacturers have, in the past, developed suitable coning or winding systems. Originally, there were basically two winding machine principles for continuous filament and spun yarns, namely the "drum winders" and "precision winders." The *drum winders*, which were originally used mainly for in-line fiber manufacturing, got its name from the fact that the yarn package is driven through frictional contact with a motor operated drum running at a constant rotational speed. This results in a yarn speed which remains constant throughout the coning operation. However, for proper yarn winding in these machines the ratio between the revolutions of the yarn package and the transversing mechanism has to change its stroke sequence per unit time with increasing yarn package diameter. *Precision winders*, on the other hand, are coning units which allow a constant ratio between spindle speed and the filament laying-on mechanism to be maintained. In most cases the yarn is laid on the cone in a defined transverse movement. The iridescent moving

Figure 158 Photograph of a split-film yarn winding station. (Courtesy of Chemiefaser Lenzing.)

filament guide is operated by a cam which is perfectly synchronized with the spindle drive. As the diameter of the filament package increases, the yarn speed has to be enhanced. The spindle speed remained in older type precision winders constant. However, with the development of modern electronic control principles it is now possible to change the spindle drive speed in accordance with increasing package diameter. This is accomplished by use of tensor controls sensitive to changes in the winding tension which are transmitted to a moving coil system. The current induced in the coil by movement resulting from changes in the yarn tension regulates the capacitor control of the variable speed motor driving the winding spindle. This system allows adjustment of the spindle drive in such a way as to slow it down in accordance with the increasing package diameter and with the constant speed of the incoming yarn.

More simple systems operate with spring-controlled tensors in combination with torque-sensitive alternate current (ac) motors, where the rotating speed of the motor driving the spindle slows down automatically with growing yarn tension due to increasing package diameter. High precision winding units are mostly built in sections, or "gangs" in which 6-30 individually driven cross-winding heads are assembled in 3-4 rows on top of each other. Each of these winding heads has its own yarn guide system. The electrical supply system for the controls and motor drives of the winding heads in the sections is designed in such a way that several of these winder sections can be assembled by plug-in connections to larger units with 120-200 winding heads, according to a building-set principle. Figure 158 shows a photograph of such a large winding machine. In-line winding operations of split- or sliced-film yarns are performed with winding speeds which are relatively moderate and rarely exceed 300 m/min. Furthermore, split- or sliced-film yarns are rarely lower in titer than 500 dtex. In most cases they are in the range of 3000-5000 dtex.

## IX. EDGE TRIM AND POLYMER WASTE RECYCLING SYSTEMS

Production units for polypropylene film, film tapes, or split-film fibers should be equipped for better efficiency with in-line edge trim recycling systems, and when large enough in capacity, also with polymer waste regranulating units.

Systems for edge trim recycling work usually on the principle of compacting the edge trim strips by twisting at elevated temperature (Chemiefaser Lenzing AG, 1975) in order to produce a compressed

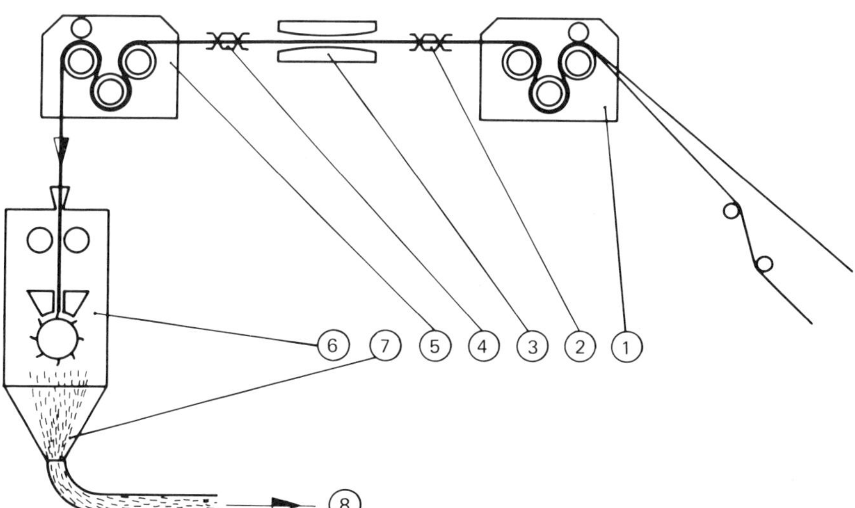


Figure 159 Functional scheme of an edge trimming recycling unit: 1, inlet trio; 2, guide and folding device; 3, compression unit; 4, twisting device; 5, takeoff trio; 6, cutting device; 7, shred container; 8, recycling line to extruder. (Chemiefaser Lenzing System.)

string which can be cut into pieces suitable for refeeding into the film extruder. Figure 159 shows a functional scheme of such an edge trimming recycling unit. Larger production units should be equipped with regranulating units consisting of a cutting or tearing machine for film, tape, or fiber waste, compactors to bring the cut or torn material into a conveyable form, a strand extruder, and a strand granulator. The recycling of granulated waste material into the product line or its use for specially designed products tolerating the somewhat inferior properties of the doubly extruded polymer results in substantially better plant economy. Such units repay themselves if the amount of waste material to be regranulated exceeds 200-300 metric tons a month and internal use is feasible.

## X. SETUP OF COMPLETE PRODUCTION LINES

In the following, the setup of production lines for uniaxially stretched film, film tapes, and split- or sliced-film yarns or fibers shall be outlined and illustrated by flow sheets.

### A. Production Lines for Uniaxially Stretched Film

The production of uniaxially stretched film from thermoplastics, such as polyolefins, can either be performed on flat film or on blown film extrusion lines. In flat film extrusion, film solidification can be achieved by water bath cooling or on chill rolls. Various setups of such film extrusion lines are illustrated in Figure 160.

Film production on a flat film extrusion line includes the following production steps: (1) plasticizing of the thermoplastic granular polymer in a single-screw extruder (double-screw extruders are only used in very special cases), (2) extrusion of the polymer melt through a flat film die, (3) cooling and solidification of the film either by water bath immersion or on chill rolls, (4) uniaxial stretching on roller stretching units (sometimes including annealing treatment allowing longitudinal shrinkage up to approx. 15%), and (5) film winding on suitable film windup equipment. In the manufacture of film on blown film extrusion lines, the process includes: (1) plasticizing of the polymer in a suitable extruder, (2) extrusion of the melt through an annular die, (3) film solidification by inflated air and sometimes with the help of cooling rings (in exceptional cases water cooling is additionally applied in downward extrusion), (4) collapsing of the film tube with the help of collapsing boards, (5) stretching of the flattened double layer film without or with annealing, and (6) winding up on suitable winders.

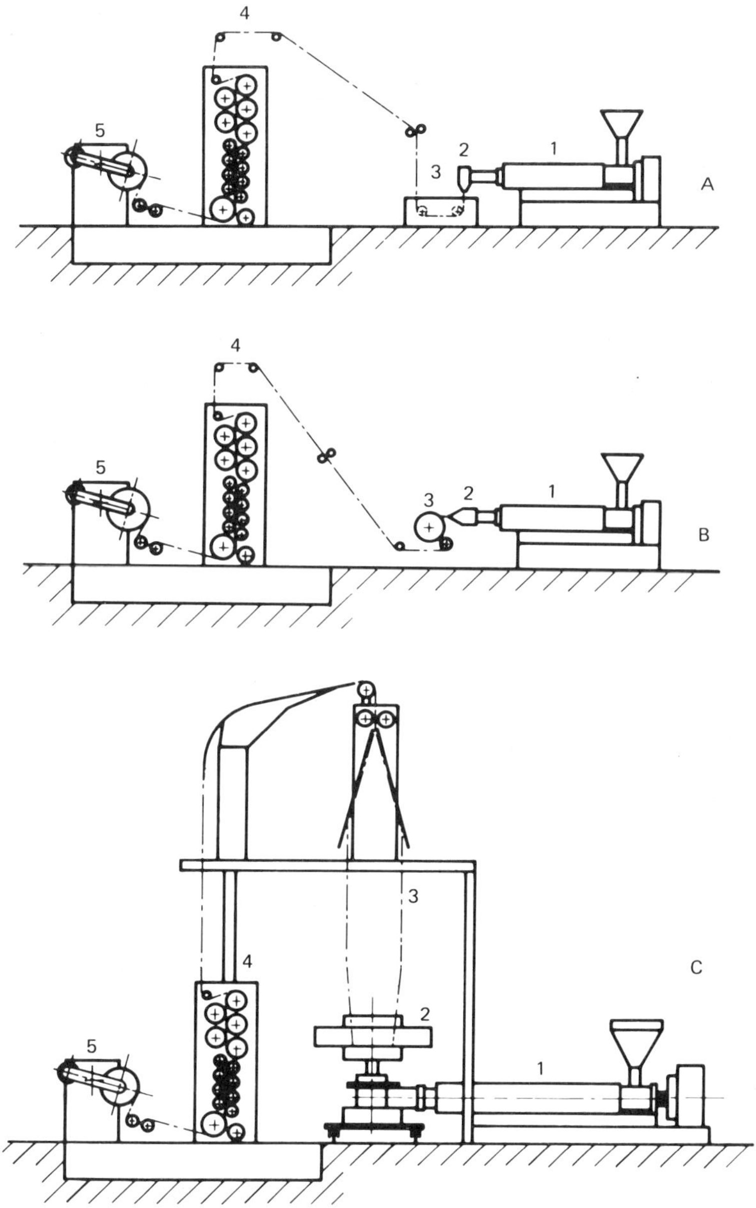


Figure 160 Scheme of various types of production lines for uniaxially stretched films: (A), with water bath quenching; (B), with chill-roll film solidification; (C), film blowing unit.

## B. Production Lines for Film Tapes

The setup of film tape production plants depends on the type and intended end use of the tapes. Normal weaving and knitting tapes can be produced on lines either using flat film extrusion with water bath or chill-roll cooling, followed by hot plate stretching or in hot air ovens, or using blown film extrusion followed by stretching on similar drawing equipment (or on roller stretching units). Flow sheets of such lines were shown in Figure 1. The production of weaving or knitting tapes for use in the manufacture of sacks or canvases seldomly calls for annealing treatments and quite often one gets away with relatively simple stretch inlet and outlet roller units. Production lines for the manufacture of tapes for ropes and twines must be built more sturdily. The lines are set up similar, but the thicker and wider tapes and the higher stretch ratios necessary to achieve optimum tensile properties demand flat film extrusion, water bath cooling, and the use of heavier stretching equipment able to supply higher stretching forces and the best possible force transmission.

Film tapes used in primary carpet-backing fabrics must have extremely high dimensional stability, particularly at elevated temperatures in order to withstand latex coating treatment during the manufacture of tufted carpets. Various manufacturers of film tape lines have developed different design concepts for their annealing units. Some of them use conventional hot air ovens in combination with multiple roller inlet and outlet equipment, the latter running at lower line speed, allowing between 5-20% longitudinal shrinkage. Other machine manufacturers use multiple roller units in which some rolls on the inlet side are heated and equipped with individual drives for each roll or sets of rolls, allowing a certain degree of creep from one roll or set of rolls to the next. Rolls on the outlet side are finally used for cooling and setting the relaxed film tapes. Figure 161 shows a photograph of such a carpet-backing tape line from its heat-setting end. Production lines for manufacturing multiple-layer film tapes or film fiber products require the use of more than one extruder feeding the various melts of the component polymers into a correspondingly designed coextrusion die head. These multicomponent extrusion units allow one to produce tapes or split-film yarns or fibers with special properties, such as thermoplastic adhesive tapes or fibers having latent bulking or crimping characteristics, and so forth.

## C. Production Lines for Split- or Sliced-Film Yarns or Fibers

Basically the setup of film yarn or film fiber production lines resembles relatively closely the design of a plant for tape manufacturing. The principle of film or tape splitting or slicing techniques determine the type and location of additional attachments or facilities to be added to a tape line.

Figure 161 Photograph of a production line for technical film tapes (carpet backing yarn) seen from its heat-setting end. (Courtesy of Windmöller & Hölscher, Lengerich/Westf.)

The purely random mechanical or chemomechanical fibrillation of highly stretched film or tape, as used in the manufacture of cords or binder twine, is mostly done by twisting, rubbing, or other types of frictional treatment. This means that the twisting, rubbing, or friction device causing the splitting is placed between the stretching unit and the final windup equipment. Lines for the production of film-to-fiber products using needle or knife roller treatment as a means for the film or tape separation have the needle or knife roller normally installed ahead of the stretching unit. Incisions introduced into the film by the needle or knife roller initiate fibrillation during the stretch. In case of staple fiber production the needle or knife roller treatment can also be performed on the film or tapes shortly before crimping or cutting procedures. Incisions introduced into the film or tapes have to be, in this case, significantly longer than the staple length.

Lines for the production of split- or sliced-film yarn using the film profiling techniques, such as used in the Barfilex process of Barmag-Barmer Maschninenfabrik AG (1971a, 1972, 1973a) and others, need no additional facilities when the film or tape splitting is achieved by high stretch or by twisting operations combined with the windup. When the film separation of the profiled film is done by a slicing or cutting operation, as in the process suggested by Hercules, Inc. (1969), the slicing attachment is normally incorporated into the line after the water bath or the chill-roll cooling system, and before the stretching unit. The same applies to the controlled film-to-fiber slicing technique invented by Chemiefaser Lenzing AG (1970), where the special film-to-fiber separation tool (see Figure 90) is positioned after the chill-roll cooling. In Figure 90, shown in Chapter 2, we have already presented the photograph of such a sliced-film yarn production unit showing the position of the film-to-fiber separation tool placed between the chill roll and the inlet roller unit of the hot table stretch.

## XI. MACHINERY FOR THE PROCESSING OF FILM TAPES, FILM YARNS, AND FILM FIBERS

The market for polyolefin film tapes, split- or sliced-film yarns or fibers has grown tremendously in the last two decades. Worldwide products made from high-density polyethylene and isotactic polypropylene film tapes or film fibers consume close to one million metric tons of these polymers today.

The main fields of application are those originally dominated by the so-called hard fibers, such as jute, hemp, and sisal. The main processing techniques to convert film tapes and split-film yarns or fibers are the *weaving* and *knitting* processes to fabrics or knits for outdoor upholstery fabrics, mats, canvases, or sacks. In many cases these tape or film fiber fabrics have to undergo *coating* or *lamination* with paper, conventional textiles, plastic foam, or metal foils. Further processing techniques for film tape or split-fiber yarns are the *tufting* into outdoor floor coverings or artificial grass substitutes. A wide field of application is in the use of uncoated and coated tape or split-fiber fabrics or knits in the *processing* of all kinds *of sacks* for a wide variety of packaging purposes.

Quite a number of machine manufacturers active in the field of film tape or film fiber manufacturing facilities have also designed special machinery for processing tapes or split-film fibers into woven or knitted fabrics, for coating or laminating such fabrics, for processing carpets, mats, or artificial grass, and for forming sacks from tape fabrics or knits. In the following paragraphs the processing of tapes or split fibers on weaving, knitting, tufting, coating, laminating, and sack forming equipment shall be discussed briefly. However, it has to be considered that this discussion can only touch these processing steps and the machinery used for them, and has to be restricted to basic considerations.

### A. Processing by Weaving and Knitting

Principally, three systems are being used for processing tapes or split-film yarns into fabrics: (1) processing on flat weaving looms,

(2) processing on circular weaving looms, and (3) processing on rashel knitting equipment.

### 1. *Processing on Weaving Looms*

Any loom can be adopted to process film tapes or split-film yarns. In main use are shuttle looms, rod gripper type looms, and Sulzer looms with gripper shuttles.

Weaving on *flat weaving looms* can be done either with a warp beam system, from a creel, or from a film reel applying the split weaving technique. In weaving from a warp beam system, part of the tapes forming the warp have to be put onto a weaving beam prior to the weaving in a special operation. The efficiency of the weaving operation and the quality of the resulting tape fabric depend to a large extent on the perfection of the warp beam. Crucial to a good beaming performance is the controlled and exactly flat delivery of tapes from the spools of the creel. An iridescent device allows lay-in of the tapes onto the beam in a laterally regular displaced pattern, facilitating even discharge in the weaving operation.

Weaving from a creel eliminates the time-consuming and work-intensive beaming procedure. In this connection the proper tension uniformity of the tapes fed from the tape packages on the creel to the loom is extremely important. Another advantage of weaving from a creel is the longer lengths of tapes or yarns on each spool. It is essential that each spool contains an equal tape or yarn length in order to diminish loss of material. Weaving from film reels (split weaving) used on the loom instead of warp beams obviates both difficult warp preparation by beaming and labor-intensive creel equipment. With the help of suitable let-off devices allowing controlled film delivery in accordance with the flauncy warp advance in weaving, film is cut on special knife bars right on the loom. Figure 162 shows a photograph of a split weaving loom from the film reel inlet side.

Normally, the width of the fabric is determined by the width of the film reels attached as warp supply to the loom. By this, the number of warp tapes per unit width is limited, unless two or more film reels are arranged one on top of the other, allowing an increase in the number of warp tapes. This split weaving technique offers the following advantages: (1) the use of film reels instead of conventional warp beams and the direct slicing of monoaxially stretched film with knife bars on the loom results in a uniform and well balanced warp tension over the whole width of the loom; (2) this technique assures a uniform laying-flat array of warp tapes, leading to a fabric giving perfect coverage at the lowest possible material consumption; (3) the changing of film reels can be performed simply by attaching the end of the empty film reel with adhesive tape to the beginning of the new film reel. The joint will easily pass through the knives of the cutting bar, allowing the weaving operation to proceed without interruption; (4) the transition to other tape widths can be achieved readily in a similar manner; and

Figure 162 Photograph of a split weaving loom seen from its film reel inlet side. (Courtesy of Chemiefaser Lenzing AG.)

(5) the omission of beaming or creel installations results in somewhat reduced floor space requirements for the split-weaving operation and in a reduced labor force.

A weakness of weaving from film reels, however, derives from the fact that in the cutting of warp tapes from film directly on the loom, any defects coming from blunted or damaged knives will be transferred into the final fabric. Careful attention, therefore, has to be given to the quality of the knives. Enhanced warp ruptures reducing the operating efficiency of the weaving and edge incisions on the tapes, resulting in reduced tensile strength of the tape fabric, will be caused by insufficient knife care or extensive knife wear. The split weaving technique is mainly applied in the manufacture of more simple woven fabrics used for articles, such as sacks, covering canvases, or packaging fabrics.

On various flat weaving looms the weft is produced by following different methods: On shuttle looms the weft material is carried through the shed together with the shuttle. The weft tape spools used in weaving on shuttle looms have to be prepared on special reeling machines guaranteeing uniform rolling takeoff performance. Working with rod gripper-type looms, the weft material is drawn-off cylindrical tape packages. On these looms it is not necessary to put the weft tape on special spools. The rod gripper carries the tape through the opened shed to the opposite side of the loom. In using Sulzer looms, weft spools are also not needed. The weft tape is drawn off cylindrical tape packages and shot through the shed by a small gripper shuttle. *Weaving with circular looms* is mainly applied wherever seamless tube fabrics offer advantages. Any kind of sacks, especially large volume sacks for holding up to one metric ton of bulk goods, are preferably manufactured on circular looms.

On these machines the warp tape can be fed to the loom either from warp beams, creels, or film reels cutting the tape with special knife bar attachments before entering the tape feed. In order to avoid overlapping and entanglement of tapes, they are directed into the tape feed unit of the loom with the help of combs. The weft can be carried through the weft opening using several shuttles simultaneously. The weaving ring controls the width of the tube fabric and warp density. The tube width can be altered by exchanging the weaving ring. The uniformity of warp tape tension is achieved by tension compensators. The weft tapes are laid down in the weaving on circular looms perfectly flat, resulting in good fabric coverage and in optimum use of the tape material. Figure 163 shows a photograph of a circular loom working with film tape. The main film tape and film fiber products manufactured on flat or circular looms are: packaging fabrics, sack tube fabrics, primary carpet-backing fabrics, woven geotextiles, filter fabrics, curtains, awnings, sun blinds, sight blinds, table cloth, etc.

### 2. *Processing on Knitting Equipment*

The processing of film tapes or split-film yarns into textile fabrics can also be performed on raschel knitting machines. Raschel knitting is a special variety of the warp knitting technique. Raschel machines can be operated either with film tapes or yarns coming from warp beams, from a creel system, or with tapes cut with knife bars right on the knitting machine. On raschel machines flat webs as well as sacks closed on three sides can be produced.

Raschel machines generally have a bigger output than weaving looms, which is mainly caused by the fact that raschel knits have an open, networklike structure requiring less tape or yarn material per unit area. The high speed of knitting sets high standards for the quality of film tapes or yarns. They must have relatively high breaking strength, reasonably high elongation at break, reduced splitting tendency, and good ability to slide. To date, such properties can only be achieved with tapes made from high-density polyethylene.

Raschel machines for the knitting of sacks where two web layers have to be produced, one on top of the other, work with two knitting devices. The two webs are joined together in longitudinal direction in knitting over a few centimeters width in distances defined by the desired sack dimensions. At regular length, depending on the desired height of the sacks, the two web layers are joined together by knitting transversely across the machine width for a few centimeters in a similar manner. The partly joined double web is then passed through a

Figure 163 Photograph of a circular weaving loom (Lenzing system) operating with film tape. (Courtesy of Chemiefaser Lenzing AG.)

heated cutting device which cuts along the longitudinal joints and subsequently through a second cutting device which separates the sacks across the transverse joints. During cutting, the sheared edges are fused together by the heated cutting device, thus preventing fraying. By changing the count and needles, the texture of the raschel fabric can be modified over a wide range. Also, the size of the sacks can be adjusted as desired. Figure 164 shows a photograph of a raschel knitting machine working from film reels.

## B. Processing on Coating and Laminating Equipment

In many applications of film tape fabrics one- or two-sided coating is required. The coating serves as a vapor or dust barrier and at the same time prevents fabric deformation and tape dislocations. For the coating of film tape fabrics, melt extrusion coating is applied, particularly with low-density polyethylene for fabrics from polyethylene tape and with polypropylene having a higher atactic content for polypropylene tape fabrics. Polymer grades used for extrusion coating usually have lower molecular weight (i.e., higher melt-flow index) than grades used in tape manufacturing.

The basic unit of a melt extrusion coating line consists of: (1) a fabric unwinding unit of roll widths between 2-5 m, equipped with a properly controllable fabric release and guide system, including an adjustable breaking mechanism allowing exact tension control; (2) a preheating unit for the fabric in order to enhance the adhesion of the coat; (3) the horizontally arranged extruder unit equipped with a wide slot die extending over the whole width of the coating machine. In many cases the extruder is mounted on a mobile undercarriage and can be separated from the unit; (4) cooling and pressure rolls, and (5) the changeable fabric winder. Figure 165 shows an operation scheme of such a melt extrusion coating unit.

The operation speeds of extrusion coating units are in the range of 50-100 m/min. The most important process requirements for achieving a well adhering coat of uniform thickness over the whole width of the fabric are: a homogeneous melt, constant extrusion rate over the whole width of the die, a well adjusted die slot, perfect alignment of the wide slot die and of the contact point between melt and material to be coated, good temperature control for all heating and cooling rolls, and a perfect fabric guide system preventing the formation of folds. There are also

**Figure 164** Photograph of a raschel knitting machine working from film reels. (Courtesy of Chemiefaser Lenzing AG.)

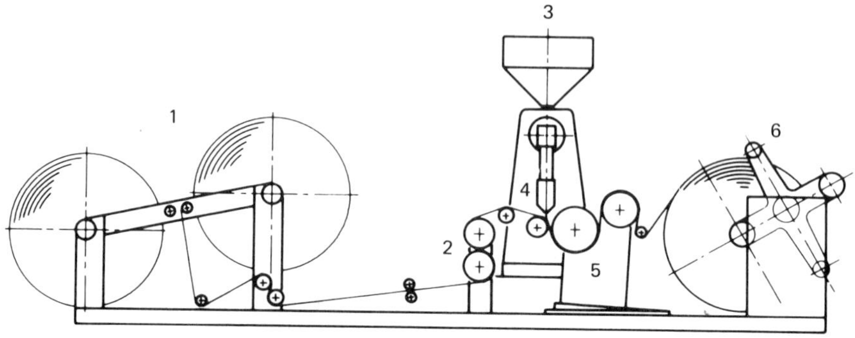


Figure 165 Operation scheme of a melt extrusion coating line: 1, fabric unreel station; 2, preheating rolls; 3, hopper for coating polymer; 4, flat coating die; 5, chill roll; 6, winding station for coated fabric.

melt extrusion coating units offered on the market which allow one to coat tube fabrics produced on circular looms in a one-step operation. Such coated tubular fabrics are used for manufacturing sacks and large volume containers for bulk goods.

Of similar design as the above-described melt extrusion coating units are the melt extrusion laminating installations used for combining tape fabrics with paper, plastic foams, metal foils, etc. Such laminates have gained increasing importance in the manufacture of bags for bulk goods, insulation material for pipes, swimming pool covers, and for ice boxes. The melt extrusion lamination unit has a second unwinding station with the necessary tension controlling release system and a pressure roll unit of heated rollers used for pressing the freshly laminated layers firmly together in order to achieve optimum adhesion. Figure 166 shows a photograph of such a melt extrusion lamination machine.

## C. Processing on Tufting Equipment

Film tapes and split-film yarns are being used to manufacture mats, carpets, and artificial grass substitute products. The processing method most frequently used to make such articles is the tufting process.

A tufted product is a pile fabric which is produced by inserting a face yarn, which may be a conventionally spun bulked continuous filament yarn, a staple fiber yarn, film tapes or split-film fiber yarns, into a backing fabric. The latter can either be a closely woven fabric made of jute, film tapes, or a spun-bonded nonwoven product. The insertion of the pile yarn is done in a sewing-type process with needles. The inserted tufts are generally locked to the backing fabric by a coat

Figure 166 Photograph of a melt extrusion laminating unit. (Courtesy of Chemiefaser Lenzing AG.)

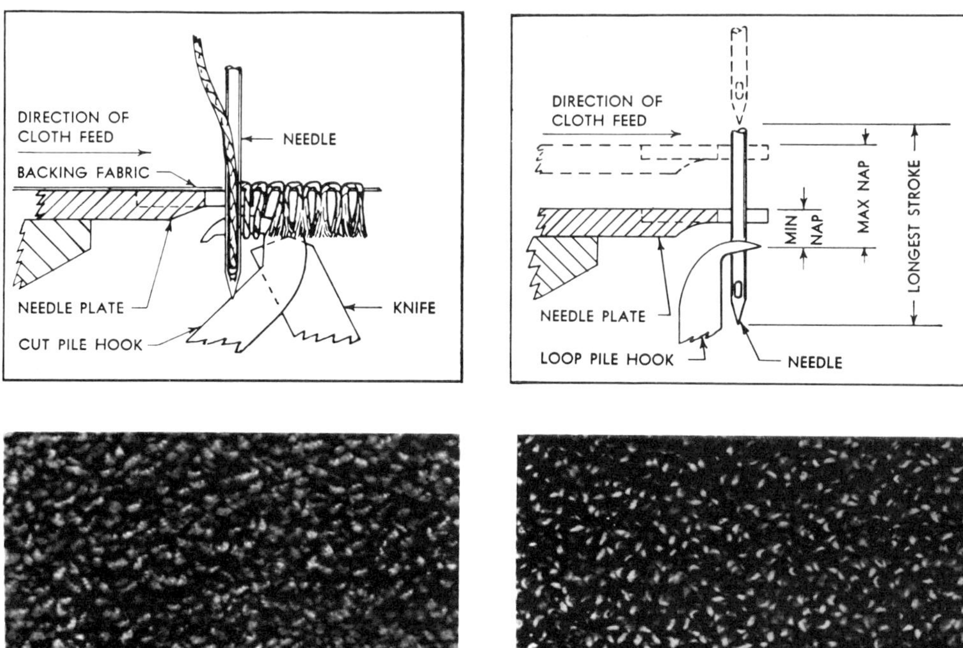


Figure 167 The operation principle of a loop pile tufting machine (right) and a cut pile tufting machine (left).

Figure 168 Photograph of a bag former unit for the manufacturing of flat sawn bags. (Courtesy of Chemiefaser Lenzing AG.)

of vulcanized latex applied to the other side of the primary backing fabric of the mat or the carpet. There are two types of tufting machines, namely those producing loop pile products and those making cut pile. The principle of these two machine types is illustrated in Figure 167.

Loop pile machines, used mainly in the manufacture of fashion styled floor coverings, have in many cases special attachments which produce patterns by controlling the amount of pile yarn fed to the needle in order to produce high and low naps. Also, patterning equipment is in use which produces contrasting areas of loop and cut pile.

## D. Processing on Bag Forming Equipment

One of the major applications of coated and uncoated film tape fabrics is the manufacture of bags for many types of packaged goods. For this purpose, some machine manufacturers have developed bag-forming equipment which works with flat or tubular tape fabrics by cutting to size, folding the bag in case of flat fabrics, and seaming the material either by welding or by sewing. Figure 168 shows a photograph of such a bag-forming unit for manufacturing sewn flat bags.

# 4

# The Prospects of Film Tape and Film Fiber Products

There are four considerations to take into account when judging the future development and prospects of film tape and film fibers with respect to their competition with conventional manmade fiber manufacturing by spinneret spinning: (1) the specific characteristics and properties of film tape and film fibers; (2) the application areas for film tape and film fibers; (3) the competitiveness of film tape and film fiber production methods with respect to technology and costs; and (4) the structure and market aspects of the fiber, textile, and related industries.

## I. PROPERTIES OF FILM TAPE AND SPLIT-FILM YARNS AND FIBERS

In film tape and film fiber processing from thermoplastic polymers, fundamentally, there exists the same flexibility as in conventional fiber production which allows "engineering" of mechanical properties, such as tensile strength, elongation at break, work potential, tendency to further fibrillation, dimensional stability, etc. of the film tapes or film fibers according to end-use requirements by thermal conditions during extrusion, the draw ratio, drawing conditions, and proper relaxation during heat setting.

In general, production modes for film tape and film fiber products allow for achieving mechanical properties similar to conventional fiber manufacturing. Properties of film tapes can be made to vary over a relatively wide range, depending on the end use they are designed for. In Table 23 the mechanical properties of film tapes for various end uses are listed and defined. In this table, taken from Hensen (1979),

**Table 23** Characteristics and Requirements for Film Tapes in Various Fields of Applications

| Application area | Quality requirement | Draw ratio | Dimensions mm (dtex) mechanical properties | | Polymer |
|---|---|---|---|---|---|
| Carpet-backing fabric | Low heat shrinkage<br>High strength<br>Temperature resistance<br>Splitting tendency<br>Dull appearance | 1:1<br>1:5 | Tape width<br>Thickness<br>Titer<br>Tenacity<br>Elongation<br>Shrinkage | 1.2/2.4<br>0.05<br>550/1100<br>4 cN/dtex<br>20%<br>1% (150°) | PP |
| Canvas cover, tarpaulin | High strength | 1:7 | Tape width<br>Thickness<br>Titer<br>Tenacity<br>Elongation | 2.4<br>0.04<br>850<br>5 cN/dtex<br>20% | PP,PE |
| Sacks, bags | High strength<br>High abrasion resistance<br>Controlled elongation | 1:7 | Tape width<br>Thickness<br>Titer<br>Tenacity<br>Elongation | 3.0<br>0.03<br>800<br>4.2 cN/dtex<br>30% | PP,PE |
| Ropes | High tensile strength<br>Controlled elongation<br>Good fibrillation tendency | 1:9<br>to<br>1:11<br>(1:15) | Tape width<br>Thickness<br>Titer | 20-60<br>0.04-0.10<br>15,000-50,000 | PP<br>PP/PE |
| Twine | High tensile strength<br>High knot strength | 1:9<br>to<br>1:11 | Tape width<br>Thickness | 30-80<br>0.03-0.06<br>14,000-30,000 | PP<br>PP/PE |
| Release fabrics | High strength | 1:7 | Tape width<br>Thickness<br>Titer | 2.1<br>0.04<br>750 | PP |
| Filter fabrics | Low heat shrinkage<br>High abrasion resistance | 1:7<br>1:5 | Tape width<br>Thickness<br>Titer | 1.0/2.0<br>0.04<br>350/700 | PP<br>PES |

Table 23. (continued)

| Application area | Quality requirement | Draw ratio | Dimensions mm (dtex) mechanical properties | | Polymer |
|---|---|---|---|---|---|
| Reinforcing fabrics | Low heat shrinkage<br>Controlled elongation<br>Heat resistance | 1:7<br>1:5 | Tape width<br>Thickness<br>Titer | 2.0<br>0.03<br>550 | PP<br>PES |
| Wallpapers, household textiles | UV resistance<br>Low static charging<br>Uniform dyeing<br>Textile handle | 1:7 | Tape width<br>Thickness<br>Titer<br>Tenacity<br>Elongation<br>Shrinkage | 1.2-3.0<br>0.035<br>350-900<br>5.5 g/dtex<br>15%<br>5% (150°) | PE |
| Outdoor carpets | Low heat shrinkage<br>Good wear resistance<br>Weathering resistance | 1:7 | Tape width<br>Thickness<br>Titer | 1.0-12.0<br>0.03-0.06<br>350-900 | PP |
| | Elastic recovery<br>Uniform dyeing | 1:5 | Tape width<br>Thickness<br>Titer | 1.0-3.0<br>0.02<br>300-1000 | PES |
| Decorative tapes | Attractive surface<br>Light weight | 1:6 | Tape width<br>Thickness<br>Titer<br>Density<br>Tenacity | 4.0-12.0<br>0.03-0.05<br>800-3000<br>0.5 g/cm$^3$<br>100 N/mm$^2$ | PP with blowing agent |
| Knitting tapes (for sacks, nets, etc.) | High knot strength<br>Low fibrillation Tendency<br>Suppleness<br>UV-resistance | 1:6<br>1:7 | Tape width<br>Thickness<br>Titer<br>Tenacity<br>Elongation | 2.4<br>0.025<br>550<br>3.5 g/dtex<br>40% | PP, PE |
| Packaging tape | High strength<br>Low fibrillation tendency | 1:9<br>1:7 | Tape width<br>Thickness<br>Tenacity<br>Elongation | 5.0-16.0<br>0.3-0.6<br>5.5 g/dtex<br>15% | PP<br>PES |

Table 23. (continued)

| Application area | Quality requirement | Draw ratio | Dimensions mm (dtex) mechanical properties | | Polymer |
|---|---|---|---|---|---|
| Nonwoven fabrics | Split-film fibers<br>Fiberlike properties | 1:7 | Starting<br>Tape width<br>Thickness<br>Split-fiber filament titer | <br>20-300<br>0.025<br>10-70 | |

*Note*: Tenacities, elongation values, and shrinkage data refer to PP tapes.

the wide variety of properties for film tapes and the wide range of end-use applications is demonstrated. Furthermore, the table indicates that mostly high-density polyethylene, isotactic polypropylene, and in a few cases also polyethyleneterephthalate are being used in tape manufacturing. The main characteristics of film tapes are the dimensions, defined by width, gauge (thickness) and titer (denier or dtex), the tensile strength and elongation at break, heat shrinkage, fibrillation tendency, etc. The conditions of drawing, the draw ratio, and thermal relaxation are the most important means to alter mechanical properties in the desired way.

The mechanical properties of split-film fiber yarns or fibers can be varied over a wide range and are, in the final products, by no means inferior to conventionally produced continuous filament yarns or staple fiber yarns, as proven by the properties of film fiber yarns produced by different methods listed in Table 24 (see Condit and Johnson, 1969; Nichols, 1971). The most apparent and basic difference between spinneret-spun fibers and film fiber products is the cross-sectional characteristic. Spinneret-spun fibers are, in most instances, circular or multilobal in cross section, whereas film fibers have rectangular, "lemon-type" or "banana-type" cross sections. In addition, film fibers are generally less regular in their cross-sectional dimensions, thus having a somewhat wider titer distribution than conventionally produced man-made fibers (see Figures 11, 101, and 170). In the cases of pin-roll fibrillation, film fiber products often have a more or less pronounced network structure and comparatively more loose fiber ends. The properties which will primarily be influenced by these differing geometric characteristics are the softness and handle, the resilience, the bulk, the pilling and shedding, and the fiber-to-fiber friction.

Nott (1971) has published some interesting considerations concerning the effect of cross-sectional difference of fibers with circular and rectangular cross sections with respect to softness or handle. He came to the conclusion that for equal softness the thickness (t) and the width (w) of film fibers should correspond to

$$t = 0.866d \qquad w = 1.047t$$

wherein d is the diameter of a circular fiber of the same titer. If the width of film fibers differs significantly from the thickness, the hand of a textile product will be caused primarily by alignment of the rectangular fibers along their larger face. Table 25, taken from Nott's paper, gives the variations of film fiber dimensions having equal denier with circular fibers for various film thicknesses, and indicates the dimensions for film fibers of equal softness and hand in a given fabric when alternately made from conventionally spun fibers. Film fibers made by uncontrolled mechanical fibrillation, having a relatively wide distributed single fiber width at a given thickness predetermined by the film thickness, have, in practice, dimensional ratios deviating significantly from Nott's rule. Therefore, they show a distinctly harsher

**Table 24** Comparison of Spinneret-Spun Polypropylene Fibers and of Polyolefin Fibers Made by Film-to-Fiber Technology

| Processing technology | Polymer | Single-filament titer (dtex) | Tenacity (gf/dtex) | Elongation (%) |
|---|---|---|---|---|
| Spinneret-spun multifilament | Polypropylene mfi 3.5 | 5 | 7.5 | 18 |
| Mechanically fibrillated | Polypropylene mfi 3.5 | 5-45 | 1.5-4.5 | 11[a] |
| Profiled film (Barfilex) | Polypropylene mfi 3.5 | 16 | 5.2 | 12 |
| Roll embossing (Shell Res.) | Polypropylene mfi 3.5 | 15 | 6.7 | 16 |
| Roll embossing (Shell Res.) | 80% Polypropylene 20% High-impact polystyrene | 23 | 5.7 | 19 |
| Stretch cutting (Chemiefaser Lenzing AG) | Polypropylene mfi 3.5 | 25 | 5.4 | 21[b] |
| Roll embossing (Shell Res.) | H.D. polyethylene mfi 0.4 Density: 0.955 | 23 | 6.5 | 21 |

[a]Depending on fibrillator type and degree of twist.
[b]Heat setting without crimp.

Table 25 Variation of Polypropylene Fibril Dimensions with Denier

| Multifil diameter (μm) | Denier | Cross-sectional area ($mm^2$) | Width of fibrils for film thicknesses of 90 (μm) | 70 (μm) | 45 (μm) | 33 (μm) | 22.5 (μm) | 16.5 (μm) | 11 (μm) |
|---|---|---|---|---|---|---|---|---|---|
| 12.5 | 1 | $0.123 \times 10^{-3}$ | 1.37 | 1.75 | 2.75 | 3.7 | 5.5 | 7.5 | $11^{x}$ |
| 17.7 | 2 | $0.25 \times 10^{-3}$ | 2.75 | 3.5 | 5.5 | 7.5 | 11 | $15^{x}$ | 22.5 |
| 25 | 4 | $0.5 \times 10^{-3}$ | 5.5 | 7 | 11 | 15 | $22^{x}$ | 30 | 45 |
| 35.4 | 8 | $1.0 \times 10^{-3}$ | 11 | 14 | 22 | $30^{x}$ | 44 | 60 | 90 |
| 50 | 16 | $2.0 \times 10^{-3}$ | 22 | 28 | $44^{x}$ | 60 | 88 | 120 | 180 |
| 75 | 36.5 | $4.5 \times 10^{-3}$ | 50 | $64^{x}$ | 100 | 135 | 200 | 270 | 400 |
| 100 | 64 | $7.87 \times 10^{-3}$ | $87.5^{x}$ | 113 | 175 | 240 | 350 | 400 | 720 |

*Note*: x, fibrils of equal softness to multifils of given denier.

handle. For film fibers made by controlled mechanical fibrillation or film cutting and slicing, much narrower width distributions are obtained. However, since the practically achievable distances of the cutting edges of slicing tools, or the distances of the tops of embossing rollers are today in the range of 100-200 μm, a lower titer limit of approximately 10-12 dtex exists for fibrillated fibers made from film, giving good textile hand in agreement with Nott's rule.

The limitation of achieving titers below 10 dtex determines the appropriate place for film fibers with respect to their end uses. In Figure 169, taken from Hensen (1974), an attempt is made to define this condition. The main areas where film tape and film fibers are of interest and have their best potential are the fields of technical fabrics, cords, twines, ropes, tows, nets, heavy upholstery fabrics, carpets, floor coverings, and nonwovens (see Sec. II). Another difference with film fibers in comparison to spinneret-spun conventional fibers is their wider titer distribution. This is particularly true for split-film fibers made by uncontrolled, random mechanical or chemomechanical fibrillation (see Figure 11). Fibers made with processes applying controlled film-to-fiber separation techniques in many cases approach the

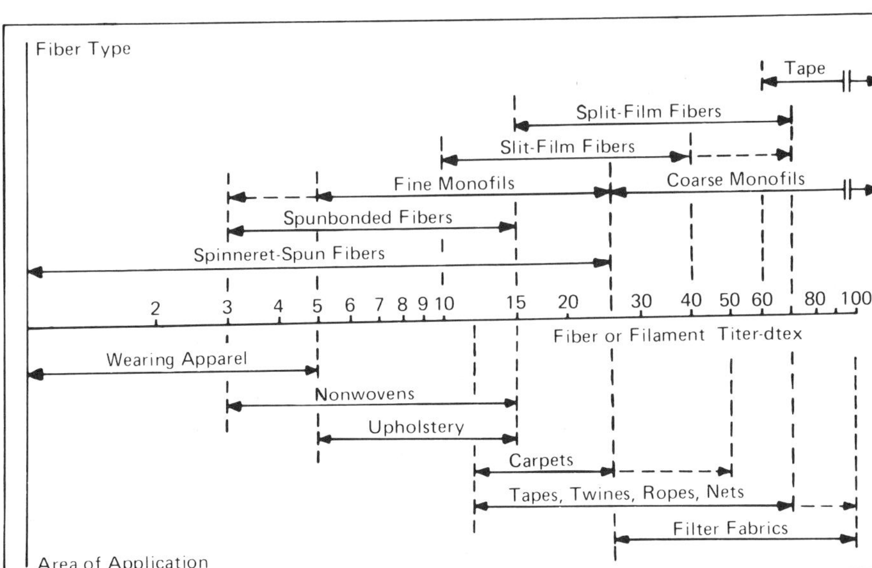


Figure 169 The application areas of various fiber types. (From Hensen, 1974.)

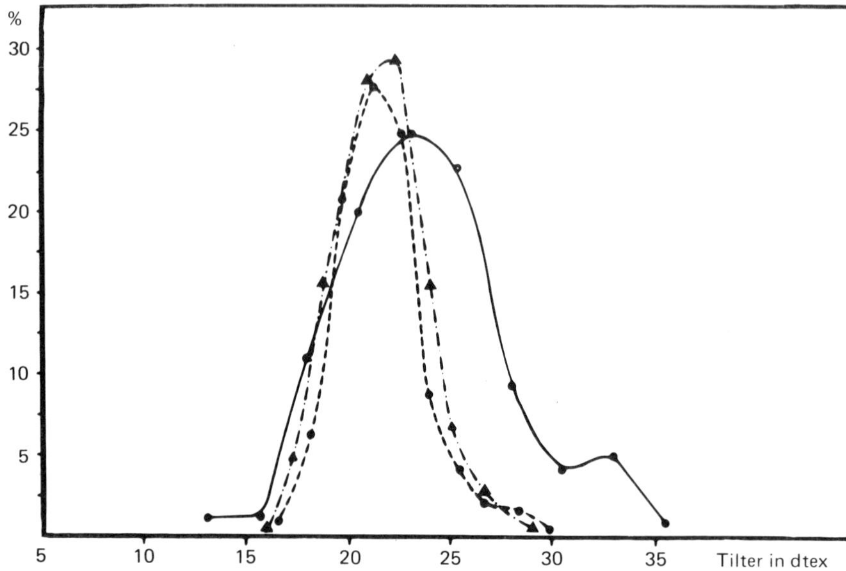


Figure 170 Comparison of the single-filament titer distribution of a slit-film polypropylene carpet yarn with those of conventionally manufactured continuous-filament polyamide and polypropylene carpet yarns. ———, polypropylene slit-film yarn (24.9 dtex); -▲--▲-, spinneret-spun polypropylene yarn (20.8 dtex); -●-●-●-, spinneret-spun polyamide yarn (21.4 dtex). (From Krässig, 1977.)

good uniformity in titer distribution characteristics known for spinneret-spun manmade fibers, as demonstrated in Figure 170.

From theoretical considerations, film fibers of rectangular cross section have a tendency to pack to higher bulk density and, therefore, tend to give less cover than fibers with circular cross section. This will be further enhanced by the wider distribution in single-filament titer. However, in practice the random arrangement of the filament's various titers in yarns, the remainders of network structures, and the presence of fibrillated loose ends will reduce the formation of an optimum bulk density. Furthermore, it is known from practical experience that spinneret-spun fibers of high "aspect ratio" (dimensional aniostropy) have better covering power than yarns made from fibers with substantially circular cross section. With respect to appearance, film fibers in many cases are designated as being "shiny" or "lustrous." In this regard they are similar in appearance to spinneret fibers of multilobal cross section. The incorporation of small amounts of dulling agents such as titanium dioxide, does not work satisfactorily partly due to the formation of shiny cleavage planes during fibrillation.

There is some evidence that multilobal spinneret fibers and fibrillated film fibers, both having a high aspect ratio, have a lower pilling tendency. On the other hand, it is known that low-denier fibers cause more pilling. This could mean that the low-denier fraction in mechanically fibrillated film may give rise to pilling. However, polypropylene constituting the major portion of fibers made by way of film apparently causes few pilling problems. Film fibers generally have a significantly higher fiber-to-fiber friction than circular spinneret fibers. This leads, in ropes or twines, to a much higher tenacity yield. However, the greater friction also gives rise to more shedding in production, more heat development on friction points in processing, and continued fibrillation in use, particularly when high draw ratios have been used to enhance film-to-fiber separation.

Generally, for the testing of mechanical properties, such as tenacity, elongation at break, knot strength, modulus, abrasion resistance, etc., the same or slightly modified methods are being used than those in the testing of spinneret-spun manmade fibers. These methods are well known and often described as basic standard methods, such as recommended by ASTM or DIN norms. Kampf (1976) and Geitel (1976) have described the more frequently used methods to characterize properties of film fibers in special publications. The major European manufacturers of polypropylene fibers made by conventional fiber spinning and by film-to-fiber separation methods are organized in EATP – (European Association of Textile Polyolefins, Paris) and are also working together to standardize test methods.

## II. AREAS OF APPLICATION FOR FILM TAPE AND FILM YARNS AND FIBERS

The application of film tapes and film fibers are numerous and manifold. In the framework of this book, it is impossible to discuss in detail the wide variety of applications of film tape and film fibers and the problems the end user encounters in processing, conversion, the use of these products, and the goods made from them. In the following, the major applications are listed without claim of completeness.

*Flat film tapes* are used in the form of woven fabrics, knits, or raschels, sometimes having plastic coats (see Copper et al., 1971; Mackenzie, 1975) or laminated with paper, films, aluminum foil, spunbondeds, or nonwovens in the following applications: bags and nets for household, agriculture, industrial packaging, transportation, or storage needs (see Burggraf, 1967; Broatch, 1968; Cook, 1968; Kaneko, 1968; Marks, 1968; Overton, 1968; Weber, 1968; Mayer, 1969; Moorwessel and Pilz, 1969; Real, 1969; Barmag-Barmer Maschinenfabrik AG, 1971b; De Monte, 1971; Kerr, 1971; Krause, 1971; *Hydrocarbon Process*, 1974; McLellan, 1975). An example for the use of raschel bags in the harvesting of agricultural goods and of various fabric constructions are given in Figure 171.

Figure 171 The use of raschel-knitted bags in the harvesting of agricultural goods (top); examples of raschel-knit fabrics (bottom). (Courtesy of Chemiefaser Lenzing AG.)

*Fabrics for baling* natural and manmade fibers and for other wrapping purposes (see Möller, 1967; Farnworth and Pemberton, 1968; Weber, 1968; Mayer, 1969; Real, 1969; McLellan, 1975). Figure 172 illustrates this interesting application of low-density polyethylene coated fabrics made from high-density polyethylene tapes.

*Covers, tarpaulins, awnings, silo liners*, etc. (see Moorwessel and Pilz, 1969; Weber, 1969; Krause, 1971). A special application is by use of roof underliners to prevent the penetration of dust and snow. Figure 173 illustrates this.

*Sun shades, wind screens, sight protections, snow fences*, etc. (see Moorwessel and Pilz, 1969; Real, 1969; Weber, 1968; Krause, 1971). Such sight protection application on a tennis court is demonstrated with Figure 174.

*Primary carpet backing* for tufted carpets or base fabrics for needle-punched floor coverings (see Möller, 1967; Broatch, 1968; Stout, 1968; Moorwessel and Pilz, 1969; Real, 1969; Welge, 1970; Badrian and Choufoer, 1971; Barmag-Barmer Maschinenfabrik AG, 1971b; Kerr, 1971; Ward, 1973; *Chemiefasern + Text.—Anwendungstech./Text. Ind.*, 1974; Hoevel, 1975).

*Upholstery fabrics, seat covers, mats, protective coverings, table covers, table sets, book bindings*, etc. (see Marks, 1968; Weber, 1968; Krause, 1971). Figure 175 illustrates the use of tape fabric laminates with plastic foams and aluminum foils as protective coverings of swimming pools.

*Hats, caps, protective work clothing*, such as aprons, wraps, etc. (see Weber, 1968). *Portfolios, pouches, suitcases*, etc. Base fabrics for nonwoven felts used as filters, insulators, etc. (see Moorwessel and Pilz, 1969; Schuur, 1969). *Base fabrics* and *pile material in artificial grass* manufacturing. *Geo-textiles* used in road construction or bank stabilization. The use of tape fabrics in this fast growing field is illustrated by the photograph in Figure 176.

*Twisted film tapes* are used whenever heavier fabrics with improved use characteristics are required in articles such as: *filter fabrics* (see Moorwessel and Pilz, 1969; Schuur, 1969). *High quality woven or knitted sacks and bags. High volume containers* for bulk goods, such as fertilizers, salts, etc., holding up to one metric ton are being used in growing numbers. This application of film tapes is illustrated in Figure 177. *Upholstery material* for garden and camping use. *Coated woven fabrics for awnings* and protective purposes (see Schuur, 1969). *Massage aids.*

*Fibrillated film yarns* are used in the manufacture of the following goods: *strings, cords*, and *laces* for various types of packaging use, *agricultural twine* for baling purposes, etc. (see *Chem. Ind.* 1966; Burggraf, 1967; Courvoisier, 1967; Möller, 1967; Broatch, 1968; Mackie, 1968; Taylor, 1968; Badrian and Choufoer, 1971; Beck, 1971; Skoroszewski, 1971; Wills, 1971). *Ropes* and *cables* for a large number of end uses (see Burggraf, 1967; Möller, 1967; Broatch, 1968; Parsey, 1968; Real, 1969; Badrian and Choufoer, 1971; Beck, 1971;

Figure 172 The application of coated tape fabrics in the baling of man-made staple fibers (top); example of a woven tape fabric (bottom). (Courtesy of Chemiefaser Lenzing AG.)

Figure 173 Illustration of the use of flame-retardant coated tape fabrics as roof underliners. (Courtest of Chemiefaser Lenzing AG.)

Figure 174 Coated tape fabrics used as sight protection on a tennis court. (Courtesy of Chemiefaser Lenzing AG.)

Figure 175 Use of tape fabric laminates with plastic foam and aluminum foil as swimming pool covers. (Courtesy of Chemiefaser Lenzing AG.)

Skoroszewski, 1971; Wills, 1971). Figure 178 shows some cord, rope, and cable products manufactured from fibrillating polyolefin film tapes. *Girdles* or *bolts*; *fish nets* and *sporting nets*; *mats* and *floor coverings* for outdoor and wet room applications. *Tufted* and *woven carpets* in the form of texturized face yarns or as chain or stuffer yarns (see Condit and Johnson, 1969; Real, 1969; Badrian and Choufoer, 1971; Nichols, 1971; Slack, 1973; Harms et al., 1973b, Krässig, 1975a, 1975b; Anderson, 1974; Cable, 1975; Ten Hoevel, 1975). In Figure 179 the application of polypropylene film tape fabrics as primary carpet backing and of fibrillated film fiber yarns as face material in tufted carpets is illustrated. *Fabrics for industrial uses*, such as filters, base fabrics in belts, felts, nonwovens, etc.

*Split- or sliced-film staple fibers* are applied in the manufacture of the following products: needle-punched or binder-bonded *nonwovens for floor coverings*, *filters* of many kinds, *insulations*, etc. (Schuur, 1969; Badrian and Choufoer, 1971; Dykes, 1971). *Spun yarns* (alone or in blends with other textile fibers), e.g., *for upholstery fabrics* or in the manufacture of knitted apparel (see Klust, 1968; Real, 1969; Schuur, 1969; Badrian and Choufoer, 1971; Nichols, 1971; Stout, 1971; McMeekin et al., 1971; Anderson, 1974). *Reinforcement of concrete* (see *Chem. Weekbl.*, 1969; Zonsveld, 1971). *Cigarette filters* (see Eastman Kodak Co., 1969, 1970).

The processing of film tapes and fibrillated film yarns or fibers in winding (Hensen, 1968; Mackie, 1968), twisting (Mackie, 1968), warp beaming (Mackie, 1968; Norton, 1968; Shenton, 1968), weaving and knitting (Phillips Petroleum Co., 1966; Burggraf, 1967; *Br. Plast.* 1968; Lennox-Kerr, 1968, Mackie, 1968; Marks, 1968; Shenton, 1968; Thomas, 1968; Zalewski, 1969; Chemiefaser Lenzing AG, 1972b; Anderson, 1974; Mackenzie, 1975, 1976; Wheatley, 1975; Waddington, 1975; Noe. 1979) have their specific problems. They were to some extent discussed in Chapter 3, Secs. VIII.B-D., XI.A., and XI.C.

Figure 176 The application of polyolefin tape fabrics as geotextiles in a bank stabilization system. (Courtesy of Patchogue Plymouth Division of Amoco Deutschland GmbH.)

Figure 177 Illustration of the use of large-volume containers from tape fabrics for the transport of fertilizers in the agriculture. (Courtesy of Windmöller and Hölscher, Lengerich/Westf.)

Figure 178 The use of film tapes and split-film yarns in cords, ropes, and cables. (Courtesy of Barmag-Barmer Maschinenfabrik AG.)

Figure 179 Illustration of the use of carpet backing tape fabrics and of split-film face yarn in the manufacture of tufted carpets. (Courtesy of Barmag-Barmer Maschinenfabrik AG.)

## III. FILM FIBER PRODUCTION METHODS AND ECONOMIC ASPECTS

Spinneret spinning of manmade fibers and the manufacture of film tape or split-film fibers are both based on the same technological processes, namely melt extrusion through a die, cooling, and quenching, orientation by drawing, heat setting, and collecting or winding. From a basic point of view, there are no additional processing steps in either the spinneret spinning or in the film-to-fiber process which could be considered specific only for one or the other of the two techniques.

The most obvious differences, however, between spinneret and film-fiber manufacturing are the following:

1. Spinneret processes by melt spinning were, until recently, usually discontinuous, whereas film tape and film-fiber processing is

performed in line. Spin drawing and spin texturizing techniques are presently changing this in the field of continuous filament yarn manufacturing.

2. Spinneret-spun filament yarns and staple fibers are generally produced on a much larger scale than film tape or film-fibers. There are still some people who consider film-fiber processing as the fiber production methods of the "poor man." They believe that even small or medium size fiber end users, such as carpet manufacturers, could competitively use film-to-fiber production methods to satisfy their own fiber needs. However, this is no longer valid, at least not in highly industrialized countries. Film-to-fiber production in these countries has to be performed on larger scale units in order to be competitive.

In spinneret fiber production it is possible to extrude and cool the unstretched primary filament much faster than one can orient, wind up, and handle the oriented fiber. This was the reason spinneret fiber production with the melt spinning technique was until now divided into two processing steps, namely fiber spinning as the first, and drawing, heat setting, and crimping as the second production steps.

Film-to-fiber production processes, on the other hand, are rarely divided into two processing steps. The extrusion speed for film production, nowadays approximately 10-50 m/min (= 30-150 ft/min), depending on film thickness, dictates the production speed ranging from 40 up to 300 m/min (120-900 ft/min), depending on the draw ratio applied. Such speeds in drawing, heat setting, and winding are easily handled with today's technical know-how. Increase in extrusion speed and consequently in line speed, however, do adversely influence film quality and film uniformity, thus lowering the quality of the resulting tapes or split-film yarns or fibers. Furthermore, higher speeds in extrusion would call for more expensive metering equipment to balance the loss otherwise experienced in uniformity. It is hard to judge whether the use of a two-step process concept, applying much higher speeds in film production followed by tape cutting, drawing, setting, and collecting in a second processing step under present line speed conditions, might be a necessary future step to assure a long-range competitive position of the film-to-fiber technology versus the novel spin-stretch and spin-texturizing technology in spinneret spinning. Surely, such a separation of the film-to-fiber production into two processing steps would eliminate major present advantages, namely the in-line character and the relative simplicity of the equipment.

A number of people have tried to make comparisons of production costs between conventional fiber spinning and film-to-fiber processing. Principally, all of them have come out with practically the same result, namely that the production costs for film tape or film fibers are lower for producing high-denier products and that they increase much more rapidly in attempting to product low-denier products than in the case of spinneret-spun fibers. This is due mainly to the fact that in spinneret spinning one can generally balance the loss in melt discharge,

Table 26 Production Costs of Polypropylene Spinneret and Film Fiber[a]

| | Film fiber | | | Spinneret fiber | | |
|---|---|---|---|---|---|---|
| Fibril/fiber denier | 15 | 6 | 3 | 15 | 6 | 3 |
| Capital ($) | 45,000 | 45,000 | 45,000 | 100,000 | 100,000 | 100,000 |
| Drawn film thickness or fiber diameter (μm) | 45 | 27.5 | 20 | 48.5 | 30.5 | 22 |
| Output (kg/h) | 90 | 55 | 40 | 90 | 85 | 88 |
| Output, tons/an (assume 5000 h/an) | 450 | 275 | 200 | 450 | 425 | 400 |
| Material cost ($/t) | 256 | 256 | 256 | 260 | 260 | 260 |
| Direct labor, (no. of workers) | 2 | $2\frac{1}{2}$ | 3 | 3 | $3\frac{1}{2}$ | 4 |
| Maintenance (no. of workers) $\frac{1}{2}$ | $\frac{1}{2}$ | 1 | $1\frac{1}{2}$ | 1 | $1\frac{1}{2}$ | 2 |
| *Cost per hour* | $ | $ | $ | $ | $ | $ |
| Raw material | 23.05 | 14.10 | 10.25 | 23.40 | 22.10 | 20.80 |
| Amortization over 4 years | 3.25 | 3.25 | 3.25 | 5.00 | 6.00 | 6.00 |
| Return on capital at 20% pa | 2.60 | 2.60 | 2.60 | 4.80 | 4.80 | 4.80 |
| Labor at $1.25/hr | 3.12 | 4.37 | 5.62 | 5.00 | 6.25 | 7.50 |
| Power and services | 0.50 | 0.50 | 0.50 | 0.50 | 0.50 | 0.50 |
| Building | 0.50 | 0.50 | 0.50 | 0.60 | 0.60 | 0.60 |
| Spares | 0.75 | 0.75 | 0.75 | 0.75 | 0.75 | 0.75 |
| Packages | 0.65 | 0.40 | 0.29 | 0.65 | 0.61 | 0.58 |
| Direct overheads and insurance | 0.75 | 0.75 | 0.75 | 0.75 | 0.75 | 0.75 |
| Total per hour | 35.17 | 27.22 | 24.51 | 42.45 | 42.36 | 42.28 |
| Production cost per kg | 0.39 | 0.50 | 0.61 | 0.47 | 0.50 | 0.53 |

[a] English pound per June 30, 1971 = 2.42 U.S. dollar.

caused by the smaller spinneret orifices, by applying spinneret assemblies with a higher number of holes. In film production, however, one is limited by the practical considerations of film handling to particular flat die widths or circular die diameters. One of the best comparisons found in literature was made by Nott (1971) and Nott and Sarwar (1975). The results of their analysis are presented in Table 26.

From the data contained in Table 26, based on 1971 currency value one can derive the fact that the production costs of film fiber production rise above those of spinneret fibers, under the assumptions made by Nott, when the titer of the film fibers is lower than 6.5 dtex (i.e., 6 den). This result is another argument that in the production of film fiber products one should not attempt to produce low-denier types. This supports the conclusion that it is likely that film fiber production will be limited in the future to products in the higher denier range.

## IV. INDUSTRY AND MARKET STRUCTURE

It is only natural that the main emphasis at the beginning of the development of new technologies is placed on improving process techniques. In this stage, product development is directed towards technologically less demanding applications. This was the reason why film tape and film fiber applications began in the field of manufacturing twines and ropes, later penetrating into markets for technical fabrics, such as primary carpet backing and base fabrics for needle-punched felts and for packaging fabrics. Only during the last decade has the attempt been made to extend into more demanding textile end uses, such as pile yarns for carpets, yarns for knitted goods, etc. However, the sometimes exaggerated expectations have been corrected to a more realistic view of the market potential of film tape and especially of film fibers.

The prevailing opinion today is that, both from a market point of view and from the structure of the fiber industry, the proper route for film fiber products may lay in the careful choice of and concentration on appropriate fields of application, where cheap fiber materials or small-volume specialties are involved. In the denier range of approximately six and higher, there are sufficient areas of potential applications for split-film yarns and fibers where their special property characteristics and lower production costs give them an advantage over spinneret-spun fibers (see Wood-Walker, 1975).

One further aspect should be considered carefully. As the film fiber moves ahead into more elaborate areas of textile use, the competition between large producers of conventionally spun manmade fibers and small film fiber manufacturers will become more actue. In more advanced areas, small film-to-fiber manufacturers may no longer be able to compete with large organizations possessing excellently equipped research departments operated by experienced research personnel.

# Bibliography

## BIBLIOGRAPHY


Achermann, A. (1982) *Lenzinger Ber.* *52* (2/1982), 35.

Akay, M. (1979) *Processing Structure of Polypropylene Films*, 2nd Int. Conference: Polypropylene for Fibers and Textiles, York, England, Sept. 26-28.

Albers, J. H. M. and Duiser, J. A. (1970) *Textil.* *29* (10), 742.

Allied Chemical Corp. (1969a) Inv.: Prevorsek, D. C., Lamb, G. E. R., and Oswald, H. J., *German Pat. 1,807,145*, June 19, 1969; U.S. Appl. Nov. 6, 1967.

Allied Chemical Corp. (1969b) Inv.: Prevorsek, D. C. and Lamb, G. E. R. *German Pat. 1,807,143*, Dec. 11, 1969; U.S. Appl. Nov. 11, 1967; *U.S. Pat. 3,539,663*, Nov. 10, 1970.

Allied Chemical Corp. (1969c) Inv.: Lamb, G. R., Prevorsek, D. C., Kotliar, A. M., and Oswald, H. J., *French Pat. 2,023,665*, Oct. 22, 1969; U.S. Appl. Nov. 20, 1968.

Allied Chemical Corp. (1972) Inv.: Lamb , G. E. R., Prevorsek, D. C., Testa, A., and Oswald, H. J., *Brit. Pat. 1,260,957*, Jan. 19, 1972; U.S. Appl. March 13, 1968.

Allied Chemical Corp. (1973) *Brit. Pat. 1,308,998*, March 7, 1973; U.S. Appl. July 10, 1969.

Anderson, P. G. (1974) *Text. Inst. Ind.* *12* (7), 208.

Andrew, W. (1975) *Textilbetrieb* *93* (6), 33.

Asahi Chemical Industries Co., Ltd. (1976) Inv.: Teraoka, T., Satake, K., and Sone, T., *Japan Kokai 76-12,273*, Feb. 12, 1976.

Asahi Dow, Ltd. (1972) *Japan Kokai 72-50-015*, Dec. 15, 1972.

Asahi Kasei Kogyo K. K. (1968) Inv.: Kaisha, Kabushiki, *U.S. Pat. 3,398,441*, Aug. 17, 1968.

Asahi Kasei Kogyo K. K. (1972) *German Pat. 1,785,412*, April 13, 1972; Japan Appl. Sept. 21, 1970.

Aspin, F. M. (1975) *Development in Roll Embossing Machines to Produce Fibers from Film*, Conference Polypropylene Fibers in Textiles, York, Sept. 20-Oct. 1, 1975.

Aspin, A. F. and Wall, D. W. (1979) *The Properties of Roll Embossed Yarns*, 2nd Conference Polypropylene Fibers and Textiles, York, Sept. 26-28, 1979.

Badische Anilin & Soda Fabrik AG (1970a) Inv.: Trieschmann, H. G., Moorwessel, D., Merkel, H., and Pilz, G., *German Pat. 1,813,652*, July 2, 1970.

Badische Anilin & Soda Fabrik AG (1970b) Inv.: Trieschmann, H. G., Moorwessel, D., Merkel, H., and Pilz, G., *German Pat. 1,813,649*, July 9, 1970.

Badische Anilin & Soda Fabrik AG (1970c) Inv.: Trieschmann, H. G., Moorwessel, D., Merkel, H., and Pilz, G., *German Pat. 1,813,650*, July 2, 1970.

Badische Anilin & Soda Fabrik AG (1970d) Inv.: Trieschmann, H. G., Moorwessel, D., Merkel, H., and Pilz, G., *German Pat. 1,813,651*, July 9, 1970.

Badische Anilin & Soda Fabrik AG (1970e) Inv.: Trieschmann, H. G., Müller-Tamm, H., Moorwessel, D., Ball, W., and Pilz, G., *German Pat. 1,917,654*, Oct. 15, 1970.

Badische Anilin & Soda Fabrik AG (1970f) Inv.: Wisseroth, K., Herbeck, R., Scholl, R., and Trieschmann, H. G., *German Offen. 1,905,835*, Sept. 3, 1970.

Badische Anilin & Soda Fabrik AG (1973) *Brit. Pat. 1,305,653*, Feb. 7, 1973; German Appl. April 5, 1969.

Badische Anilin & Soda Fabrik AG (1974) Inv.: Trieschmann, H. G., Müller-Tamm, H., Moorwessel, D., Pilz, G., and Rau, W., *German Offen, 2,034,024*, Aug. 8, 1974.

Badische Anilin & Soda Fabrik AG (1976) Inv.: Schneider, P., Schick, H., Müller-Tamm, H., and Hennenberger, P., *German Offen. 2,441,541*, April 15, 1976.

Badonsky, J. (1960) *SEP-Journal 16*, 1303.

Badrian, W. H. J. and Choufoer, J. H. L. (1971a) *Text. Production 1971* (8), 27.

Badrian, W. H. J. and Choufoer, J. H. L. (1971b) *Text. Manuf. 97* (7) 263.

Badrian, W. H. J. (1974) *Mod. Text. Mag. 55* (7), 58.

Bakelite Xylonite, Ltd. (1968) *Brit. Pat. 1,104,694*, Feb. 28, 1968.

Balk, H. (1978) *Chemiefasern + Text.-Anwendungstech. /Text. Ind. 28* (6), 527.

Balslev, J. (1971) *Process and End Product Comparisons Between Film Production Methods of Extrusion, Cooling, Stretching and Annealing*, Conference Textiles from Film II, Manchester, July 6-7, 1971.

Baltá-Calleja, F. J. and Peterlin, A. (1971) *Makromol. Chem. 141*, 91.

Barbe, G. and Sieroff, M. (1977), *Lenzinger Ber. 42* (Jan.), 158.

Barmag-Barmer Maschinenfabrik AG (1971a) *German Pat. 1,660,230*, April 29, 1971; *Austrian Pat. 314,063; U.S. Pat. 3,594,870.*

Barmag-Barmer Maschinenfabrik AG (1971b) Inv.: Schippers, H. and Hessenbruch, R., *German Offen. 1,660,231*, July 1, 1971.

Barmag-Barmer Maschinenfabrik AG (1971c) *Int. Text. Bull. /Spinnerei 3/1971*, 305.

Barmag-Barmer Maschinenfabrik AG (1972) *German Pat. 2,127,792*, Dec. 14, 1972.

Barmag-Barmer Maschinenfabrik AG (1973a) *PRT Polym. Age 4* (10), 370.

Barmag-Barmer Maschinenfabrik AG (1973b) Inv.: Schippers, H., Jensen, F., Koslowski, G., and Czerwon, D., *German Pat. 2,328,639*, June 6, 1973.

Barmag-Barmer Maschinenfabrik AG (1974) *German Pat. 1,660,223*, Oct. 31, 1974.

Bash, D. P. (1978) *Fiber Producer 1978* (4), 24.

Beck, T. (1971) *Magy. Tech. 1971* (1), 34.

Becker, H. (1962) *German Pat. 1,127,535*, April 12, 1962.

Bemberg AG (1970) *French Pat. 1,599,551*, Aug. 21, 1970; German Appl. Jan. 3, 1968.

Bemberg AG (1971) *German Pat. 1,710,608*, March 11, 1971; *Brit. Pat. 1,246,052.*

Berger, W. and Kammer, H.-W. (1974) *Faserforsch. Textiltech. 25* (4), 140.

Berger, W. and Kammer, H.-W. (1978) *Textiltechnik 28* (6), 344.

Berger, W. and Kammer, H.-W. (1979) *Textiltechnik 29* (2), 93.

Berger, W. and Michels, Ch. (1975) *Faserforsch. Textiltech. 26* (7), 340.

Berger, W. and Schmack, G. (1974) *Textiltechnik 24* (1), 36.

Berger, W., Grossmann, H., and Schmack, G. (1976) *Faserforsch. Textiltech. 27* (2), 81.

Bergonzoni, A. and Discresce, A. J. (1965) *SPE Techn. Papers 11*, VI 1-2.

Botros, R. (1979) *Nickel Chelating Dyestuffs for Printing Polypropylene*, 2nd Int. Conference Polypropylene Fibers and Textiles, York, Sept. 26-28, 1979.

Bouriot, P., Hagege R., and Sotton, M. (1976) *Bull. Sci. Inst. Text. Fr. 5* (18), 139.

Brändle, K. (1977) *German Offen. 2,551,393*, May 18, 1977.

Brändle, K. (1978) *German Offen. 2,725,348*, Dec. 7, 1978.

Brandrup, J. and Immergut, E. H. (1966) *Polymer Handbook*, Interscience, New York.

Bransom Instruments, Inc. (1967) Inv.: Soloff, R. S., *U.S. Pat. 3,331,719*, July 18, 1967.

Braun, S. (1975) *New Installations for the Production of Low Shrinkage Film Tape*, Conference Polypropylene Fibers in Textiles, York, Sept. 30-Oct. 1, 1975.

Briston, J. (1972) *Converter 9* (7), 14.

Briston, J. H. and Katan, L. L. (1974) *Plastic Films*, Iliffe Books, London.

Brittain, D. J. (1968) *Text. Inst. Ind.* *6* (10), 271.

Broatch, W. N. (1968) *Polypropylene Tapes for Stretched Tape Outlets; Cord and Fabric Performance*, Conference Textiles from Film, Manchester, April 4-5, 1968.

*Br. Plast.* (1967) *40* (9), 92.

*Br. Plast.* (1968) *41* (2), 165.

Brydson, J. A. (1970) *Flow Properties of Polymer Melts*, Iliffe Books, London.

Burggraf, K. (1967) *Kunststoffe 57* (2), 78.

Cable, J. (1975) *Carpet Backing Yarns in Woven Carpets*, Conference Polypropylene Fibres in Textiles, York, Sept. 30-Oct. 1, 1975.

*Cah. Text. Cenatra* (1971) *8* (5), 43.

Campbell, I. A. and Skoroszewski, W. H. (1972) *Text. Inst. Ind.* *5* (6), 170.

*Can. Text. J.* (1975) *92* (3), 36.

Cansfield, D. L. M., Capaccio, G., and Ward, I. M. (1976) *Polym. Eng. Sci.* *16* (11), 721.

Celanese Corp. (1969) *Brit. Pat. 1,165,934*, Oct. 1, 1969; U.S. Appl. Sept. 12, 1966.

Celanese Corp. (1971a) *German Pat. 1,660,259*, April 8, 1971; U.S. Appl. Sept. 12, 1966.

Celanese Corp. (1971b) Inv.: Chopra, S. N. and Turmel, H. M., *U.S. Pat. 3,576,931*, April 27, 1971.

Celanese Corp. (1971c) Inv.: Keuchel, H. W., *Brit. Pat. 1,231,866*, May 12, 1971.

Celanese Corp. (1971d) *Brit. Pat. 1,242,346*, Aug. 11, 1971; U.S. Appl. March 1, 1968.

Chemiefaser Lenzing AG (1970) Inv.: Sasshofer, F. and Krässig, H., *Belg. Pat. 744,984*, March 31, 1970; Austrian Appl. Jan. 28, 1969; *Brit. Pat. 1,287,874; French Pat. 2,029,593; Austrian Pat. 290,710; Swiss Pat. 525,300; U.S.S.R. Pat. 344,652* and *393,837.*

Chemiefaser Lenzing AG (1972a) Inv.: Eichler, J. A. and Plammer, A. *Austrian Pat. 302,634*, Oct. 25, 1972.

Chemiefaser Lenzing AG (1972b) Inv.: Eichler, J. A. and Plammer, A. *Austrian Pat. 302,214*, Oct. 10, 1972; *Brit. Pat. 1,202,035; U.S. Pat. 3,645,299.*

Chemiefaser Lenzing AG (1972c) Inv.: Sasshofer, F. and Krässig, H., *U.S. Pat. 3,676,243*; Austrian Appl. Nov. 21, 1969; *Brit. Pat. 1,326,772; French Pat. 2,068,572; Swiss Pat. 529,248.*

Chemiefaser Lenzing AG (1973) Inv.: Eichler, J. A. and Plammer, A. *Austrian Pat. 311,666*, Nov. 26, 1971; *Brit. Pat. 1,275,866; French Pat. 2,088,978.*

Chemiefaser Lenzing AG (1975) Inv.: Plammer, A., *Austrian Pat. 330,332.*

Chemiefaser Lenzing AG (1977) Inv.: Plammer, A., *Austrian Pat. 340,335*, Dec. 12, 1977.

Chemiefaser Lenzing AG (1979) Inv.: Seyfriedsberger, F., *Austrian Pat. 354,335*, May 15, 1979.

Chemiefasern (1970) *20* (6), 468.

*Chemiefasern + Text.-Anwendungstech./Text. Ind.* (1972) *22* (3), 239.

*Chemiefasern + Text.-Anwendungstech./Text. Ind.* (1974a) *24* (7), 566.

*Chemiefasern + Text.-Anwendungstech./Text. Ind.* (1974b) *24* (9), 720.

*Chemiefasern + Text.-Anwendungstech./Text. Ind.* (1974c) *24* (8), 607.

Chemie Linz AG (1976) Inv.: Haltrich. H. and Blauhut, E., *Austrian Pat. 330,334*, June 15, 1976.

*Chem. Ind.* (1966) *18* (4), 214.

*Chem. Weekbl.* (1969) *65* (3), 9.

Chevron Research Co. (1968) *Brit. Pat. 1,128,274,* Sept. 25, 1968; U.S. Appl. March 31, 1966; *German Pat. 1,704,532.*

Chevron Research Co. (1970) Inv.: Salmela, J. M., *Brit. Pat. 1,213,269,* Nov. 25, 1970; U.S. Appl. July 3, 1968; *Austrian Pat. 315,351.*

Chevron Research Co. (1973a) Inv.: Salmela, J. M., *Brit. Pat. 1,327,459,* Aug. 22, 1973; U.S. Appl. Jan. 26, 1970.

Chevron Research Co. (1973b) Inv.: Guenther, L. M., *U.S. Pat. 3,756,484*, Sept. 4, 1973.

Chori Co., Ltd. (1972a) *German Pat. 1,779,290*, April 27, 1972; Japan Appl. July 26, 1967.

Chori Co., Ltd. (1972b) *German Pat. 1,660,253*, Dec. 14, 1972.

Choufoer, J. H. L. (1975) *Recent Developments in Mechanical Fibrillation of Polypropylene Film*, Conference Polypropylene in Textiles, York, Sept 30-Oct. 1, 1975.

Clark, A., Hogan, J. P., Banks, R. L., and Lanning W. C. (1956) *Ind. Eng. Chem. 48,* 1152.

Clark, A., Hogan, J. P., Banks, R. L., and Lanning, W. C. (1976) *Chemtechnol. 1976* (11), 694.

Clayton, C. H. (1976) *Text. Asia 7* (6), 42.

Coen, A. and Zilio, G. (1970) *Mat. Plast. Elast. 36* (6), 633.

Condit, P. C. and Johnson, G. B. (1969) *Mod. Text. Mag. 50* (8), 23.

Consolidated Bathwirt, Ltd. (1973) Inv.: Blais, B., *German Offen. 2,232,758*, Jan. 25, 1973.

Continental Linoleum Union (1971) Inv.: Berglund, P. H. S., Fanto-Kurtos, P. R., and Mattson, N. A. U., *French Pat. 2,072,315*, Oct. 29, 1971; Swed. Appl. Nov. 24, 1969; *Brit. Pat. 1,329,426.*

Cook, B. (1968) *Sacks for the Post Office*, Conference Textiles from Film, Manchester, April 4-5, 1968.

Cooper, M. L., Groom, S., Hopper, D., and Viney, M. J. (1971) *The Coating of Woven Polypropylene Fabrics*, Conference Textiles from Film II, Manchester, July 6-7, 1971.

Cooper, M. L. (1975) *Low Shrinkage Polypropylene Yarns*, Conference Polypropylene Fibers in Textiles, York, Sept. 30-Oct. 1, 1975.

Courtaulds, Ltd. (1968) *German Pat. 1,435,407*, Dec. 12, 1968; Brit. Appl. May 14, 1962 and March 1, 1963.

Courtaulds, Ltd. (1971) Inv.: Adams, G. T., *Brit. Pat. 1,234,782*, June 9, 1971.

Courvoisier, J. D. (1967) *Plast. Mod. Elastomers 19* (5), 68, 111.

Darragh, J. L. and Johnson, G. B. (1968) *Mod. Text. Mag. 49* (9), 24.

Dawczynski, H., Sattler, W., Michels, Ch., Franz, H., and Markel, K. (1978) *Lenzinger Ber. 45*, 132.

Deanin, R. D. and Sansone, M. F., (1978) *Polymer Reprints 19* (No. 1/March), 211.

De Monte, N. T. H. (1971) *Text. Inst. Ind. 9* (5), 15.

Diedrich, B. (1975) *Appl. Polymer Symp. 26*, 1.

Domininghaus, H. (1960) *Kunststoff-Rdsch. 7*, 474.

Domininghaus, H. (1969) *Monoaxial verstreckte Folienbändchen*, Chapter 4.1.3 in *Kunststoff-Handbuch*, Vol. IV (Edit.: Vieweg, R., Schley, A., and Schwarz, A.) Carl Hanser Verlag, München.

Dow, J., Patchell, A. G., and Piskozub, Z. T. (1972) *Plast. Polym. 40* (146), 80; see *Newer Routes from Film to Fiber I*, Conference Textiles from Film II, Manchester, July 6-7, 1971.

Dowd, L. D. (1963) *SPE-Journal 19*, 451.

Dow Chemical Company (1955) Inv.: Smith, C. P., Le Boeuf, E. W., and McIntire, O. R., *U.S. Pat. 2,707,805*, May 10, 1955.

Dow Chemical Company (1956) Inv.: Annesser, R. J., *U.S. Pat. 2,728,950*, Jan. 3, 1956.

Dow Chemical Company (1958) Inv.: Costa, J. E., Lefevre, L. E., and Le Boeuf, E. W., *U.S Pat. 2,853,741*, Sept. 30, 1958.

Dow Chemical Company (1961) Inv.: Costa, J. E., Le Boeuf, E. W., and Lefevre, L. E., *U.S. Pat. 2,980,982*, April 25, 1961.

Dow Chemical Company (1965) Inv.: Faircloth, C. R., *German Pat. 1,207,543*, Dec. 23, 1965.

Dowgielewicz, S. (1974) *Technik Wlokienniczy 28* (2), 38.

Droghanova, E. L. and Martynova, L. A. (1976) *Tekst. Prom. (Moscow) 36* (9), 36.

Dunlop Rubber Co., Ltd. (1971) *German Pat. 1,660,268*, April 22, 1971; Brit. Appl. Sept. 8, 1965.

du Pont de Nemours, E. I. & Co. (1966a) *U.S. Pat. 3,242,035*, March 22, 1966; *German Pat. 1,175,385*.

du Pont de Nemours, E. I. & Co. (1966b) Inv.: Blades, H. and White, J. R., *U.S. Pat. 3,227,784*, Jan. 4, 1966.

Dykes, E. M. (1971) *The Use of Staple Fibers in Needle-Punched Floorcoverings*, Conference Textiles from Film II, Manchester, July 6-7, 1971.

Eastman Kodak Co. (1967) Inv.: Hagemeyer, H. J. Jr. and Marvin, B. E., *U.S. Pat. 3,304,295*, Feb. 14, 1968.

Eastman Kodak Co. (1968) Inv.: Touey, G. P. and Mumpower, R. C., *U.S. Pat. 3,393,120*, July 16, 1968.

Eastman Kodak Co. (1969) Inv.: Johnson, W. L. and Harkleroad, J. K., *German Pat. 1,921,678*, Nov. 13, 1969; U.S. Appl. April 29, 1968.
Eastman Kodak Co. (1970) Inv.: Touey, G. P. and Mumpower, R. C., *U.S. Pat. 3,501,361*, March 17, 1970.
Elfers, H. and Schlegel, E. (1974a) *German Pat. 2,360,461*, June 12, 1974.
Elfers, H. and Schlegel, E. (1974b) *German Pat. 2,360,462*, June 12, 1974.
Evans, M. E. (1971) *The Influence of Additives on Processing and Properties of Polypropylene Film Fibers*, Conference Textiles from Film II, Manchester July 6-7, 1971.
Farbwerke Hoechst AG (1970) *German Pat. 1,921,182*, Nov. 12, 1970.
Farbwerke Hoechst AG (1971) Inv.: Rust, K. and Schrott, E., *German Pat. 1,940,329*, Feb. 25, 1971.
Farbwerke Hoechst AG (1974) *German Pat. 2,238,377*, Feb. 14, 1974.
Farnworth, A. J. and Pemberton, G. M. (1968) *Woolpacks from Woven Plastics*, Conference Textiles from Film, Manchester, April 4-5, 1968.
Feuerböther, D. (1976) *Tech. Textilien 19* (4), 98.
Fiber Industries, Inc. (1972) Inv.: Gibbon, J. D., *U.S. Pat. 3,695, 025*, Oct. 3, 1972.
Fibron, Inc. (1977) Inv.: Gustavson, K. C. and Chastain, E. L., *U.S. Pat. 4,065,538*, Dec. 27, 1977.
Fischer, P. (1969) Z. *Gesamte Textilind. 71* (11), 758.
Fitton, S. L. and Gray, J. (1971) *Factors Affecting the Light Stability of Polyolefin Stretched Films*, Conference Textiles from Film II, Manchester, July 6-7, 1971.
Fitzgerald, W. E. and Knudsen, J. P. (1967) *Text. Res. J. 37* (6), 447.
Fletcher, R. (1971) *Mechanical Fibrillation of Film*, Conference Textiles from Film II, Manchester, July 6-7, 1971.
Ford, J. E. (1965) *Brit. Plast. 38*, 157.
Ford, J. E. (1966) *Text. Manuf. 92*, 477.
Ford, J. E. (1967) *Text. Inst. Ind. 5* (6), 171.
Ford, J. E. (1968) *Introduction to Textiles from Film*, Conference Textiles from Film, Manchester, April 4-5, 1968.
Ford, J. E. (1970) *Textiles from Film*, Shirley Seminar, Manchester, Oct. 1970.
Ford, J. E. and Govier, T. H. H. (1971) *Future Applications for Film Textiles*, Conference Textiles from Film II, Manchester, July 6-7, 1971.
Foster, B. W. (1975) *Yarns Produced by the Hot-Melt Embossing Technique*, Conference Polypropylene Fibres in Textiles, York, Sept. 30-Oct. 1, 1975.
Franz, H. and Michels Ch. (1975) *DDR Pat. 186,198*, May 15, 1975.
Franz, H. and Michels, Ch. (1978) *Textiltechnik 28* (7), 404.
Freedman, A. R. (1970) *Mod. Plast. 47* (6), 109.
Freedman, A. R. (1975) *Roll Embossed Fibres–A Systems Approach*, Conference Polypropylene Fibres in Textiles, York, Sept. 30-Oct. 1, 1975.
Friedman, M. C. and Vlasov, S. V. (1974) *Plast. Massy 1974*, 27.

Friedrich, H.-J. (1971) *German Pat. 1,685,662*, April 15, 1971.

Galanti, A. V. and Mantell, C. L. (1965) *Polypropylene Fibers and Films*, Plenum Press, New York.

Geitel, K.-H. (1976) *Tech. Textilien 19* (5), 128.

Geitel, K.-H. (1977) *Textiltechnik 27* (4), 218.

Gouw, L. H. and Skoroszewski, W. H. (1968) *Film Production Techniques and Stretching Conditions and their Influence on Fiber Properties*, Conference Textiles from Film, Manchester, April 4-5, 1968.

Grilli, E. R. (1975) *The Interaction between Polypropylene and Jute/ Sisal*, Conference Polypropylene Fibres in Textiles, York, Sept. 30-Oct. 1, 1975.

Gropper, H., Birnkraut, H.-W., Payer, W., Scheible, P., Dittmann, W., Plenikowski, J., Goldbach, G., and Immel, W. (1960) *Polyolefine*, in *Ullmanns Enzyklopädie der technischen Chemie*, 4th Edition, Vol. 19, Verlag Chemie, Weinheim, pp. 167-226.

Hagen, H. and Domininghaus, H. (1961) *Polyäthylen und andere Polyolefine*, Verlag Brunke Garrels, Hamburg.

Hajamasy, T. and Alberti, J. (1976) *Magy. Textiltech. 30* (2), 65.

Hajamasy, T., Winkler, I., Alberti, J., and Kalmar, Z. (1972) *Magy. Textiltech. 24* (12), 657.

Hanson, D. E. and Reed, F. H. (1975) *Factors Governing the Orientation Process in Polypropylene*, Conference Polypropylene Fibres in Textiles, York, Sept. 30-Oct. 1, 1975.

Harms, J., Krässig, H., and Sasshofer, F. (1971a) *Fäden und Fasern aus Folien*, X. International Man-Made Fiber Conference, Dornbirn, Austria, Sept. 21-23, 1971.

Harms, J., Krässig, H., and Sasshofer, F. (1971b) *Lenzinger Ber. 32*, 81.

Harms, J. Krässig, H., and Sasshofer, F. (1973a), *Melliand Textilber. Int. 54* (5), 439.

Harms, J., Krässig, H., and Sasshofer, F. (1973b) *Chemiefasern + Text.-Anwendungstech. /Text. Ind. 23* (9), 845; *23* (10), 979.

Heggs, T. G. (1973) in *Block Copolymers* (Edit.: Allport, D. C. and Janes, W. H.), Chapter 4, pg. 105-149 and Chapter 80, pg. 493-529, Halsted Press, New York.

Hensen, F. (1968) *Winding of Monoaxially Stretched Film Tapes*, Conference Textiles from Film, Manchester, April 4-5, 1968.

Hensen, F. (1969) *Ind. Anzeiger 91*, 77; *91*, 1883.

Hensen, F. (1973) *Chemiefasern + Text.-Anwendungstech. /Text. Ind. 23* (7), 617.

Hensen, F. (1974) *Neue Anlagen zur Herstellung von Fasern aus Polyolefinen*, Intercarpet, Baden/Vienna/Austria, April 24-26.

Hensen, F. (1978) *Chemiefasern + Text.-Anwendungstech. /Text. Ind. 28* (1), 36.

Hensen, F. (1979) *Anlagen zur Herstellung von Folienbändchen und Monofilamenten/Plants for the Production of Film Tapes and Monofilaments* in *Folien, Gewebe, Vliesstoffe aus Polypropylene/Films, Woven and Nonwoven Fabrics made from Polypropylene* (Edit.: Committe "Extrusion" of the VDI-German Society of Plastics Technology), VDI-Verlag, Düsseldorf (published in German and English).

Hensen, F. and Braun, S. (1978a) *Industrial & Production Eng. 1978* (3), 40.

Hensen, F. and Braun, S. (1978b) *Kunststoffe 68* (4), 221.

Hensen, F. and Klawonn, G. W. (1967a) *Melliand Textilber. 48* (4), 379.

Hensen, F. and Klawonn, G. W. (1967b) *Text. Prax. Int. 22* (5), 329; *22* (6), 410.

Hercules, Inc. (1969) Inv.: Kim, C. W. and Samluk, S. D., *U.S. Pat. 3,470,285*, Sept. 30, 1969.

Hercules, Inc. (1970) *Brit. Pat. 1,210,644*, Oct. 10, 1970; U.S. Appl. Jan. 8 and July 22, 1968.

Hercules, Inc. (1971a) *French Pat. 2,044,770*, April 2, 1971; U.S. Appl. May 6, 1969.

Hercules, Inc. (1971b) *German Pat. 1,760,075*, Dec. 12, 1971; U.S. Appl. March 30, 1967.

Hercules, Inc. (1972) *French Pat. 2,115,463*, Aug. 11, 1972; U.S. Appl. Nov. 23, 1970; *Brit. Pat. 1,340,587.*

Hercules, Inc. (1973) *Brit. Pat. 1,333,474*, Oct. 10, 1973; U.S. Appl. May. 4, 1970.

Hoechst AG (1976) Inv.: Janocha, S. and Porrmann, H., *German Pat. 1,660,436*, June 24, 1976.

Hogan, J. P. and Witt, D. R. (1979) *Preprints Div. Petrol. Chem./ Amer. Chem. Soc. 24* (2) 377.

Hossack, D. C. (1971) *Warp Sheet*, Conference Textiles from Film II, Manchester, July 6-7, 1971; see also Hossack, D. C., *Tape Yarns*, Merrow Publishing Co., Ltd., Watford, Hertsh, England, 1971.

*Hydrocarbon Process.* (1974) *53* (2), 115.

*Hydrocarbon Process.* (1979) *58* (11), 223.

Idemitsu Kosan Co., Ltd. (1975a) Inv.: Fujimoto, N., Mugino, Y., Sugi, N., Kawachi, T., Sasaki, N., Kawasaki, K., Kozakura, S., and Tomikawa, M., *Japan Kokai 75-65,571*, June 3, 1975.

Idemitsu Kosan Co., Ltd. (1975b) Inv.: Fujimoto, N., Mugino, Y., Sugi, N., Kawachi, T., Sasaki, N., Kawasaki, K., Kozakura, S., and Tomikawa, M., *Japan Kokai 75-65,572*, June 3, 1975.

Idemitsu Kosan Co., Ltd. (1975c) Inv.: Fujimoto, N., Mugino, Y., Sugi, N., Kawachi, T., Sasaki, N., Kawasaki, K., Kozakura, S., and Tomikawa, M., *Japan Kokai 75-69,322*, June 10, 1975.

Idemitsu Kosan Co., Ltd. (1975d) Inv.: Fujimoto, N., Mugino, Y., Sugi, N., Kochi, T., Sasaki, N., Kawasaki, K., Kozakura, S., Kaji, H., and Tomikawa, M., *Japan Kokai 75-41,968*, April 16, 1975.

I. G. Farbenindustrie AG (1938) Inv.: Jacque, H., *German Pat. 667,234,* Nov. 7, 1968; *U.S. Pat. 2,185,789*; *Brit. Pat. 479,202.*

Imperial Chemical Industries, Ltd. (1937) Inv.: Fawcett, E. W., Gibson, R. O., Perrin, M. W., Paton, J. G., and Williams, E. G., *Brit. Pat. 471,590*, Sept. 6, 1937; *U.S. Pat. 2,210,774*; *Indian Pat. 23,709.*

Imperial Chemical Industries, Ltd. (1969) *German Pat. 1,807,957*, June 19, 1969.

Imperial Chemical Industries, Ltd. (1970) Inv.: Mascia, L., *S. African Pat. 68-07,254*, May 8, 1970; Brit. Appl. Nov. 9, 1968.

Imperial Chemical Industries, Ltd. (1971a) *Austrian Pat. 294,298*, March 15, 1971; Brit. Appl. Oct. 12, 1964.

Imperial Chemical Industries, Ltd. (1971b) Inv.: Blacker, J. G., *Brit. Pat. 1,235,121,* June 9, 1971.

Imperial Chemical Industries, Ltd. (1971c) Inv.: Ostrowski, H. S., Roberts, J. F. L., and Blacker, J. G., *Brit. Pat. 1,235,122*, June 9, 1971; *Austrian Pat. 311,534*; *German Pat. 1,942,166.*

Imperial Chemical Industries, Ltd. (1971d) Inv.: Blacker, J. G., *Brit. Pat. 1,253,036*, Nov. 10, 1971.

Imperial Chemical Industries Australia, Ltd. (1978) Inv.: Keith, D. G., *Japan Kokai 78-15,192*, May 23, 1978.

Inf. Chim. (1979) *189* (5), 255.

Institute Textile de France (1976) *Neth. Pat. 76-03,690*, Oct. 12, 1976.

Isvoranu, C. (1972) *German Pat. 2,040,702*, Feb. 24, 1972.

Ivanyukov, D. V. Konysheva, A. V., and Amerik, V. V. (1972) *Plast. Massy 1972* (10), 26.

Ivett, R. W. (1968) *Coloring of Polypropylene Fibers*, Conference Textiles from Film, Manchester, April 4-5, 1968.

Jakob, W. and Michels, Ch. (1974) *Faserforsch. Textiltech. 25* (6), 229.

Jezl, J. L. and Honeycutt, E. M. (1969) "*Propylene Polymers*", in *Encyclopedia of Polymer Sci. and Technol.* (Edit.: Mark, H. F., Gaylord, N. G., and Bikales, N. M.) Vol. II, Interscience, p. 597-619.

Johnson & Johnson (1970) Inv.: Kawaites, F., *U.S. Pat. 3,515,325*, June 2, 1970.

Käufer, H. (1981) *Arbeiten mit Kunststoffen,* 2nd Ed., Vol. 2 *Verarbeitung*, Springer Verlag, Berlin.

Kalle AG (1968) Inv.: Janocha, S., Seifried, W., and Gehler, H., *German Pat. 1,779,150*, July 12, 1968.

Kammer, H.-W. (1978) *Faserforsch. Textiltech. 29* (7), 459.

Kampf, R. (1976) *Techn. Textilien 19* (10), 104.

Kaneko, T. (1968) *Polyolefin Woven Sacks as Containers for Rice and Wheat in Japan*, Conference Textiles from Film, Manchester, April 4-5, 1968.

Kerr, A. A. (1971) *Present and Future Applications for Film Fibers in the United States*, Conference Textiles from Film II, Manchester, July 6-7, 1971.

Kim, C. Y., Evans, J., and Goring, D. A. J. (1971) *J. Appl. Polymer Sci. 15*, 1365.

Klust, G. (1968) *Text. Prax. Int. 23* (6), 361.

Knappe, W. (1975) "*Kunststoffverarbeitung*", in *Kunststoff-Handbuch*, Vol. I, Chapter 5 (Edit.: Schaab, H. and Stoeckhert, K.), Carl Hanser Verlag, München-Wien.

Koninklijke Shell Plastics Laboratorium (1971) *Symposium on Polypropylene Yarns and Film*; see *Tex 30* (11), 50.

Krämer, A. (1979) "Extruder und Extrusionsanlagen" in *Kunststoff-Maschinenführer* (Edit.: Schaab, H. and Stoeckhart, K.), Carl Hanser Verlag, München-Wien.

Krässig, H. (1977) Review: "Film to Fiber Technology" in *J. Polymer Sci. Macromolecular Reviews* Vol. 12 Wiley, New York, p. 321.

Krässig, H. (1972) *Dtsch. Textiltech. 22* (10), 641.

Krässig, H. (1975a) *Faserforsch. Textiltech. 26* (3), 135.

Krässig, H. (1975b) *The Manufacture of a Continuously Separated Filament Yarn or Staple Fiber Product by a Novel Slitting Technique*, Conference Polypropylene Fibers in Textiles, York, Sept. 30-Oct. 1, 1975; *Plastics & Rubber Processing 1976* (12), 143.

Krause, M. (1971) *Application of Raschel Knitted Fabrics*, Conference Textiles from Film II, Manchester, July 6-7, 1971.

Krejci, M. and Stepankowa, V. (1973) *Textil Praque 26* (9), 326.

Kresser, Th. O. J. (1960) *Polypropylene*, Rheinhold, New York.

Krueter H. and Diedrich, B. (1974) *Chem. Eng. 1974* (Aug. 5), 62.

*Kunststoffe* (1966) *56* (11), 798.

*Kunststoffe* (1969) *59* (5), 296.

Kuranray Co. (1972) Inv.: Watanabe, P., *Japan Kokai 72-49,783*, Dec. 14, 1972.

La Cellophane S. A. (1971) Inv.: Vivien, A., *U.S. Pat. 3,611,700*, Oct. 12, 1971.

Lagowski, W. (1967) *Przegl. Wlok. 21* (6), 299.

Lambeg Industrial Research Assoc. (1971) Inv.: McMeekin et al., *Brit. Pat. 1,232,738*, Nov. 10, 1971.

Lambeg Industrial Research Assoc. (1972) Inv.: Ruddel, J. and Todd, H. A. C., *U.S. Pat. 3,636,185*, Jan. 18, 1972.

Lambeg Industrial Research Assoc. (1973a) *German Pat. 2,261,182*, June 20, 1973; Brit. Appl. Dec. 18, 1971.

Lambeg Industrial Research Assoc. (1973b) Inv.: Ruddel, J. N. and McMeekin, S., *Brit. Pat. 1,315,306*, May 2, 1973.

Lambeg Industrial Research Assoc. (1975) Inv.: Ruddel, J. N. and McMeekin, S., *Brit. Pat. 1,404,753*, Dec. 3, 1975.

Larsen, L. (1932) *U.S. Pat. 1,862,687*, June 14, 1932.

Laus, Th. (1975) *The Influence of Various Parameters at the Artificial Weathering of Polypropylene Fiber Grades*, Conference Polypropylene Fibres in Textiles, York, Sept. 30-Oct. 1, 1975.

Lawrence, A. M. (1971) *Recovery of Oriented Polypropylene Waste*, Conference Textiles from Film II, Manchester, July 6-7, 1971.

Lee, J. G. (1975) *Dyeing of Polypropylene Fiber Including the Printing of Nickel Modified Polypropylene*, Conference Polypropylene Fibres in Textiles, York, Sept. 30-Oct. 1, 1975.

Lennox-Kerr, P. (1968) *Text. Ind. 70* (10), 713.

Lenz, J. (1977) *Lenzing Plastic Information 1.*

Lenz, R. W. and Stein, R. S. (1973) *Structure and Properties of Polymer Films*, Plenum Press, New York.

Leu, K. W., Linhart, H., and Müller, H. (1975) *Recent Developments in UV-Stabilization of Polypropylene Fibers*, Conference Polypropylene Fibres in Textiles, York, Sept. 30-Oct. 1, 1975.

Leykam-Josefsthal AG (1970) Inv.: Senger, F., *Austrian Pat. 286,764,* April 15, 1970.

Lurex N. V. (1974) Inv.: Shoots, B., *U.S. Pat. 3,837,893*, Sept. 24, 1974.

Mackenzie, J. P. (1975) *Polyolefines and their Processing into Woven Fabrics*, Conference Polypropylene Fibres in Textiles, York, Sept. 30-Oct. 1, 1975.

Mackenzie, J. P. (1976) *Text. Inst. Ind. 14* (1), 15.

Mackie, G. (1968) *Aspects of Plant for Conversion of Film Fibers*, Conference Textiles from Film, Manchester, April 4-5, 1968.

Mackie, James & Sons, Ltd. (1971) *German Pat. 2,045,756*, April 15, 1971; Brit. Appl. Sept. 22, 1969; *French Pat. 2,062,492.*

Mackie, James & Sons, Ltd. (1973a) Inv.: Mackie, J. K. P., *Brit. Pat. 1,318,347*, May 31, 1973.

Mackie, James & Sons, Ltd. (1973b) *German Pat. 2,305,053*, Aug. 16, 1973; Brit. Appl. Feb. 8, 1972; *French Pat. 2,171,289.*

Maddock, B. H. (1960) *SPE-Journal 16*, 375.

Maillefer, C. (1963) *Mod. Plast. 40* (5), 132.

Malcomesz, H.-P. and Blechschmidt, D. (1977) *Techn. Textilien 20* (2/3), 48.

Mark, H. F., Atlas, S. M., and Cernia, E. (1968) *Man-Made Fibers: Science and Technology*, Interscience, New York.

Marks, R. H. (1965) *U.S. Pat. 3,214,943*, Nov. 2, 1965.

Marks, R. H. (1968) *Knitting Sheet from Polypropylene and other Plastic Materials*, Conference Textiles from Film, Manchester, April 4-5, 1968.

Martinova, L. A. Zverev, M. P., and Usenko, V. A. (1979) *Khim. Volokna 1979* (1), 18.

Mattler, P. (1933) *U.S. Pat. 1,901,250*, March 14, 1933.

Mayer, W. F. (1969) *Plastverarbeiter 20* (5), 323.

McIllhagger, R. (1975) *Some Aspects of Research on Polypropylene in Textiles*, Conference Polypropylene in Textiles, York, Sept. 30-Oct. 1, 1975.
McKelvey, J. M. (1962) *Polymer Processing*, Wiley, New York.
McLellan, W. I. (1975) *Industrial Fabric Packaging*, Conference Polypropylene Fibres in Textiles, York, Sept. 30-Oct. 1, 1975.
McMeekin, S. (1968) *Aspects of Staple Fiber*, Conference Textiles from Film, Manchester, April 4-5, 1968.
McMeekin, S., Ruddel, J. N., and Todd, H. A. (1971) *Manufacture of Spun Staple Yarns from Film Fibres*, Conference Textiles from Film II, Manchester, July 6-7, 1971.
Mehta, H. (1971) *Multicomponent Film Fiber Developments*, Conference Textiles from Film II, Manchester, July 5-6, 1971.
Meskat, W. (1955) *Kunststoffe 45,* 87.
Messerschmidt-Boelkow-Blohm GmbH (1978) Inv.: Holzer, J., Herzer, K., and Schöner, F., *German Offen. 2,657,506,* June 22, 1978.
Michels, Ch. (1976) *Faserforsch. Textiltech. 27* (12), 639.
Michels, Ch. and Franz, H. (1977) *Faserforsch. Textiltech. 28* (4), 165.
Michels, Ch. and Franz, H. (1978) *Textiltechnik 28* (6), 348.
Michels, Ch., Geitel, K.-H., Franz, H., and Eberhart, H. (1975) *Textiltechnik 25* (6), 375.
Michels, Ch., Franz, H., and Reuter, K. (1977) *Textiltechnik 27* (4), 213.
Mirsky, A. (1967) *Brit. Pat. 1,067,514,* May 3, 1967.
Mitsubishi Rayon Co. (1970a) *Brit. Pat. 1,207,733,* Oct. 7, 1970; Japan Appl. May 24, 1967.
Mitsubishi Rayon Co. (1970b) Inv.: Yokouchi, N., Achikawa, A., Yasui, A., and Kuwabara, K., *Japan Kokai 70-40,439*, Dec. 18, 1970.
Mitsubishi Rayon Co. (1971a) *Brit. Pat. 1,233,713*, May 26, 1971; Japan Appl. Dec. 11, 1967 and Feb. 7, 1968.
Mitsubishi Rayon Co. (1971b) Inv.: Kiyoshim S. and Nobuaki, N., *Japan Kokai 71-36,890*, Oct. 29, 1971.
Mitsubishi Rayon Co. (1971c) Inv.: Okumura, M., Nishikawa, T., and Nishikawa, Y., *Japan Kokai 71-11,059*, March 20, 1971.
Mitsubishi Rayon Co. (1971d) Inv.: Shimoto, E., Ueno, K., Okumura, M., and Nishikawa, Y., *Japan Kokai 71-30,456*, Sept 4, 1971.
Mitsubishi Rayon Co. (1971e) Inv.: Okumura, M., Nishikawa, T., Echi, M., and Kagawa, K. *Japan Kokai 71-41,461*, Dec. 7, 1971.
Mitsubishi Rayon Co. (1971f) Inv.: Miyanoki, T., Mikita, S., Hirose, K., and Kagawa, K., *Japan Kokai 71-42,624*, Dec. 12, 1971.
Mitsuibishi Rayon Co. (1971g) Inv.: Masahiro, O., Tatsuaki, N., Masaharu, F., and Sachio, M. *Japan Kokai 71-43,265*, Dec. 21, 1971.
Mitsubishi Rayon Co. (1972a) Inv.: Okumura, M., Nishikawa, T., and Nishikawa, Y., *Japan Kokai 72-03,863,* Feb. 2, 1972.

Mitsubishi Rayon Co. (1972b) Inv.: Shirosawa, K. and Sugai, J. *Japan Kokai 72-06,570*, Feb. 24, 1972.

Mitsubishi Rayon Co. (1972c) Inv.: Okumura, M., Nishikawa, T., and Kajawa, K., *Japan Kokai 72-08,142*, March 8, 1972.

Mitsubishi Rayon Co. (1972d) Inv.: Yoshihiki, M., Masahiro, O., Katsuaki, H., and Tatsuaki, N., *Japan Kokai 72-04,726*, Feb. 9, 1972.

Mitsubishi Rayon Co. (1973a) Inv.: Okumura, M. and Nishikawa, T., *Japan Kokai 73-15,223*, May 12, 1973.

Mitsubishi Rayon Co. (1973b) Inv.: Okumura, M., Nishikawa, T., and Nishikawa, N., *Japan Kokai 73-15,233*, May 12, 1973.

Mitsubishi Rayon Co. (1974) Inv.: Okumura, M., Nishikawa, T., and Nishikawa, N., *German Pat. 1,768,444*, Nov. 28, 1974; Japan Appl. May 24, 1967.

Mitsubishi Rayon Co. (1975) Inv.: Shimoto, E., *Japan Kokai 75-03,822*, Feb. 10, 1975.

Mitsubishi Rayon Co. (1976) Inv.: Okumura, M., Nishikawa, T., and Nishikawa, N., *Japan Kokai 76-41,908*, Nov. 12, 1976.

Mitsui Petroleum Industries, Ltd. (1968) *Brit. Pat. 1,126,759*, Sept. 11, 1968; Japan Appl. Nov. 27, 1965.

Mitsui Petrochemical Industries, Ltd. (1974) Inv.: Toyota, A., Kashiwa, N., *Japan Kokai 74-12,860*, Feb. 1, 1974; *German Offen. 2,504,036*.

*Mod. Plast.* (1967) *44* (8), 82.

*Mod. Plast. Int.* (1974) *4* (9), 27.

*Mod. Plast. Int.* (1980) Jan., 22.

*Mod. Text. Mag.* (1968) *49* (9), 27.

*Mod. Text. Mag.* (1973) (3), 56.

Möller, H. (1967) *Melliand Textilber. 48* (10), 1168.

Monsanto Chemicals, Ltd. (1968) *French Pat. 1,543,891*, Oct. 25, 1968; Brit. Appl. Nov. 17, 1966.

Monsanto Chemicals, Ltd. (1969) *German Pat. 1,816,125*, Oct. 16, 1969; Brit. Appl. Dec. 21, 1967.

Monsanto Chemicals, Ltd. (1971) *German Pat. 2,029,761*, April 19, 1971; Brit. Appl. June 23, 1969.

Monsanto Chemicals, Ltd. (1971b) *German Pat. 2,050,294*, April 22, 1971; Brit. Appl. Oct. 14, 1969.

Monsanto Chemicals, Ltd. (1973) Inv.: Changani, P. D. and Penny, M. T., *Brit. Pat. 1,336,811*, Nov. 14, 1973.

Montecatini-Societa Generale per l'Industria Mineraria e Chimica (1955) Inv.: Natta, G., *Italian Pat. 535,712*, Nov. 17, 1955; Inv.: Ziegler, K., *Brit. Pat. 810,023*, March 11, 1959.

Montecatini Edison S.p.A. (1974) Inv.: Giannini, U., Cassata, A., and Longi, P. *German Pat. 2,347,577*, May 2, 1974; Italian Appl. Sept. 26, 1972.

Moore, P. F. (1979) *Polypropylene Lubrication–Fibers–Tapes and Fabrics*, 2nd Int. Conference Polypropylene Fibres in Textiles, York, Sept. 26-28, 1979.

Moorwessel, D. and Pilz, G. (1969) *Kunststoffe 59* (4), 205.

Moorwessel, D. and Pfirrmann, G. (1975) *Self-Extinguishing Polypropylene Fiber Material*, Conference Polypropylene Fibres in Textiles, York, Sept. 30-Oct. 1, 1975.

Mould, J. D. (1975) *Woven Polypropylene for Use in the Tufted Carpet Industry*, Conference Polypropylene Fibres in Textiles, York, Sept. 30-Oct. 1, 1975.

Muzzy, J. D., Day, C., and Levy, A. C. (1976) *Non-Woven Technol. Challenges. Achiev.* 71.

Naphtachimie (1974) Inv.: Avaro, M. and Mangin, P., *German Offen. 2,346,714*, March 28, 1974; *Brit. Pat. 1,414,883*.

Naphtachimie (1975) Inv.: Dormenval, R., Havas, L., and Mangin, P., *U.S. Pat. 3,922,322*, Nov. 25, 1975; *French Pat. 2,207,145* (Prior. Nov. 17, 1972); *Brit. Pat. 1,413,613*; *Japan Kokai 75-46,784*; *Neth. Pat. 7,315,652*; German Offen. 2,357,120.

Natta, G. (1955) *J. Polymer Sci. 16*, 143; *Makromol. Chem. 16*, 213; *Angew. Chem.* (1965) *68*, 393; (1957) *69*, 213.

Nau, G. (1968) *Chemiefasern + Text.-Anwendungstech./Text. Ind. 18* (1), 31.

Nawlicka, J. (1977) *Text. Inst. Lodz, Polym. Pr. Int. Wlok. 1977* (27), 51.

Naylor, R. C. (1975) *The Extrusion and Weaving of Carpet Backing Fabrics*, Conference Polypropylene Fibers in Textiles, York, Sept. 30-Oct. 1, 1975.

Neveu, J. L. (1975) *Ind. Textile 1975* (10), 677.

Nichols, G. J. (1971) *Newer Routes to Fibers from Film*, Conference Textiles from Film II, Manchester, July 6-7, 1971.

Nichols, G. J. (1972) *Plast. Polym. 40* (146), 84.

Nightingale, R. J. (1970) *Brit. Plast. 43* (11), 121.

Nippon Ekika Seikei K. K. (1973) Inv.: Suzuki, S., *Japan Kokai 73-25,073*, April 2, 1973.

Nippon Steel Chemical Corp., Ltd. (1975a) Inv.: Kawaguchi, N. et al. *Japan Kokai 75-111,320*, May 2, 1975.

Nippon Steel Chemical Corp., Ltd. (1975b) Inv.: Yamase, Y., Kawaguchi, N., and Shibata, H., *Japan Kokai 75-156,722*, Dec. 16, 1975.

Noe, M. (1979) The Manufacture of Raschel-Knit Fabrics, Flat and Circular Woven Fabrics in *Films, Woven and Non-Woven Fabrics Made from Polypropylene* (Edit.: Committee "Extrusion" of the VDI-German Society of Plastics Technology), VDI-Verlag, Düsseldorf.

Nojiri, A., Morinoto, H., and Ishizuka, O. (1967) *Kobunshi Kagaku 24* (No. 264), 250.

Norton, J. (1968) *Problems of Beaming Polypropylene and other Synthetic Tapes*, Conference Textiles from Film, Manchester, April 4-5, 1968.

Nott, R. E. (1971) *Spinneret Fibers versus Film Fibers*, Conference Textiles from Film II, Manchester, July 6-7, 1971.

Nott, R. E. and Sarwar, G. (1975) *Spinneret Fibers versus Film Fibers*, Conference Polypropylene Fibers in Textiles, York, Sept. 30-Oct. 1, 1975.

Nozawa, M. (1966) *Brit. Plast. 39* (11), 642.

Ohta, K. (1977) *Japan Kokai 77-82,978*, July 11, 1977.

Oike & Co., Ltd. (1973) Inv.: Kohzo, O. and Kenichi, M., *Japan Kokai 73-15,701*, May 16, 1973.

Okabe, K. and Chino, T. (1967) *Plast. Age. 13* (8), 39.

Overton, B. W. (1968) *The Use of Woven Plastics for Heavy Duty Packaging*, Conference Textiles from Film, Manchester, April 4-5, 1968.

Owens, D. K. (1975) *J. Appl. Polym. Sci. 19*, 265.

Pakleppa, G. (1966) *Brit. Pat. 1,035,227*, July 6, 1966; S. African Appl. May 18, 1962.

Parsey, M. R. (1968) *Some Properties of Ropes and Fishnet Twines from Fibrillating Tapes*, Conference Textiles from Film, Manchester, April 4-5, 1968.

Paton, J. B., Squires, P. H., Darnell, W. H., Cash, F. H., and Carley, J. F. (1959) Extrusion, in *Processing of Thermoplastic Materials* (Edit.: Bernhardt, E. C.), Reinhold, New York.

Peters, E. F., Zletz, A., and Evering, B. L. (1957) *Ind. Eng. Chem. 49*, 1879.

Peuker, H. (1968a) *Herstellung, Weiterverarbeitung, Verwendung und Entwicklungstendenzen von Flachfasern und Splitfäden*, VIIth Int. Man-Made Fiber Conference, Dornbirn, Austria.

Peuker, H. (1968b) *Lenzinger Ber. 26*, 133.

Peuker, H. (1968/69) *Textilind. 70* (11), 774; *70* (12), 867; *71* (1), 10; *71* (2), 82.

Peterlin, A. (1972) *Textile Res. J. 42* (1), 20.

Peterlin, A. and Baltá-Calleja, F. J. (1969) *J. Appl. Phys. 40* (11), 4238.

Phillips Petroleum Co. (1966) Inv.: Bottomley, A. and Bottomley, C., *U.S. Pat. 3,283,788*, Nov. 8, 1966.

Phillips Petroleum Co. (1967) Inv.: Greene, M. E., *U.S. Pat. 3,302,501*, Feb. 7, 1967.

Phillips Petroleum Co. (1969) Inv.: Hughes, J. K. and Williams, J. E., *German Pat. 1,925,608*, Nov. 20, 1969; U.S. Appl. May 20, 1968.

Phillips Petroleum Co. (1970a) *Brit. Pat. 1,207,412*, Sept. 30, 1970; U.S. Appl. Jan. 26, 1967.

Phillips Petroleum Co. (1970b) *U.S. Pat. 3,511,901*, May 12, 1970.

Phillips Petroleum Co. (1971a) *German Pat. 1,604,676*, July 15, 1971.

Phillips Petroleum Co. (1971b) Inv.: Brown C. V., *U.S. Pat. 3,596,816*, Aug. 3, 1971.

Phillips Petroleum Co. (1971c) *German Pat. 1,660,579*, March 11, 1971; U.S. Appl. Dec. 6, 1965.

Plasticisers, Ltd. (1967) *Brit. Pat. 1,073,741*, June 28, 1967; *Austrian Pat. 257,029*.

Plasticisers, Ltd. (1971) *German Pat. 1,785,509*, March 18, 1971.

Poelchau, W. (1966) *Verpack. Rundsch. 18* (12), 1678.

Polovikhina, L. A. and Zverev, M. P. (1969) *Khim. Volokna 1969* (5), 12.

Polovikhina, L. A. and Zverev. M. P. (1972) *Khim. Volokna 1972* (4), 56.

Polovikhina, L. A. and Zverev, M. P. (1977) *Vysokolmol. Soedin. Ser. B 19* (7) 513.

Polovikhina, L. A., Shimko, J. G., and Zverev, M. P. (1978) *Khim. Volokna 1978* (1), 12.

Polymer Processing Research Institute, Ltd. (1970) *German Pat. 1,957,998*, May 27, 1970, Japan Appl. Nov. 18, 1968; *Brit. Pat. 1,250,855.*

Polymer Processing Research Institute, Ltd. (1972) *German Pat. 1,660,565*, Dec. 21, 1972; Japan Appl. June 5, 1965; *U.S. Pat 3,739,053.*

Pozleb, A. (1974) *Hem. Vlakna 1974* (4), 22.

Raff, R. A. V. and Doak, K. W. (1964/65) *Crystalline Olefin Polymers*, Vol. I and II, Interscience, New York.

Raff, R. A. V., Campbell, P. E., Jones, R. V., Caldwell, E. D., Friedlander, H. N., and Canterino, P. J. (1967) Ethylene Polymers, in *Encyclopedia of Polymer Science and Technology* (Edit.: Mark, H. F., Gaylord, N. G., and Bikales, N. M.), Interscience, New York, pp. 275-454.

Rasmussen, O. B. (1960) *U.S. Pat. 2,954,587*, Oct. 4, 1960; Dan. Appl. May 29, 1954.

Rasmussen, O. B. (1961) *U.S. Pat. 3,003,304*, Oct. 10, 1961.

Rasmussen, O. B. (1963a) *Textilind. 65* (9), 708.

Rasmussen, O. B. (1963b) *Belg. Pat. 637,119*, Dec. 31, 1963; Dan. Appl. Sept. 7, 1962.

Rasmussen, O. B. (1963c) assigned to Kontor for Industrial Eneret Ved Svend Schoenning, *Dan. Pat. 95,894*, July 29, 1963; *German Pat. 1,195,428* (assigned to Felix Schleuter).

Rasmussen, O. B. (1964) *Swiss Pat. 380,289*, July 31, 1964; Dan. Appl. June 25, 1959; *U.S. Pat. 3,165,563*; *German Pat. 1,237,058.*

Rasmussen, O. B. (1965) *Austrian Pat. 246,324*, Aug. 15, 1965; Dan. Appl. May 11, 1962.

Rasmussen, O. B. (1966a) *Neth. Pat. 66-02,387*, Aug. 24, 1966; Brit. Appl. No. 7870/65.

Rasmussen, O. B. (1966b) assigned to Phillips Petroleum Co., *U.S. Pat. 3,233,029*, Feb. 1, 1966; Dan. Appl. June 9, 1961.

Rasmussen, O. B. (1966c) *Neth. Pat. 66-06,062*, Nov. 7, 1966; Brit. Appl. May 5, 1965.

Rasmussen, O. B. (1966d) *Neth. Pat. 66-04,943*, Oct. 17, 1966; Brit. Appl. April 15, 1965.

Rasmussen, O. B. (1967a) *Austrian Pat. 251,751*, Jan. 25, 1967; Dan. Appl. May 15, 1966.

Rasmussen, O. B. (1967b) assigned to Phillips Petroleum Co., *U.S. Pat. 2,323,978*, June 6, 1967.
Rasmussen, O. B. (1969a) *French Pat. 1,568,499*, May 23, 1969; Brit. Appl. May 30, 1967.
Rasmussen, O. B. (1969b) assigned to Metal Containers, Ltd., *U.S. Pat. 3,454,455*, July 8, 1969; Brit. Appl. Nov. 12, 1963.
Rasmussen, O. B. (1969c) *German Pat. 1,494,732*, Sept. 4, 1969; Brit. Appl. Aug. 10, 1963 and April 16, 1964.
Rasmussen, O. B. (1970a) *Split Fiber Techniques*, 2nd Shirly Seminar, Manchester, Oct. 1970.
Rasmussen, O. B. (1970b) *U.S. Pat. 3,511,742*, May 12, 1970; Brit. Appl. Aug. 28, 1963 and April 16, 1964.
Rasmussen, O. B. (1971a) *Brit. Pat. 1,229,793*, April 28, 1971.
Rasmussen, O. B. (1971b) *Brit. Pat. 1,230,991*, May 5, 1971.
Rasmussen, O. B. (1971c) *Austrian Pat. 298,658*, Sept. 15, 1971; Brit. Appl. June 8, 1965.
Rasmussen, O. B. (1971d) *Brit. Pat. 1,243,512*, Aug. 18, 1971; *German Pat. 1,785,094*; *Austrian Pat. 315,350.*
Rasmussen, O. B. (1972a) *Amer. Chem. Soc. (Div. Org. Coatings Plast. Chem. Pap.) 32* (1), 264.
Rasmussen, O. B. (1972b) *German Offen, 1,785,094*, Jan. 12, 1972; Brit. Appl. Aug. 9, 1967.
Rasmussen, O. B. (1976a) *Austrian Pat. 336,759*, Sept. 15, 1976.
Rasmussen, O. B. (1976b) *Austrian Pat. 338,404*, Dec. 15, 1976.
Rasmussen, O. B. (1979) *German Pat. 2,166,931*, Nov. 11, 1979; Dan. Prior. March 3, 1970.
Real, W. J. (1969) *Can. Text. J. 86* (11), 53.
Reifenhäuser KG (1972) *German Pat. 2,129,001*, Dec. 28, 1972.
Reifenhäuser KG (1974) Inv.: Reifenhäuser, F., *German Pat. 2,316,484*, Oct. 24, 1974.
Reinshagen, J. H. and Dunlap, R. W. (1975) *J. Appl. Polym. Sci. 19* (4), 1037.
Reuter, K. and Kampf, R. (1978) *Methods and Results of Fineness Investigations on Split Film Fibers*, 3rd Int. Colloquium on Split Fibers, Dresden.
Rheinstahl Henschel AG (1971a) *Brit. Pat. 1,238,120*, July 7, 1971; German Appl. Aug. 24, 1968; *Austrian Pat. 317,407.*
Rheinstahl Henschel AG (1971b) *German Gebrauchsmuster* (Registered Design) *7,044,132*, March 4, 1971.
Roberts R. H. (1971) *Dyeing and Printing of Film Fibers*, Conference Textiles from Film II, Manchester, July 6-7, 1971.
Roff, W. J. and Scott, J. R. (1971) *Fibres, Film, Plastics and Rubbers*, Butterworths, London.
Russek, W. and Glowacka, St. (1978) *Production, Properties and Application of Bicomponent Film Fibers and Yarns*, 3rd Int. Colloquim on Split Film Fibers, Dresden.

Rusznak, J., Bertalan, G., Berek, A., Trezl. L., and Zaoui A. (1973) *Magy. Textiltech.* *25* (5), 235; *26* (4), 176.

Rusnak, J. and Huszar, A. (1975) *Period. Polytech. Chem. Eng.* *19* (1/2), 201.

Ruta, D. (1971) *Textilia 47* (2), 9.

Samuels, R. J. (1968) *J. Polymer Sci. Part A-2 6*, 1101.

Samuels, R. J. (1970) *J. Macromol. Sci. Phys. B 4* (3), 701.

Schaab, H. and Stoeckhert, K. (1979) *Kunststoff-Maschinenführer*, Carl Hanser Verlag, München-Wien,

Scheiner, L. L. (1977) *Plast. Technol.* *23* (2), 53.

Schenkel, G. (1963) *Kunststoff Extrudertechnik*, Carl Hanser Verlag, München.

Schmack, G., Berger, W., and Kammer, H.-W. (1975) *Faserforsch. Textiltech.* *26* (6), 284.

Schrader, H. (1968) *Textilind.* *70* (3), 135.

Schubert, F. R. (1970) *Textilind.* *72* (3), 219.

Schumann, W., Eichler, R., Ebert, Ch., and Güther, W. (1978) *Textiltechnik 1978* (Nr.28), 492.

Schuur, G. (1966) *Kolloid Z. & Z. Polym.* *208*, 123.

Schuur, G. (1969) *Text. Prax. Int.* *24* (2), 65.

Schuur, G. and Van der Vegt, A. K. (1975) *Structure and Properties of Oriented Polymers* (Edit.: Ward, I. M.), Applied Science Publishers, London, pp. 413-453.

Schwenkedel, S. (1967) *Textilind.* *69* (11), 835.

Scragg, Ernest & Sons, Ltd. (1969) *German Pat. 1,922,695*, Nov. 13, 1969; Brit. Appl. May 3, 1968.

Scragg, Ernest & Sons, Ltd. (1970) Inv.: Waterhouse, G., *Brit. Pat. 1,214,543*, Dec. 2, 1970.

Scragg, Ernest & Sons, Ltd. (1971a) Inv.: Waterhouse G., *German Pat. 2,117,632*, Oct. 21, 1971; Brit. Appl. April 10, 1970, *Brit. Pat. 1,333,089*; *U.S. Pat. 3,768,938.*

Scragg, Ernest & Sons, Ltd. (1971b) *German Pat. 1,435,661*, Feb. 4, 1971; Brit. Appl. March 9, 1962.

Scragg, Ernest & Sons, Ltd. (1972) Inv.: Waterhouse, G., *French Pat. 2,123,349*, Sept. 8, 1972; Brit. Appl. Jan. 26, 1971; *U.S. Pat. 3,801,252.*

Scragg, Ernest & Sons, Ltd. (1973a) Inv.: Waterhouse, G., *Brit. Pat. 1,333,089*, Oct. 10, 1973.

Scragg, Ernest & Sons, Ltd. (1973b) Inv.: Waterhouse, G., *Brit. Pat. 1,331,751*, Sept. 26, 1973.

Scragg, Ernest & Sons, Ltd. (1974) Inv.: Waterhouse, G., *Brit. Pat. 1,349,249*, Apr. 3, 1974.

Seifert, K. (1968) *German Pat. 1,266,441*, Nov. 21, 1968; *Austrian Pat. 298,659.*

Sekusui Chemical Co., Ltd. (1971) Inv.: Yoshihara, R., *Japan Kokai 71-30,455*, Sept. 4, 1971.

Sekusui Jushi Co., Ltd. (1978) Inv.: Sugita, M. and Nagae, S., *Japan Kokai 78-98,371*, Aug. 28, 1978.

Seymour, D. E. and Dow, J. (1968) *Plastic Net: A New Product*, Conference Textiles from Film, Manchester, April 4-5, 1968.

Sheehan, W. C. (1965) *Text. Res. J. 35*, 636.

Sheehan, W. C. and Cole, T. B. (1964) *J. Appl. Polym. Sci. 8*, 2359.

Sheehan, W. C., Wellmann, R. E., and Cole, T. B. (1964) *Relationship Between Structural Parameters and Tenacity of Polypropylene Monofilaments*, Symposium on Polypropylene Fibers, Southern Research Institute, Sept. 1964.

Shell Internationale Research Maatschappij N. V. (1966) *Neth. Pat. 64-14,962*, June 23, 1966; *Austrian Pat. 282,799.*

Shell Internationale Research Maatschappij N. V. (1967a) Inv.: Gouw, L. H. and Voncken, J. F., *Neth. Pat. 65-11,455*, March 3, 1967; *German Pat. 1,660,540*; *U.S.S.R. Pat. 235,654.*

Shell Internationale Research Maatschappij N. V. (1967b) *Belg. Pat. 694,039*, Feb. 14, 1967; Neth. Appl. Feb. 16, 1966; *Brit. Pat. 1,120,872*; *French Pat. 1,511,362.*

Shell Internationale Research Maatschappij N. V. (1969a) *Neth. Pat. 69-05,427*, Oct. 14, 1969; Brit. Appl. March 26, 1969.

Shell Internationale Research Maatschappij N. V. (1969b) *Austrian Pat. 282,799*, Nov. 15, 1969; Neth. Appl. Dec. 22, 1964.

Shell Internationale Research Maatschappij N. V. (1970a) *Neth. Pat. 69-15,490*, April 17, 1970.

Shell Internationale Research Maatschappij, N. V. (1970b) Inv.: Badrian, W. H. J., *German Offen. 2,025,493*, Dec. 3, 1970.

Shell Internationale Research Maatschappij N. V. (1971a) *French Pat. 2,075,257*, Nov. 12, 1971; Brit. Appl. Jan. 8, 1970.

Shell Internationale Research Maatschappij N. V. (1971b) *French Pat. 2,046,406,* April 9, 1971; Brit. Appl. April 25, 1969.

Shell Internationale Research Maatschappij N. V. (1971c) *French Pat. 2,051,249*, May 5, 1971.

Shell Internationale Research Maatschappij N. V. (1971d) *Neth. Pat. 70-18,840*, July 1, 1971.

Shell Internationale Research Maatschappij N. V. (1972a) Inv.: Skoroszewski, W. H., *German Pat. 2,159,142*, May 31, 1972; Brit. Appl. Nov. 30, 1970; *French Pat. 2,115,473.*

Shell Internationale Research Maatschappij N. V. (1972b) Inv.: Vernon, B. J. and Skoroszewski, W. H., *Brit. Pat. 1,260,836*, Jan. 19, 1972.

Shell Internationale Research Maatschappij N. V. (1972c) Inv.: Vernon, B. J. and Skoroszewski, W. H., *Brit. Pat. 1,260,837*, Jan. 19, 1972.

Shell Internationale Research Maatschappij N. V. (1972d) Inv.: Ruddel, J. N. and McMeekin, S., *German Offen, 2,118,593*, Oct. 19, 1972.

Shell Internationale Research Maatschappij N. V. (1973a) Inv.: Stones, D. S., *Brit. Pat. 1,317,558*, May 23, 1973.

Shell Internationale Research Maatschappij N. V. (1973b) Inv.: Skoroszewski, W. H., *Brit. Pat. 1,306,214*, May 23, 1973.

Shell Internationale Research Maatschappij N. V. (1974) Inv.: Gouw, L. H. and Badrian, W. H. J., *German Pat. 2,413,714*, Sept. 26, 1974.

Shell Internationale Research Maatschappij N. V. (1975a) Inv.: Badrian, W. H. J., *German Offen. 2,457,277*, June 12, 1975.

Shell Internationale Reserach Maatschappij N. V. (1975b) Inv.: Mehta, H., *German Offen. 2,457,373*, June 12, 1975.

Shell Internationale Research Maatschappij N. V. (1978) *Brit. Pat. 1,497,540*, Jan. 12, 1978.

Shell Internationale Research Maatschappij N. V. (1979) Inv.: Skoroszewski, W. H. and Stones, D. S., *German Pat. 2,031,338*, Feb. 1, 1979; Brit. Appl. May 26, 1970.

Shell, Den Haag (1977) *Austrian Pat. 367,418*, May 25, 1977.

Shell Oil Co. (1970a) *U.S. Pat. 3,542,909*, Nov. 24, 1970.

Shell Oil Co. (1970b) Inv.: Dekker, J. and Schuur, G., *U.S. Pat. 3,500,517*, March 17, 1970.

Shell Oil Co. (1971) Inv.: Schuur, G., *U.S. Pat. 3,528,418*, June 1, 1971.

Shell Oil Co. (1975) Inv.: Vernon, B. J. and Skoroszewski, W. H., *U.S. Pat. 3,907,478*, Sept. 23, 1975.

*Shell Rev.* (1970) *1970* (2), 16.

Shirley Institute, Manchester (1964), *Man-Made Textiles 41* (486), 27.

Showa Yuka K. K. (1974) *Japan Kokai 74-75,667*, July 22, 1974.

Showa Yuka K. K. (1975a) Inv.: Taka, T. and Toda, H., *Japan Kokai 75-13,631*, Feb. 13, 1975.

Showa Yuka K. K. (1975b) Inv.: Taka, T., Miyazono, Y., and Haruta, Y., *Japan Kokai 75-143,847*, Nov. 19, 1975.

Skoroszewski, W. H. (1971) *Parameters Affecting Processing of Polymers and Polymer Blends*, Conference Textiles from Film II, Manchester, July 6-7, 1971.

Slack, P. T. (1968) *Machinery for the Extrusion, Slitting and Drawing of Fibers*, Conference Textiles from Film, Manchester, April 4-5, 1968.

Slack, P. T. (1973) in *Carpet Substrates*, Ellis, Ed., Textile Trade Manchester, pp. 51-63.

Smith & Nephew Plactics, Ltd. (1979) Inv.: Lloyd, R. and Patchell, A., *German Offen. 2,830,089,* March 1, 1979.

Smith & Nephew Polyfabrik, Ltd. (1970) Inv.: Lemmonier, A., *French Pat. 2,030,319*, Nov. 13, 1970; Brit. Appl. Feb. 5, 1969.

Smith & Nephew Polyfabrik, Ltd. (1971) Inv.: Dow, J., Lloyd, R., and Patchell, A. G., *German Pat. 2,031,686*, April 8, 1971; Brit. Appl. June 27, 1969.

Smith & Nephew Research, Ltd. (1971) Inv.: Dow, J. and Jones, B. M., *Brit. Pat. 1,232,282*, May 19, 1971.

Smith & Nephew Research, Ltd. (1972) *German Pat. 2,222,274*, Nov. 16, 1972; Brit. Appl. May 7, 1971.

Société Normande de Matière Plastiques (1966) Inv.: Rigaud, J., *Belg. Pat. 670,558*, Jan. 30, 1966; French Appl. Oct. 9, 1964.

Société Rhodiaceta (1965) *French Pat. 1,426,288*, Dec. 20, 1965; Neth. Appl. 65-15,900; Austrian Appl. 11284/65.

Société Rhodiaceta (1966) *Neth. Pat. 66-01,471*, Aug. 15, 1966.

Soko Co., Ltd. (1970) *German Pat. 2,016,359*, Oct. 8, 1970; Japan Appl. April 5, 1969; *Swiss Pat. 509,434.*

Soko Co., Ltd. (1971) *German Pat. 2,021,986*, Oct. 14, 1971; *Swiss Pat. 509,434.*

Soko Co., Ltd. (1973) Inv.: Gomi, M., *Swiss Pat. 534,232*, Feb. 28, 1973.

Solomon, D. H. (1967) *The Chemistry of Organic Film Formers*, Wiley, New York.

Standard Oil Co. (1978) Inv.: Haas, P. and Koblischke, K., *German Pat. 2,655,198*, Nov. 9, 1978.

Stewart, W. R. & Sons, Ltd. (1971) *French Pat. 2,037,850*, Feb. 5, 1971; Brit. Appl. March 10, 1969.

Stewart, W. R. & Sons, Ltd. (1976) Inv.: Steward, W. R., *Brit. Pat. 1,421,324*, Jan. 14, 1976.

Stout, H. P. (1968) *Woven Polypropylene Tapes for Tufted Carpet Backing*, Conference Textiles from Film, Manchester, April 4-5, 1968.

Stout, H. P. (1971) *Film Fibers in Floor Coverings*, Conference Textiles from Film II, Manchester, July 6-7, 1971.

Taylor, R. I. (1968) *Packing and Agricultural Twines*, Conference Textiles from Film, Manchester, April 4-5, 1968.

TBA Industrial Products, Ltd. (1976) Inv.: Edleston, J., *German Pat. 2,556,130*, June 24, 1976.

Teijin, Ltd. (1970) Inv.: Kaneko, H., *Japan Kokai 70-39,477,* Dec. 11, 1970.

Teijin, Ltd. (1973) Inv.: Matsukura, Y. and Washida, K., *Japan Kokai 73-04,387*, Feb. 7, 1973.

Ten Hoevel, B. (1975) *The Market for Polypropylene in Carpets*, Conference Polypropylene Fibres in Textiles, York, Sept. 30-Oct. 1, 1975.

*Text. Inst. Ind.* (1973) *11* (12), 325.

*Text. Rec.* (1966) *84* (9), 88.

*Text. World* (1971) *122* (6) 57.

Thomas, G. O. (1971) *Text. Manuf.* *97* (1158), 239.

Thomas, I. H. (1968) *Weaving Methods for Polypropylene Film Fibers*, Conference Textiles from Film, Manchester, April 4-5, 1968.

Toray Industries, Inc. (1971) Inv.: Keiichi, A., Mitsunori, O., and Koji, W., *Japan Kokai 67-02,492*, Jan. 24, 1971.

Toyobu Co., Ltd. (1974) Inv.: Isaka, T., *Japan Kokai 74-10,276.*

Toyo Kayaku Co., Ltd. (1976) Inv.: Sakai, K., Tajuchi, T., Ozawa, M., Ito, M. and Matsuzawa, S., *Japan Kokai 76-12,878.*

Toyo Stauffer Chemical Co., Ltd. (1978) Inv.: Takahaski, Y., Tomiyasu, S., and Takitani, M., *Belg. Pat. 870,106,* Aug. 1978; *Japan Kokai 79-38,291*; *French Pat. 2,401,983*; *German Offen. 2,838,013.*

*Tufting Needling News Bull.* (1973) *1973* (39), 11.

Tweedale, A. (1974) *Text. Inst. Ind. 12* (3), 76.

Ube Industries, Ltd. (1973) Inv.: Tokiura, M., Giwara, S., Iakamoto, Y., and Okuda, H., *Japan Kokai 72-50,333*, Dec. 18, 1973.

Union Carbide Australia, Ltd. (1970) Inv.: Miller, A. R., *Australian Pat. 445,455*, Oct. 2, 1970.

Van Boskirk, R. L. (1969) *Mod. Plast. 46* (7), 8.

Van Schuppen, D. S. (1974) *Chemiefasern + Text.-Anwendungstech./ Text. Ind. 24* (12), 984.

Van Tilburg, J. (1975) *Brit. Pat. 1,408,262*, Oct. 1, 1975.

VDI-Gesellschaft für Kunststofftechnik (1978) *Films, Woven and Nonwoven Fabrics made from Polypropylene*, VDI-Verlag, Düsseldorf.

Veba Chemie AG (1978) Inv.: Buhren, D., Klimmek, U., and Bleyel, E., *German Pat. 2,314,052*, Nov. 16, 1978.

VEB Textilkombinat Cottbus (1973) Inv.: Heger, A, Pässler, H., and Patitz, E., *German Offen. 2,132,932*, Jan. 25, 1973.

VEB Textilkombinat Cottbus (1977) *Austrian Pat. 338,405*, Aug. 25, 1977.

Velluire, M. (1972) *Melliand Textilber. Int. 53* (2), 123.

Volans, P. and Changani, P. D. (1968) *Faserforsch. Textiltech. 19* (12), 590.

Waddington, C. (1975) *Processing of Synthetic Tapes from Package to Weavers Beam*, Conference Polypropylene Fibres in Textiles, York, Sept. 30-Oct. 1, 1975.

Ward, D. (1973) *Mod. Text. Mag. 54* (3), 46.

Ward, I. M., Capaccio, G., and Cansfield, D. L. M. (1975) *The Preparation of Ultra-High Modulus Polypropylene Films and Fibers*, Conference Polypropylene Fibres in Textiles, York, Sept. 30-Oct. 1, 1975.

Weber, E. (1967) *Verpack. Rundsch. 1967* (5), 576.

Weber, E. (1968) *The Place of High-Density Polyethylene in Woven Fabrics*, Conference Textiles from Film, Manchester, April 4-5, 1968.

Welge, W. E. (1970) *Chemiefasern 20* (2), 113.

Wheatley, B. (1975) *Raschel Knitting of Polyolefine Yarns and Tapes*, Conference Polypropylene Fibres in Textiles, York, Sept. 30-Oct. 1, 1975.

Whitehead, Ch. A. (1975) *Polypropylene Fiber Usage in the United States*, Conference Polypropylene Fibres in Textiles, York, Sept. 30-Oct. 1, 1975.

Wills, A. E. (1971) *Development in Applications of Film Yarns in Ropes and Twines*, Conference Textiles from Film II, Manchester, July 6-7, 1971.

Windmüller & Hölscher (1978) Inv.: Bosse, F., *German Offen, 2,652,011*, May 24, 1978.

Windmöller & Hölscher/Starlinger (1981) *Chemiefasern + Text.-Anwendungstech. /Text. Ind.* *31* (5), 386.

Wood-Walker, C. P. (1975) *The Current Use and Future Potential of Polypropylene in Textiles*, Conference Polypropylene Fibres in Textiles, York, Sept. 30-Oct. 1, 1975.

Zalewski, E. (1969) *Chemiefasern 19* (3), 188.

*Z. Gesamte Textilind.* (1969) *71* (3), 198.

Ziegler, K. (1955) *Angew. Chem. 67*, 426; see also Ziegler, K. and Hoberg, H., *Kunststoffe-Plastics. Solothurn 6*/1, 39.

Ziegler, K., Holzkamp, E., Breil, H., and Martin, H. (1955) *Angew. Chem. 67*, 541.

Zonsveld, J. J. (1971) *The Application of Staple Film Fibers in Concrete*, Conference Textiles from Film II, Manchester, July 6-7, 1971.

# Index